Integrated Power Devices and TCAD Simulation

Devices, Circuits, and Systems

Series Editor
Krzysztof Iniewski
CMOS Emerging Technologies Research Inc.,
Vancouver, British Columbia, Canada

PUBLISHED TITLES:

Atomic Nanoscale Technology in the Nuclear Industry
Taeho Woo

Biological and Medical Sensor Technologies
Krzysztof Iniewski

Building Sensor Networks: From Design to Applications
Ioanis Nikolaidis and Krzysztof Iniewski

Circuits at the Nanoscale: Communications, Imaging, and Sensing
Krzysztof Iniewski

Electrical Solitons: Theory, Design, and Applications
David Ricketts and Donhee Ham

Electronics for Radiation Detection
Krzysztof Iniewski

**Embedded and Networking Systems:
Design, Software, and Implementation**
Gul N. Khan and Krzysztof Iniewski

Energy Harvesting with Functional Materials and Microsystems
Madhu Bhaskaran, Sharath Sriram, and Krzysztof Iniewski

**Graphene, Carbon Nanotubes, and Nanostuctures:
Techniques and Applications**
James E. Morris and Krzysztof Iniewski

High-Speed Photonics Interconnects
Lukas Chrostowski and Krzysztof Iniewski

Integrated Microsystems: Electronics, Photonics, and Biotechnology
Krzysztof Iniewski

Integrated Power Devices and TCAD Simulation
Yue Fu, Zhanming Li, Wai Tung Ng, and Johnny K.O. Sin

Internet Networks: Wired, Wireless, and Optical Technologies
Krzysztof Iniewski

Low Power Emerging Wireless Technologies
Reza Mahmoudi and Krzysztof Iniewski

Medical Imaging: Technology and Applications
Troy Farncombe and Krzysztof Iniewski

FORTHCOMING TITLES:

3D Circuit and System Design: Multicore Architecture, Thermal Management, and Reliability
Rohit Sharma and Krzysztof Iniewski

Circuits and Systems for Security and Privacy
Farhana Sheikh and Leonel Sousa

CMOS: Front-End Electronics for Radiation Sensors
Angelo Rivetti

Gallium Nitride (GaN): Physics, Devices, and Technology
Farid Medjdoub and Krzysztof Iniewski

High Frequency Communication and Sensing: Traveling-Wave Techniques
Ahmet Tekin and Ahmed Emira

High-Speed Devices and Circuits with THz Applications
Jung Han Choi and Krzysztof Iniewski

Labs-on-Chip: Physics, Design and Technology
Eugenio Iannone

Laser-Based Optical Detection of Explosives
Paul M. Pellegrino, Ellen L. Holthoff, and Mikella E. Farrell

Metallic Spintronic Devices
Xiaobin Wang

Mobile Point-of-Care Monitors and Diagnostic Device Design
Walter Karlen and Krzysztof Iniewski

Nanoelectronics: Devices, Circuits, and Systems
Nikos Konofaos

Nanomaterials: A Guide to Fabrication and Applications
Gordon Harling, Krzysztof Iniewski, and Sivashankar Krishnamoorthy

Optical Fiber Sensors and Applications
Ginu Rajan and Krzysztof Iniewski

Organic Solar Cells: Materials, Devices, Interfaces, and Modeling
Qiquan Qiao and Krzysztof Iniewski

Power Management Integrated Circuits and Technologies
Mona M. Hella and Patrick Mercier

Radio Frequency Integrated Circuit Design
Sebastian Magierowski

Semiconductor Device Technology: Silicon and Materials
Tomasz Brozek and Krzysztof Iniewski

Soft Errors: From Particles to Circuits
Jean-Luc Autran and Daniela Munteanu

FORTHCOMING TITLES:

VLSI: Circuits for Emerging Applications
Tomasz Wojcicki and Krzysztof Iniewski

Wireless Transceiver Circuits: System Perspectives and Design Aspects
Woogeun Rhee and Krzysztof Iniewski

Integrated Power Devices and TCAD Simulation

Yue Fu • Zhanming Li
Wai Tung Ng • Johnny K.O. Sin

CRC Press
Taylor & Francis Group
Boca Raton London New York

CRC Press is an imprint of the
Taylor & Francis Group, an **informa** business

CRC Press
Taylor & Francis Group
6000 Broken Sound Parkway NW, Suite 300
Boca Raton, FL 33487-2742

First issued in paperback 2017

Version Date: 20140225

ISBN 13: 978-1-4665-8381-8 (hbk)
ISBN 13: 978-1-138-07185-8 (pbk)

Library of Congress Cataloging-in-Publication Data

Fu, Yue (Electrical engineer)
  Integrated power devices and TCAD simulation / authors, Yue Fu, Zhanming Li, Wai Tung Ng, Johnny K.O. Sin.
      pages cm. -- (Devices, circuits, and systems)
  Includes bibliographical references and index.
  ISBN 978-1-4665-8381-8 (hardback)
  1. Power semiconductors--Computer-aided design. 2. Three-dimensional imaging. I. Li, Zhanming. II. Ng, Wai Tung. III. Sin, Johnny K. O. IV. Title.

TK7871.85.F794 2014
621.3815--dc23   2013047329

Visit the Taylor & Francis Web site at
http://www.taylorandfrancis.com

and the CRC Press Web site at
http://www.crcpress.com

We would like to thank our family members

Holly Li, Angelina Fu, Vivian Zhou, Wailyn Chung, Casey Ng, Erin Ng, Susan Ho, Andrea Sin, and Benjamin Sin

for their understanding and support.

Contents

Preface

Ever since the invention of the transistor in the late 1940s, the development of transistors followed in two major directions: device miniaturization and performance improvement. One of the key parameters for performance improvement is obviously the power rating of the transistor, with development that resulted in the area of power semiconductors. Because all electrical devices require a supply of power and power management electronics for proper operation, power semiconductors are an important area of transistor development over the past decades.

In recent years, device miniaturization has allowed the minimum feature size to approach the nanoscale, and present ultra-large scale integration (ULSI) technology is capable of putting billions of transistors onto a single chip. This causes serious problems in powering up the chips. Furthermore, higher power efficiency required due to environmental issues also puts a heavy burden on the power management and power electronics part of the system. These and other related issues spur continued research into the area of power semiconductor devices and technology.

In the area of power semiconductor development, the focus was first on discrete power devices for high-power ratings. Typical structures were power bipolar transistors and thyristors. Due to the slow switching speed and high switching loss of these devices, a fast switching device such as the vertical double-diffused MOS (VDMOS) transistor was invented, and finally for overall lower power loss, the insulated-gate bipolar transistor (IGBT) was created. As integrated circuit (IC) technology became more popular, there was a push to integrate power transistors with control IC for low-cost, compact, and high-performance applications. To accomplish this, lateral double-diffused MOS (LDMOS) transistors and lateral insulated-gate bipolar transistors (LIGBTs) were developed. That was the golden age for the development of the power IC (PIC) technology, and various kinds of bipolar-CMOS-DMOS (BCD) technologies were developed.

With today's well-developed ULSI and PIC technologies, it is envisaged that development of a power system-on-a-chip (PowerSOC) for future consumer and industrial applications will be a very promising direction. Of course, to make that a reality, various other technologies for achieving high-performance monolithic passive components such as efficient passive and IC integration and effective power dissipation will also be needed.

Both high-performance lateral power transistors and process technologies are necessary to develop PIC technologies. For efficient design of semiconductor devices and process technologies, technology computer-aided design (TCAD) tools are commonly used in the industry. Some books have already been published on power device design and technology development, but none particularly focus on how to use TCAD tools to design and develop power devices and PICs. It is the purpose of this book to cover this need and especially to give engineers who are new to the field of power semiconductors a quick start on device and technology design and development using TCAD tools.

The book adopts a top-down approach to introduce new engineers to the field. It begins with basic power electronics systems and an introduction to power ICs and guides the reader to explore the semiconductor industry before getting into smart power IC technology. It then goes on to explain the basics of TCAD modeling for process and device simulation. Model calibration for specific fabrication facilities for accurate and reliable simulation results are also discussed. Details on how to use TCAD tools for power IC process development and power device designs are then introduced. This includes many simulation examples on TCAD methodology and procedures for industrial design of practical power devices and process technologies. More than 300 diagrams effectively illustrate the key concepts and techniques for power device and technology design. Finally, a brief introduction to GaN power devices is presented with TCAD simulations to give readers, especially those with a silicon-only background, a head start in this field.

In the course of writing this book, the authors have received substantial help and support from many individuals. We would like to acknowledge and extend our heartfelt gratitude to each of them for all their generous help and support. In particular, we would like to thank Michel Lestrade from Crosslight Software, who has made significant contributions to the review and proofreading; Professor Maggie Xia and Dr. Yuanwei Dong from the University of British Columbia, for their reviews of the process simulations in Chapter 7 and other chapters; Professor Gang Xie from Zhejiang University, for his contributions to the GaN device simulations and review in Chapter 10; Robert Taylor, chief executive officer of Mega Hertz Power Systems Ltd., and Dr. Roumen Petkov from Greecon Technologies Ltd., for their reviews and suggestions of Chapter 1; Dr. Gary Dolny from Fairchild Semiconductor (USA) and Professor John Shen of the Illinois Institute of Technology for their useful initial reviews and suggestions.

Finally, we would like to give our special thanks to Nora Konopka, Michele Smith, Kathryn Everett, Iris Fahrer, and Theresa Delforn from Taylor & Francis for their professional assistance and kind help.

About the Authors

Yue Fu, PhD earned his PhD from the University of Central Florida and BS from Zhejiang University. He is currently the vice president of Crosslight Software, Inc., Canada. Dr. Fu is a senior member of the IEEE with more than 10 years of industry and academic experience in power semiconductor devices and power electronics. He holds multiple US patents and has authored or coauthored many peer-reviewed papers.

Zhanming (Simon) Li, PhD, earned his PhD from the University of British Columbia in 1988. He was with the National Research Council of Canada (NRCC) from 1988 to 1995, where he developed semiconductor device simulation software. He founded Crosslight Software, Inc. in 1995 with simulation technology transferred from the NRCC. Since then, Dr. Li has been the chief designer of many semiconductor process and device simulation software packages. He has been actively involved in research of TCAD simulation technology and has authored or coauthored more than 70 research papers.

Wai Tung Ng, PhD, earned his BASc, MASc, and PhD in electrical engineering from the University of Toronto in 1983, 1985, and 1990, respectively. He was a member of the technical staff at Texas Instruments in Dallas from 1990 to 1991. His started his academic career at the University of Hong Kong in 1992. He returned to the University of Toronto as a faculty member in 1993 and is currently a full professor. His research is focused in the areas of power management integrated circuits, integrated DC–DC converters, smart power integrated circuits, power semiconductor devices, and fabrication processes.

Johnny Kin-On Sin, PhD, earned his PhD in electrical engineering from the University of Toronto in 1988. He was a senior member of the research staff of Philips Laboratories, New York, from 1988 to 1991. He joined the ECE Department, the Hong Kong University of Science and Technology in 1991 and is currently a professor there. He holds 13 patents and has authored more than 270 technical papers. His research interests include novel power semiconductor devices and power system-on-chip technologies. Dr. Sin is a fellow of the IEEE for contributions to the design and commercialization of power semiconductor devices.

1 Power Electronics, the Enabling Green Technology

This book adopts a top-down approach so that readers can start with a glimpse of system-level applications and then proceed to the integrated circuit (IC) chips level and finally down to the semiconductor devices (elements) level. This chapter is a brief introduction of power electronics and power management systems for power device engineers. Power electronics engineers who are specifically focused on power device physics can safely skip this chapter.

1.1 INTRODUCTION TO POWER ELECTRONICS

Since the discovery of electricity, electronic appliances are found almost everywhere (now even on Mars). Power delivery is one of the most important and often neglected requirements within electronic equipment. Direct utilization of the alternating current (AC) line voltage or available battery power without any power conversion is rare. Most of the time electric energy required for electronic systems is provided by either internal or external power supplies.

Enviromental protection advocates are now promoting green awareness; thus, power efficient technologies are under consideration as an important design criteria for future applications. Various forms of clean energy such as solar and wind energy need to be converted and stored for use. Efficient energy conversion is the key to cost reduction and better utilization of these natural resources.

Power electronics that use solid-state devices to control and convert electric power are considered as the enabling technology for a greener future. The United States Department of Energy has estimated that approximately 40% of all the energy consumed is first converted into electricity. In the transporation sector the growing use of electric and plug-in hybrid cars and high-speed rail transportation may increase this to even 60% [1].

Figure 1.1 shows how power electronics systems are applied to an electric vehicle. In terms of voltage conversion, power electronic converters can be divided into four types: alternating current/direct current (AC/DC) rectifier; AC/AC converter; DC/DC converter; and DC/AC inverter. Figure 1.2 illustrates these four types of converters. A comprehensive treatment of power electronics is out of the scope of this book, so we focus on DC/DC converters.

A complete power electronics system contains three parts (Figure 1.3). The power converter topology governs how the power elements should be connected and controlled. The topology can be regarded as the "backbone" of the power electronic system. Controllers, which automatically monitor and control the power switches according to input and output conditions, are the "brains" of the system. Power elements such as power devices, transformers, inductors, and capacitors are the basic building blocks for a power supply. They can be viewed as the "muscles" of the power electronic system.

Unfortunately, the best power devices will not necessarily dominate the global marketplace. Successful implementation and adoption of power devices and their controllers within power electronic designs can be subjected to many factors such as economics, market conditions, intellectual property rights, and industry supply relationships. Additional technical factors such as PCB layout, magnetic design, and optimal filtering components all play a vital role.

Figure 1.4 is a black-box illustration of the power electronics systems. Power input is converted to power output by the power converter main circuitry. The controller senses signals from both the input/primary side and output/secondary side and compares it with the reference signal. A control

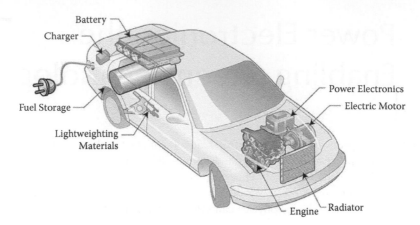

FIGURE 1.1 Electrical power system found in a plug-in hybrid electric vehicle. (Photo courtesy of Argonne National Laboratory [2].).

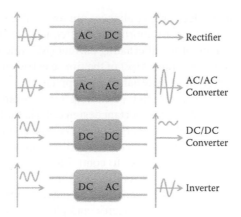

FIGURE 1.2 The four different types of power conversions.

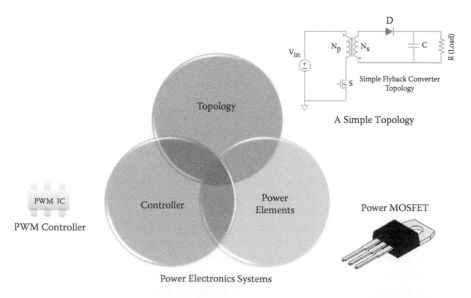

FIGURE 1.3 Power electronic systems encompassing topology, controller, and power elements such as power devices, transformers, inductors, and capacitors.

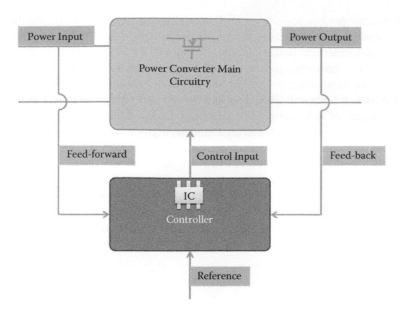

FIGURE 1.4 A black-box illustration of a power electronics system.

signal is fed into the main power transistors. This structure is for the analog (A) control method. A digital (D) control method needs A/D and D/A converters with a microprocessor or DSP.

1.2 HISTORY OF POWER ELECTRONICS

The power electronics industry continues to evolve [3]. A selected history of power electronics innovations since the invention of the transistor is listed in Table 1.1. It is safe to say that every milestone of power electronics development comes from the innovations and improvements of

TABLE 1.1

Selected Historical Timeline of Power Electronics [3] [5] [6]

Timeline	Events
1947	First transistor invented at Bell Labs
1957	GE introduces silicon-controlled rectifier (SCR)
1959	Transistor oscillation and rectifying converter power supply system U.S. Patent 3,040,271 is filed
1960	MOSFET first introduced by Atalla and Khang of Bell Labs
1972	Hewlett-Packard's first pocket calculator is introduced with transistor switching power supply
1976	Power MOSFET becomes commercially available
1976	Switched mode power supply U.S. Patent 4,097,773 is filed
1977	Apple II is designed with a switching mode power supply
1980	High-power GTOs are introduced in Japan
1980	The HP8662A synthesized signal generator went with a switched power supply
1981	The present renaissance in soft switching for power converters appears in a paper [7]
1982	Insulated gate bipolar transistor (IGBT) is introduced
1983	Space vector PWM is introduced

power semiconductor devices, topologies, and controlling methodologies. Among them, one of the most important contributions has resulted from the advancement in power semiconductor devices, which are the basic elements of power electronics. From BJT to DMOS, from SCR to IGBT, power semiconductor devices have changed the face of the power electronics industry.

Today novel compound materials such as SiC and GaN, are attracting more and more attention in power electronics. Silicon will likely continue to dominate the lower power region in the foreseeable future because of its low cost and mature technology. SiC and GaN devices may gradually encroach and eventually replace silicon in high-voltage, high-power applications. Other innovations like soft switching (e.g., zero voltage, zero current switching), which boomed in the 1990s, provide much higher switching frequency and efficiency. Digital control with microprocessors, DSPs, or programmable logic devices have been pervasive in applications like motor drives and three-phase power converters for utility interfaces [4]. With the advances in power semiconductor and digital VLSI technologies, digital control will benefit from a much wider power electronics applications in the future. Further development in power electronics on system miniaturization will be possible with the additional advancement in circuit topologies, control technologies, and integrated passive devices.

1.3 DC/DC CONVERTERS

As mentioned before, there are four main types of power converters: AC/DC, DC/DC, DC/AC, and AC/AC. This book focuses on DC/DC converters since they are found within substages of all switching type power converters and are abundantly applied in mobile power mangement systems, which is the main application area of integrated power devices. There are three basic types of DC/DC converters: linear regulators, charge pumps, and switching mode regulators.

1.4 LINEAR VOLTAGE REGULATORS

Linear voltage regulators were the mainstay of power conversion until the late 1970s when the first commercial switch mode became available [8]. The linear voltage regulator can be considered as a variable resistor connected in series with the load as depicted in Figure 1.5.

A control circuit monitors the voltage at the load (V_{out}) and continuously adjusts the value of R_{reg} such that V_{out} is constant, regardless of changes in V_{in} and R_{load}. In a practical implementation, the

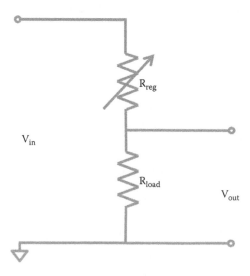

FIGURE 1.5   A typical linear voltage regulator invalid source specified.

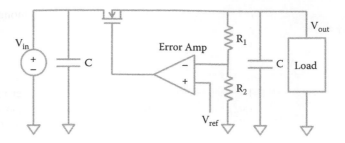

FIGURE 1.6 Simple schematic of a linear regulator.

series element is usually a power transistor (BJT or MOSFET). A typical linear regulator with feedback control is shown in Figure 1.6 [9]. A feedback loop is provided by sensing the output voltage and comparing it with the reference voltage. The gate control signal for the series element is then generated by the error amplifier. In this example, an NMOS is used as the series element to reduce the power loss and achieve small drop-out voltage (the voltage difference between V_{in} and V_{out}). The ideal power conversion efficiency (neglecting the power consumed by the feedback circuit) can be calculated as

$$P_{total} = (V_{in} - V_{out}) \cdot I_{out} + V_{out} \cdot I_{out} \tag{1.1}$$

In this case, the input current is assumed to be the same as the output current. The first term is the power dissipated in the series element (power MOSFET), and the second term is the power delivered to the load. Depending on the difference between input and output voltages, the power conversion efficiency can be quite poor. For example, a 10 V to 1 V conversion will lead to an efficiency of no more than 10%!

In most portable battery-powered applications, the battery voltage will change drastically from fully charged to near depletion (e.g., a lithium-ion battery can change from 4.7 V to 2.7 V). If the output voltage is required to be near 2.7 V, it will be difficult for the linear regulator to operate as a minimum voltage between the series element is required to maintain proper operation.

To address this issue, a special type of linear voltage regulator called low drop-out (LDO) is available. LDOs are widely accepted in many applications where the output voltage can be very close to the input voltage. The difference bewteen a LDO and a standard linear voltage regulator is the pass element and amount of drop-out voltage [9]. The standard linear voltage regulator, such as the popular LM340/LM78xx series [10], uses Darlington NPN or PNP as the pass elements. These regulators require a minimum drop-out voltage of as high as 2 V with a load current of 1 Amp. They are acceptable with applications where a 5 V input needs to drop to a 3 V output. LDOs, on the other hand, can handle much smaller drop-out. The pass elements in a LDO are usually power NMOS and PMOS; they can have drop-out voltages of less than 100 mV [9]. For example, LT3026 from Linear Technology has the minimum dropout voltage of 100 mV at $I_{OUT} = 1.5$ A [11].

1.5 SWITCHED CAPACITOR DC/DC CONVERTERS (CHARGE PUMPS)

Switched capacitor circuit or charge pump is a simple DC/DC converter that uses capacitors as energy storage elements to create a step-up or step-down voltage output. Compared with LDOs, charge pumps usually achieve a higher efficiency of above 90%. Modern charge pumps can double, triple, or halve and invert voltages. Charge pumps fill a niche in the performance spectrum between LDOs and switching regulators and offer a nice alternative to designs that may be inductor-averse. Compared to LDOs, charge pumps require an additional capacitor (a flying capacitor) to operate but do not require inductors. They are generally more costly, have higher output noise levels, and

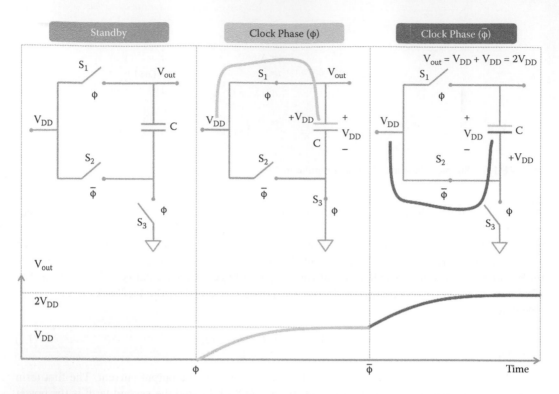

FIGURE 1.7   The operation and switching of waveforms for a simple voltage doubler [12].

lower output current capability [12]. However, a high current version of switched capacitor DC/DC converters does exist [13]. A simple voltage doubler charge pump, consisting of one capacitor and three switches, is shown in Figure 1.7. During clock phase ϕ, switches S_1 and S_3 are closed, and the capacitor is charged to the supply voltage V_{DD} (charge current flow is shown in the middle figure). During the next clock phase $\bar{\phi}$, switch S_2 is closed, and the bottom plate of the capacitor assumes a potential of V_{DD} while the capacitor maintains its charge $Q = V_{DD} \cdot C$ from the previous phase. This means that during clock phase $\bar{\phi}$, we have [12]:

$$Q = (V_{out} - V_{DD}) \cdot C = V_{DD} \cdot C \tag{1.2}$$

or

$$V_{out} = 2 \cdot V_{DD} \tag{1.3}$$

The output voltage level is qualitatively illustrated in Figure 1.7.

Charge pumps are widely used in EEPROM, flash memory, and solid-state drives (SSD) that require a high-voltage pulse to program and erase the stored data. They are also popular for LED drivers, where a higher bias voltage than the battery voltage is needed [14].

In this type of switching circuit, the power transistors operate as simple "on" and "off" switches. Ideally, they do not dissipate power in either state. In reality, leakage current and finite on-resistance will lead to certain power losses. Nonetheless, this type of circuit has the potential to provide very close to 100% power efficiency because no resistive component is used. However, switched capacitors converters have several drawbacks. It is difficult to provide arbitrary output voltage without having a large number of ratioed capacitors and switches. The amount of output current and voltage

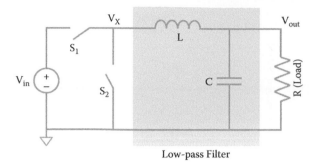

FIGURE 1.8 An ideal buck converter with ideal on–off switches and a low pass filter.

handling capability is dependent on the size of the capacitors and the fabrication technology. The charging current path and output current path usually involve multiple power transistors in series, resulting in large conduction and gate drive losses. Although experimental prototypes have demonstrated output current capacity in the ampere range with high switching frequency and multiphase operation, the size of the silicon die remains large and costly [13].

A more practical implementation of DC/DC power conversion is the use of switched inductor converters. Their operation and power semiconductor device requirements will be discussed in the next section.

1.6 SWITCHED MODE DC/DC CONVERTERS

As mentioned before, the biggest problem with linear regulators or LDOs is low efficiency. This is especially true when the input/output voltage difference is large and high current/high power are needed. Charge pumps have better efficiency and are capable of step up or down. However, the output current from the integrated charge pumps is typically very low.

Switched mode converter is an alternate method to regulate power. The input DC voltage is chopped by a switching network followed by a low pass filter as shown in Figure 1.8.

The switching network toggles the potential at the switching node, V_x alternately between V_{in} and ground. An LC low pass filter is used to smooth out the chopped voltage waveform V_x such that a constant voltage that is equivalent to the average value V_x is available at V_{out}. Figure 1.9 illustrates

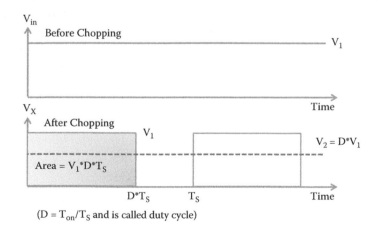

FIGURE 1.9 Idealized PWM switching waveform for a DC/DC buck converter.

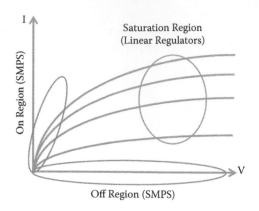

FIGURE 1.10 Power MOSFETs operation regions used in linear regulators and switched mode power supplies.

the voltage waveforms at the input and at the switching node. The fraction of time per period that the switching node is connected to the input is called the duty cycle. The average output voltage is equal to the area under the curve for V_x spreaded over one switching period, T_s, and is expressed as

$$V_{out} = V_{in} \cdot \frac{D \cdot T_s}{T_s} \qquad (1.4)$$

$$V_{out} = D \cdot V_{in} \qquad (1.5)$$

This is an important relationship as the output voltage can be adjusted by simply changing the fraction of time (D). This switching scheme is called pulse width modulation (PWM).

The switches used in this type of power converter basically operate either in fully on or fully off states. In either case, the power dissipation across the switches is zero because during the off-state the switch conducts no current and during the on-state the switch conducts current but with no voltage drop due to zero on-resistance. In practice, the switches are implemented using power MOSFETs. The operating modes that correspond to fully on and fully off states are the triode region and the cut-off state as shown in Figure 1.10. If power MOSFETs with fast switching speed and low on-resistance (1/ slope of the I-V curve in the triode region) is available, very low power loss can be achieved, leading to near 100% power conversion efficiency. By comparion, the power MOSFETs in linear voltage regulators operate in the saturation region and in the triode region (for LDOs). In this case, the power MOSFET conducts current (I_{DS}) while supporting a finite source to drain voltage (V_{DS}). This leads to an appreciable power dissipation, which is the main cause for the low power converson efficiency in linear voltage regulators.

The characteristics of power MOSFETs and how they affect the performance of switched mode DC/DC converters will be described in detail later.

1.7 COMPARISON BETWEEN LINEAR REGULATORS, CHARGE PUMPS, AND SWITCHED REGULATORS

A comparison is made to give the reader a better view of selection between linear regulators, charge pumps, and switched regulators [12] as listed in Table 1.2.

TABLE 1.2

Comparison of Different Technologies for DC/DC Converters

Features	Linear Regulators (LDOs)	Charge Pumps	Switched Regulators
Design complexity	Low	Moderate	Moderate to high
Cost	Low	Moderate	Moderate to high
Noise	Very Low	Low	Moderate
Efficiency	Low	Moderate	High
Thermal management	Poor	Good	Very good
Output current	Moderate	Low	High
Magnetics required	No	No	Yes
System on chip	Very likely	Likely	Not available now

1.8 TOPOLOGIES FOR NONISOLATED DC/DC SWITCHED CONVERTERS

For power electronics system designers, topology is one of the first things to consider. Depending on the applications, a power electronics system can be either nonisolated or isolated. Here isolation means whether a magnetic transformer is used to isolate the input and output. For the past decades, many topologies are proposed and utilized. Table 1.3 shows the typical topologies used for nonisolated switched mode converters.

1.8.1 BUCK CONVERTER

A simple buck converter schematic is shown together with the basic operation principle in Figure 1.11. The buck converter consists of a power transistor, a diode, an inductor, and a capacitor. Let us

TABLE 1.3

Topologies Overview for Nonisolated Switched Mode Converters

Nonisolated Topologies	Features
Buck converter	$V_{out} < V_{in}$ Same voltage polarity for input and output Output current ripple is low due to series connected inductor at the output
Boost converter	$V_{out} > V_{in}$ Same voltage polarity for input and output Input current ripple is low due to series connected inductor at the input
Buck–boost converter	$V_{out} < V_{in}$ or $V_{out} > V_{in}$ Inverting polarity for input and output Input and output current ripples are high
Ćuk converter	$V_{out} < V_{in}$ or $V_{out} > V_{in}$ Inverting polarity for input and output Input and output current ripples are low due to series connected inductors
SEPIC	$V_{out} < V_{in}$ or $V_{out} > V_{in}$ Same voltage polarity for input and output Input current ripple is low due to series connected inductor at the input

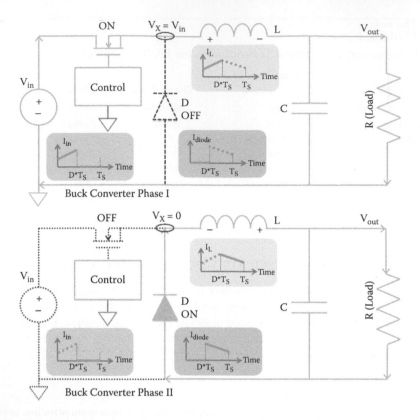

FIGURE 1.11 Simplified operation and schematic of a nonsynchronous buck DC/DC converter.

assume that the converter is working in the steady state and all the power elements are ideal (e.g., no voltage drop when the transistor and the diode are on; no leakage current when they are off). There are two phases of operation during one switching cycle, T_s. The power transistor is fully on during phase I and fully off during phase II. The time lapse during phase I is $D \cdot T_s$, where D is the duty cycle, and is the ratio between the transistor on-time to the total switching period: $D = T_{on}/T_s$. The time lapse during phase II is therefore $(1 - D) \cdot T_s$.

During phase I, the power transistor is fully on, and current flows from the input power source to the output via the power transistor and the inductor, as shown in Figure 1.11. The voltage at diode's cathode V_x equals to V_{in}. Since the diode's anode is grounded, the diode is reverse biased, no current flows through the diode. For an ideal inductor, the current I_L cannot change instantaneously. Instead, it increases linearly during phase I. The voltage polarity on the two terminals of the inductor is as shown in Figure 1.11. Since the output voltage V_{out} has small ripples, the average value is used in the steady state. The change in the inductor current ΔI_{L_on} during phase I is calculated as

$$V_L = V_x - \bar{V}_{out} = L \cdot \frac{dI_L}{dt} = L \cdot \frac{\Delta I_{L\_on}}{T_{on}} \tag{1.6}$$

$$\Delta I_{L\_on} = \frac{1}{L} \cdot \left(V_x - \bar{V}_{out}\right) \cdot T_{on} = \frac{1}{L} \cdot \left(V_{in} - \bar{V}_{out}\right) \cdot T_{on} = \frac{1}{L} \cdot \left(V_{in} - \bar{V}_{out}\right) \cdot D \cdot T_s \tag{1.7}$$

The subscript "on" indicates the period when the power transistor is on and the subscript "off" means that the power transistor is off. In phase II, the power transistor is turned off. Since the inductor current I_L and its direction cannot change immediately, it has to force its way through the diode. The polarity of the inductor voltage is reversed, and $V_x = 0$. The change in the inductor current, ΔI_{L_off} during phase II can be calculated as

$$\Delta I_{L\_off} = \frac{1}{L} \cdot (V_x - \bar{V}_{out}) \cdot T_{off} = \frac{1}{L} \cdot (-\bar{V}_{out}) \cdot T_{off} = -\frac{1}{L} \cdot \bar{V}_{out} \cdot (1 - D) \cdot T_s \tag{1.8}$$

In steady state, the average current should stay the same during one switching cycle. This means that the total current changes should sum up to 0:

$$\Delta I_{L\_on} + \Delta I_{L\_off} = 0 \tag{1.9}$$

$$\frac{1}{L} \cdot (V_{in} - \bar{V}_{out}) \cdot D \cdot T_s - \frac{1}{L} \cdot \bar{V}_{out} \cdot (1 - D) \cdot T_s = 0 \tag{1.10}$$

The average output voltage, $\bar{V}_{out}$ can be derived as

$$\bar{V}_{out} = D \cdot V_{in} \tag{1.11}$$

This result indicates that the output voltage of the buck converter equals to the input voltage multiplied by the duty cycle. In fact, this conclusion can be understood in another way. The ideal inductor is an energy storage element, which only stores and transfers energy. The energy consumption on the ideal inductor should be 0. In the steady state, current flows through the inductor either continuously or discontinuously. The average voltage over an ideal inductor should be zero. This means the average voltage on the cathode of the diode $\bar{V}_x$ has to be equal to the average $\bar{V}_{out}$. Since we know the average value of $\bar{V}_x$ is just a "chopped" version of V_{in}, the input and output voltage relationship can also be derived instantly:

$$\bar{V}_x = \bar{V}_{out} = D \cdot V_{in} \tag{1.12}$$

1.8.2 BOOST CONVERTER

The boost converter provides an output that is a step up from the input voltage. Similar to the buck converter, the boost converter consists of a power transistor, an inductor, a diode, and a capacitor. In the analysis, we assume ideal conditions for all the components. There are two phases of operation. The power transistor is fully on during phase I and is fully off during phase II. The time lapse during phase I is $D \cdot T_s$, where D is the duty cycle, while the time lapse during phase II is $(1 - D) \cdot T_s$.

During phase I, the power transistor is fully on. Current flows from the input power source to the inductor and power transistor as shown in Figure 1.12. The voltage at the diode's anode V_x equals to 0. Since the diode's cathode is connected to the output, the diode is reverse biased with no current flow. For an ideal inductor, the current I_L cannot change instantaneously. Instead, it increases linearly during phase I. The voltage polarity on the two terminals of the inductor is as shown in Figure 1.12. Since the output voltage has small ripples, the average value is used in the steady state. The change in the inductor current ΔI_{L_on} during phase I can be calculated as

$$V_L = V_{in} - V_x = L \cdot \frac{dI_L}{dt} = L \cdot \frac{\Delta I_{L\_on}}{T_{on}} \tag{1.13}$$

$$\Delta I_{L\_on} = \frac{1}{L} \cdot (V_{in} - V_x) \cdot T_{on} = \frac{1}{L} \cdot V_{in} \cdot T_{on} = \frac{1}{L} \cdot V_{in} \cdot D \cdot T_s \tag{1.14}$$

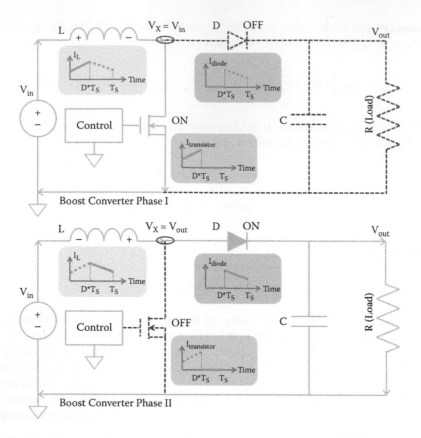

FIGURE 1.12 Simplified operation and schematic of a nonsynchronous boost DC/DC converter.

During phase II, the power transistor is turned off. Since the inductor current cannot change immediately, it forces its way through the diode. The polarity of the inductor voltage is reversed, and $V_x = V_{out}$. Therefore, the current through the inductor can be expressed as

$$\Delta I_{L\_off} = \frac{1}{L} \cdot (V_{in} - \overline{V}_{out}) \cdot T_{off} = \frac{1}{L} \cdot (V_{in} - \overline{V}_{out}) \cdot (1 - D) \cdot T_s \tag{1.15}$$

In the steady state, the average current should stay the same during each switching cycle. This means that the total inductor current change should equal to 0:

$$\Delta I_{L\_on} + \Delta I_{L\_off} = 0 \tag{1.16}$$

$$\frac{1}{L} \cdot V_{in} \cdot D \cdot T_s + \frac{1}{L} \cdot (V_{in} - \overline{V}_{out}) \cdot (1 - D) \cdot T_s = 0 \tag{1.17}$$

The average output voltage $\overline{V}_{out}$ can be expressed as

$$\overline{V}_{out} = \frac{1}{1 - D} \cdot V_{in} \tag{1.18}$$

Similar to the buck converter, this conclusion can be understood in another way. The average voltage over an ideal inductor should be zero at steady state during one switching cycle. This means that the average $\overline{V}_x$ has to be equal to the input voltage V_{in}. If we consider that the average value of

$\bar{V}_x$ is a "chopped" version of $\bar{V}_{out}$, with an "on" period of $(1-D)\cdot T_s$ in phase II, then the input and output voltage relationship can be derived as

$$\bar{V}_x = V_{in} = (1-D)\cdot\bar{V}_{out} \tag{1.19}$$

$$\bar{V}_{out} = \frac{1}{1-D}\cdot V_{in} \tag{1.20}$$

1.8.3 BUCK–BOOST CONVERTER

A buck–boost converter is convenient as the output voltage can be either lower or higher than the input voltage. Similar to the buck and the boost converters, a buck–boost converter has a power transistor, an inductor, a diode, and a capacitor, but with a different circuit topology. Again, the buck–boost converter has two phases of operation. The power transistor is fully on during phase I and fully off during phase II. The diode is conducting during phase II to provide a continuous path for inductor current. The basic operation is similar to the buck and boost converters, but with a very different input and output voltage relationship—again, assuming that all the power elements are ideal and the converter is working in continuous conduction mode (CCM). The schematic and the phases of operation for a buck–boost converter are as shown in Figure 1.13.

During phase I, the power transistor is turned on. Current flows through the transistor and inductor. This is very similar to the boost converter except that the position of the power transistor and inductor is swapped. The voltage at the cathode of the diode should be equal to V_{in}. Note that in the steady state a buck–boost converter has an inverted output voltage, meaning that the anode of

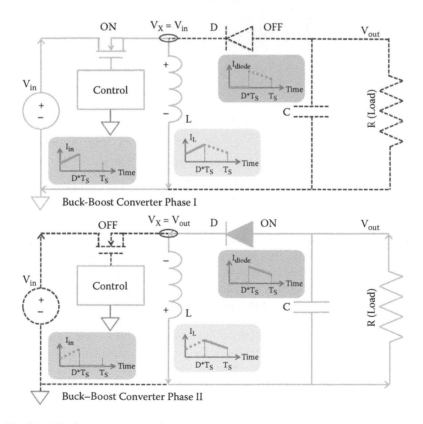

FIGURE 1.13 Simplified operation and schematic of a nonsynchronous buck–boost DC/DC converter.

the diode is negative and the diode is reverse biased. The inductor current follows the following relationships:

$$V_L = V_x = V_{in} = L \cdot \frac{dI_L}{dt} = L \cdot \frac{\Delta I_{L\_on}}{T_{on}} \tag{1.21}$$

$$\Delta I_{L\_on} = \frac{1}{L} \cdot V_{in} \cdot T_{on} = \frac{1}{L} \cdot V_{in} \cdot D \cdot T_s \tag{1.22}$$

During phase II, the power transistor is turned off by a gate signal; due to the need for continuous flow of the inductor current, the diode is subsequently turned on. The capacitor is charged with positive voltage on the lower plate and negative on the upper plate. In the buck–boost converter, the current flow during phase II is counterclockwise (vs. clockwise for the buck converter and the boost converter). The lower plate of the capacitor becomes positively charged, which brings a negative polarity on the output voltage. The voltage at the cathode of the diode is equal to the output voltage ($V_x = \bar{V}_{out}$) in phase II. During steady state, the change in the inductor current can be expressed as

$$\Delta I_{L\_off} = \frac{1}{L} \cdot V_x \cdot T_{off} = \frac{1}{L} \cdot \bar{V}_{out} \cdot (1 - D) \cdot T_s \tag{1.23}$$

The average current should stay the same during each switching cycle, meaning that the total current change should sum up to 0:

$$\Delta I_{L\_on} + \Delta I_{L\_off} = 0 \tag{1.24}$$

$$\frac{1}{L} \cdot V_{in} \cdot D \cdot T_s + \frac{1}{L} \cdot \bar{V}_{out} \cdot (1 - D) \cdot T_s = 0 \tag{1.25}$$

The average output voltage is found to be

$$\bar{V}_{out} = -\frac{D}{1 - D} \cdot V_{in} \tag{1.26}$$

The negative sign indicates that the polarity of the output is opposite to the input voltage. This may cause certain inconvenience in many applications.

1.8.4 Ćuk Converter

The Ćuk converter is named after its inventor, Dr. Slobodan Ćuk, a professor from the California Institute of Technology [15]. Unlike the previous topologies, the Ćuk converter has one power transistor, two inductors, two capacitors, and a diode. Similar to the buck–boost converter, the Ćuk converter can achieve an output voltage that is higher or lower than the input with reversed polarity. The schematic of the Ćuk converter is shown in Figure 1.14. One of the advantages of the Ćuk converter is its ability to reduce the current ripple at both input and output terminals.

The derivation of the input–output relationship is omitted in this book. Interested readers can refer to a specialized power electronics textbook for a detailed explanation. Like the buck–boost converter, the Ćuk converter has the same output and input voltage relationship:

$$\bar{V}_{out} = -\frac{D}{1 - D} \cdot V_{in} \tag{1.27}$$

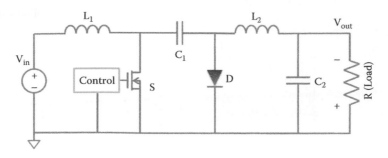

FIGURE 1.14   Schematic of a Ćuk DC/DC converter.

1.8.5   ADDITIONAL TOPICS IN NONISOLATED CONVERTERS

1.8.5.1   Nonideal Power Elements

In the previous sections, we treat all the power elements in the power converters as ideal components. In practice, power transistors have finite on-voltage drop and behave like a small resistor in the on-state. The power diodes have a voltage drop between 0.7 and 1.1 V when fully on. The inductors and capacitors have equivalent series resistance (ESR) that can also dissipate power.

1.8.5.2   Synchronous Rectification

In the buck converter example, suppose that the duty cycle D is small; then the transistor would be on for a very short period of time while the diode would be on most of the time. We know that diodes have an inherent voltage drop (even Schottky diodes could have a relatively large voltage drop). The power dissipated by the diode can be calculated as

$$P_{diode} = I_{diode} \cdot V_{diode} \cdot (1-D) \cdot T_s \tag{1.28}$$

where V_{diode} is the voltage drop between the anode and the cathode terminals. When D is small, the diode power dissipation dramatically reduces the total power conversion efficiency. This is especially true when the output voltage is very low (< 1.5 V) and output current is very high (e.g., voltage regulator module for modern CPUs). To mitigate this problem, a power transistor, usually called a "Sync-FET" is used to replace the diode, as shown in Figure 1.15. Power transistors such as power MOSFETs have much lower on-resistance when fully on. Therefore, the voltage drop from the power transistor is much smaller compared with that of the diode. The "sync" transistor is working in the third quadrant, meaning current is flowing from source to drain for N-type MOSFETs (N-type MOSFETs are preferred over P-type MOSFETs for lower on-resistance with the same device sizes). Both power transistors can be controlled by a single controller with almost complementary gate signals. A dead time is always necessary between the turning off of one transistor and the turning on of

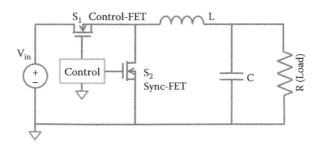

FIGURE 1.15   Buck converter with synchronous rectification.

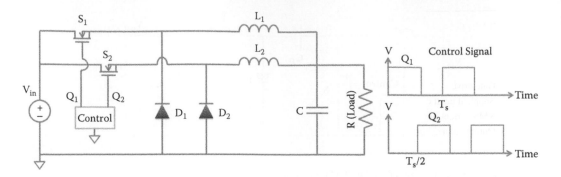

FIGURE 1.16   Two-stage interleaving buck converter topology.

the other. If both transistors are on, the input will be short-circuited, causing a significant unwanted shoot-through current between V_{in} and ground.

1.8.5.3   Continuous Conduction Mode (CCM) and Discontinuous Conduction Mode (DCM)

CCM or DCM refers to whether the inductor current is interrupted (goes to zero) or not. Until now, we considered only CCM. DCM happens when the converter is connected to a light load with small output current. If the current through the inductor cannot reverse when the average inductor current drops below a critical value, the inductor current will be interrupted until the power transistor (or the control-FET in synchronous rectification) turns on again.

When the inductor current goes to zero, there is no power drawn from the input power source and no energy in the inductor to transfer to the output [1]. This results in increased device stresses and the ratings of the passive components. DCM also results in higher noise and electromagnetic interference (EMI), although the diode reverse recovery problem is minimized [1].

1.8.5.4   Interleaving

Interleaving is achieved by paralleling and phase-shifting two pairs of power transistors, diodes and inductors, to reduce the current ripple on both input and output current. Figure 1.16 illustrates a two-stage interleaving topology for a buck converter. The phase-shift of the two stages is $T_s/2$. Multiple-staged interleaving is possible with even better ripple reduction, with the phase-shift of T_s/n, where n is the number of stages.

1.8.5.5   Ripples

Ripples are always present in switched mode power converters. Different technologies yield different magnitudes of voltage and current ripples at the input and the output terminals. Generally speaking, inductors are capable of suppressing current ripples, while large capacitors are used to suppress voltage ripples. As an example, the boost converter topology is a popular choice for the front stage of a power factor correction (PFC) circuit due to the series connection of an inductor to the input, while a buck converter is excellent in reducing root mean square (RMS) current in the output capacitor, thanks to the series connected inductor at the output side.

1.8.5.6   Comparison of Nonisolated Topologies

Table 1.4 is a comparison of three nonisolated topologies [16] in terms of gate drive methods, continuity of the input current, RMS, and output voltage.

TABLE 1.4

Comparison of Nonisolated Topologies [16]

	Buck Converter	Boost Converter	Buck–Boost Converter
Gate drive	Floating	Grounded	Floating
Supplied input current	Discontinuous	Continuous	Discontinuous
RMS current in C_{out}	Low	High	High
Output voltage	$V_{in} \cdot D$	$\dfrac{V_{in}}{1-D}$	$-V_{in} \cdot \dfrac{D}{1-D}$

1.9 TOPOLOGIES FOR ISOLATED SWITCHING CONVERTERS

For switched mode power supplies, numerous applications require electrical isolation. One primary reason for this is safety concerns. The secondary side with low DC voltage should be isolated from the AC supply lines. Also, for very large or very small input–output voltage ratios, nonisolated topologies are not practical since the duty cycle will be either too large or too small.

In this book, five types of isolated switching converters will be discussed: flyback converter; forward converter; full-bridge converter; half-bridge converter; and push–pull converter.

1.9.1 FLYBACK CONVERTERS

Flyback converters are derived from buck–boost converters. They are common in low-power applications (< 100 W). As shown in Figure 1.17, the flyback converter has the same power elements as the buck–boost converter except that the inductor is replaced by a transformer. The transformer is designed in such way that the current entering the dotted terminals produces aiding fluxes (right-hand rule: the current flowing into the dotted terminal of either windings will produce core flux in the same direction). The transformer in the flyback converter can be viewed as two inductors sharing a common core. Note that there is an air gap in the transformer for energy storage purposes and that the dots on the primary and secondary are at different positions.

Similar to the buck–boost converter, the flyback converter can be analyzed using two phases (states). Figure 1.18 shows the operation schematic. During phase I, the power switch is fully turned on, and current flows through the primary side of the transformer and the transistor. The current

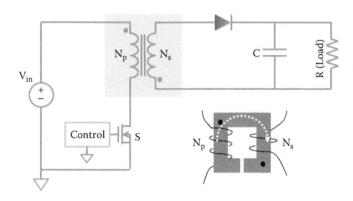

FIGURE 1.17 Flyback converter topology with detailed view of the transformer.

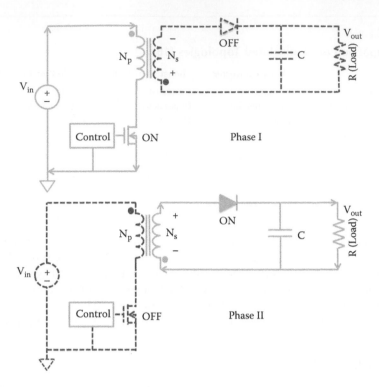

FIGURE 1.18 Operation schematic of a flyback converter.

in the primary winding and the magnetic flux increases and energy is stored in the air gap of the transformer. Due to the negatively induced voltage on the secondary winding, the diode is reverse biased. It remains off during phase I. The analytical equations are very similar to buck–boost converter and will not be repeated here.

During phase II, the power transistor is turned off, the current in the primary winding and the magnetic flux decreases. The polarity of the voltage on the secondary winding is reversed, forward biasing the diode. Energy is released from the transformer to the load.

The voltage relationship between the input and the output is similar to the buck–boost converter; the only difference is the added multiplication of a transformer's turns ratio and polarity:

$$\bar{V}_{out} = \frac{D}{1-D} \cdot \frac{N_s}{N_p} \cdot V_{in} \tag{1.29}$$

where N_s is the number of turns on the secondary side, and N_p is the number of turns on the primary side of the transformer. The transformer turns ratio N_s/N_p extends the design flexibility for the isolated buck–boost converters.

1.9.2 Forward Converters

The forward converters are derived from buck converters. They are widely used in low (< 10 W) to high-power applications of less than 1000 W. Compared with a flyback converter, the transformer in a forward converter doesn't store energy. Instead, energy is passed from the primary side to the secondary side simultaneously. The transformer design is a little bit different from flyback converters, with no air gap, and the dot position of the primary side and secondary side are at the same position. Figure 1.19 illustrates the schematic of a simplified forward converter.

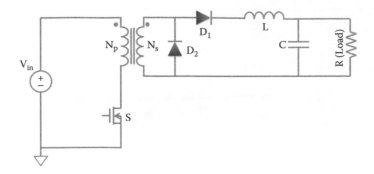

FIGURE 1.19 Schematic of a simplified forward converter.

Since the mode of operation is very similar to the buck converter, detailed analysis is omitted. The input–output voltage relationship is the same as the buck converter except that it is multiplied by the transformer's turns ratio:

$$\bar{V}_{out} = D \cdot \frac{N_s}{N_p} \cdot V_{in} \tag{1.30}$$

In practice, a third winding is always necessary with a third diode, as shown in Figure 1.20, because during the transistor's off state the energy stored in the transformer needs to be dissipated. With the third winding, a current will be forced through the diode, D_3, until the stored energy is fully dissipated before the transistor is turned on again for the next switching cycle.

1.9.3 FULL-BRIDGE CONVERTERS

Full-bridge converters are derived from buck converters, similar to the forward converters. They are mainly used in high-power applications (750 W and up) [17]. In our previously discussed converters, only one power transistor is used; full-bridge converters use four power transistors on the primary side. At lower power levels, the forward converter is preferred over the full-bridge converter because of its lower parts count. The transformer is designed in such a way that the secondary side has a winding with a centered tap. The schematic of a simplified full-bridge converter is shown in Figure 1.21.

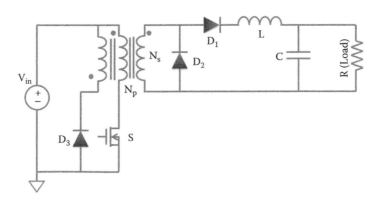

FIGURE 1.20 A more practical forward converter schematic.

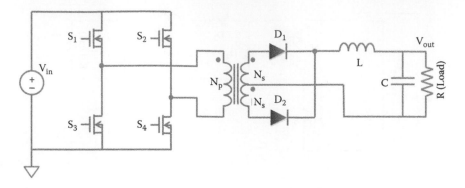

FIGURE 1.21   Simplified schematic of a full-bridge converter topology.

Operating a full-bridge converter is a little bit more complicated than forward or flyback converters. The diagonal switches S_1, S_4 and S_2, S_3 are turned on in combination during each half-cycle of the converter, and the transformer primary side winding is provided with the full input voltage. The polarity of the transformer reverses in each half cycle [18]. Figure 1.22 illustrates the operation principle of the full-bridge converter.

The input–output voltage relationship can be derived as [17]

$$\bar{V}_{out} = D \cdot \frac{N_s}{N_p} \cdot V_{in} \tag{1.31}$$

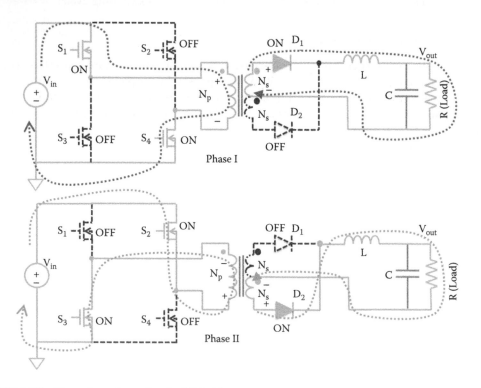

FIGURE 1.22   Operation schematic of a full-bridge converter.

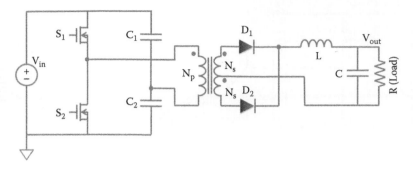

FIGURE 1.23 Simplified schematic of a half-bridge converter.

1.9.4 HALF-BRIDGE CONVERTERS

Half-bridge converters are similar to full-bridge converters, except that S_2 and S_4 are replaced by two large capacitors, as shown in Figure 1.23. The voltage at the center point between the capacitors is half V_{in}. When S_1 or S_2 is turned on, only $V_{in}/2$ is applied to the primary side of the transformer. Compared with the full-bridge converter, for the same power delivery half-bridge doubles the input current. This impairs the efficiency of power conversion at full load. Nonetheless, the half-bridge topology is useful since the cost is lower than the full-bridge version. The input–output relationship for the half-bridge is changed to

$$\bar{V}_{out} = 0.5 \cdot D \cdot \frac{N_s}{N_p} \cdot V_{in} \tag{1.32}$$

1.9.5 PUSH–PULL CONVERTERS

Similar to the half-bridge converters, the push–pull converters are derived from the buck converters. Two transistors are used in push–pull converters, as shown in Figure 1.24. One of the merits of the push–pull converters is that the controller is easy to implement, since both transistors are

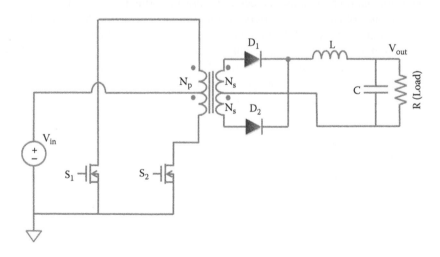

FIGURE 1.24 Simplified schematic of a push–pull converter.

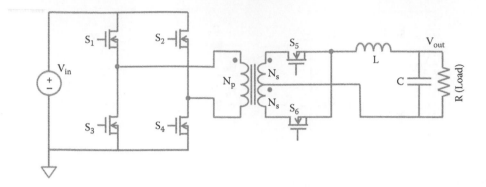

FIGURE 1.25 Schematic of a full-bridge converter with synchronous rectification.

operated at the low side (i.e., power MOSFETs source/body terminals are grounded). The input–output relationship can be expressed as

$$\bar{V}_{out} = 2 \cdot D \cdot \frac{N_s}{N_p} \cdot V_{in} \qquad (1.33)$$

1.9.6 Additional Topics in Isolated DC/DC Converters

1.9.6.1 Synchronous Rectification of Isolated DC/DC Converters

Although we used diodes D_1 and D_2 in previous isolated topologies on the secondary side, in real application synchronous rectification with active MOSFET transistors are used to increase the power conversion efficiency, as seen in Figure 1.25.

1.9.6.2 Power MOSFET Parallel Body Diode Conduction

Power MOSFET has an inherent parasitic body diode (which is not the case with IGBT or LIGBT), formed by the drain and body regions. As seen in Figure 1.26, this intrinsic body diode profoundly affects the performance of DC/DC converters. As an example, during the dead time of a full-bridge converter, the only current flow path is through the diodes, as seen in Figure 1.27. When S_2 and S_3 are turned off and before S_1 and S_4 are turned on, the current cannot change immediately due to the leakage inductance of the transformer. The current flow is then diverted through the body diodes of S_1 and S_4 until S_1 and S_4 power transistors are turned on (with very low drain to source voltage) at the onset of the next switching cycle.

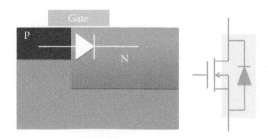

FIGURE 1.26 Intrinsic body diode in a power MOSFET.

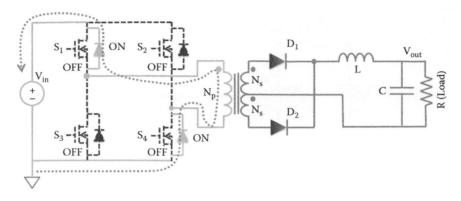

FIGURE 1.27   The current flows through the MOSFET's body diodes during dead time.

1.9.6.3   Transformer Utilization

While a detailed analysis of transformers is out of the scope of this book, a brief mentioning of transformer utilization in different isolated DC/DC converter topologies is still necessary. As we have seen, the flyback and forward converters use only one coil in the secondary side, which means that flux swings in only one quadrant of the B–H curve. This implementation is called single-ended. For the full-bridge, half-bridge, and push–pull converters, two quadrants of the B–H curve are used. These converters are then named as double-ended. For a given set of requirements, a double-ended topology requires a smaller core than a single-ended topology and does not require an additional reset winding [19]. Therefore, higher power density can be achieved with double-ended topologies.

1.9.6.4   Voltage Stress on the Active Power Transistor

During the operation of power converters, the power transistors could suffer from voltage stress. The stress level is different from topology to topology and is an important criterion of selecting the right power transistors. Table 1.5 illustrates the comparison between topologies. As we can see, full-bridge and half-bridge converters suffer the lowest voltage stress.

TABLE 1.5

Comparison of Isolated DC/DC Converter Topologies

Topologies	Power Range (W)	Transformer Utilization	Active Switch Number	Output Voltage	Voltage Stress	Cost
Flyback	< 100	Single-ended	1	$\bar{V}_{out} = \dfrac{D}{1-D} \cdot \dfrac{N_s}{N_p} \cdot V_{in}$	$> V_{in} + \dfrac{N_s}{N_p} \cdot V_{out}$	Lowest
Forward	50~200	Single-ended	1	$\bar{V}_{out} = D \cdot \dfrac{N_s}{N_p} \cdot V_{in}$	$> 2V_{in}$	>Flyback
Push–pull	100~500	Double-ended	2	$\bar{V}_{out} = 2 \cdot D \cdot \dfrac{N_s}{N_p} \cdot V_{in}$	$> 2V_{in}$	>Flyback
Half-bridge	100~500	Double-ended	2	$\bar{V}_{out} = 0.5 \cdot D \cdot \dfrac{N_s}{N_p} \cdot V_{in}$	$>= 0.5V_{in}$	>Push–Pull
Full-bridge	> 500	Double-ended	4	$\bar{V}_{out} = D \cdot \dfrac{N_s}{N_p} \cdot V_{in}$	$>= V_{in}$	>Half-bridge

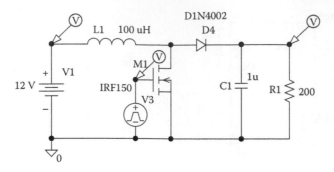

FIGURE 1.28   PSPICE schematic of a boost DC/DC converter.

1.9.7   COMPARISON OF ISOLATED DC/DC CONVERTER TOPOLOGIES

A comparison of the isolated DC/DC converter topologies is given in Table 1.5 [19]. A flyback converter uses only one power switch and no inductor at the secondary side, which makes it the easiest topology to implement and with the lowest cost. However, the problem with flyback is the poor transformer utilization (single-ended) and high voltage stress on the active power transistor. The forward converter is somewhat similar to the flyback converter and is single-ended with lower power density. Push–pull, half-bridge, and full-bridge are all double-ended converters. The push–pull converter suffers from high voltage stress. The half-bridge converter has a much lower peak voltage at the primary side. The full-bridge converter has the benefits of double-ended design, with relatively low voltage stress on the transistors. It also has a reduced input current when compared with the half-bridge converter. The issue with the full-bridge converter is the complicated control for four transistors. It easily becomes the most costly converter of the group.

1.10   SPICE CIRCUIT SIMULATION

PSPICE [20] can be used to simulate power converters. As an example, a boost converter is simulated using PSPICE with schematic and simulation windows as shown in Figures 1.28 and 1.29. Here, the MOSFET is controlled by a pulse generator, with a pulse width = 5 µs, period = 10 µs, and a fixed duty cycle $D = 0.5$. From basic calculation, we know the output voltage is related to input voltage with

$$\bar{V}_{out} = \frac{1}{1-D} V_{in} \qquad (1.34)$$

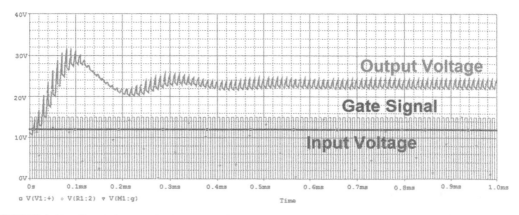

FIGURE 1.29   Simulation result of a boost DC/DC converter.

With input voltage of 12 V, the output voltage is found to be 24 V at steady state. The fluctuation of output voltage at the beginning reflects the time needed to establish steady state.

The main focus on this book is on the design and operation of the power devices using TCAD simulation tools. SPICE circuit simulations represent the next design step in the development process. This topic is widely covered by numerous books and will not be repeated here.

1.11 POWER MANAGEMENT SYSTEMS FOR BATTERY-POWERED DEVICES

Power electronics typically applies to board-level switched-mode power converters where discrete power elements such as power transistors, power diodes, capacitors, inductors, transformers, resistors, and controller ICs are packed and connected on a PCB board. This implementation is simple and flexible, capable of converting large amounts of power. With higher switching frequencies, the volume of the magnetic components can be made smaller, allowing the power density to increase. However, for very low-power applications, such as the battery-operated mobile devices, the bulky form factor is not convenient for users. Except for battery chargers (AC adapters), the conversion of battery power in a mobile device to supply the various components such as a CPU and LCD screen typically relies on integrated solutions. We refer to these integrated solutions as power management systems for battery-powered devices.

Power management systems can be regarded as a group of power converter circuits. They are becoming more diversified to keep up with the demands of modern mobile electronics systems. As an example, the power management system in a modern 3G smartphone is illustrated in Figure 1.30. It consists of many functioning units, each of which requires a different dedicated power supply [21]. Figure 1.31 is a more detailed illustration of the power management architecture of a battery-operated system with various functional blocks [22] [23]. Step-down DC/DC converters (e.g., buck converters) are used for DSP and microprocessors that require a lower voltage input, while inductorless DC/DC charge pumps are used for noise-sensitive RF circuitry since switched mode power converters typically bring some current and voltage ripples.

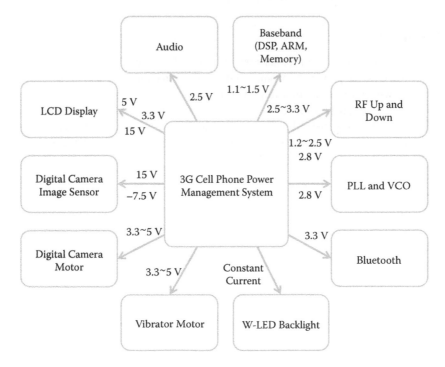

FIGURE 1.30 A typical example of 3G smartphone power management system.

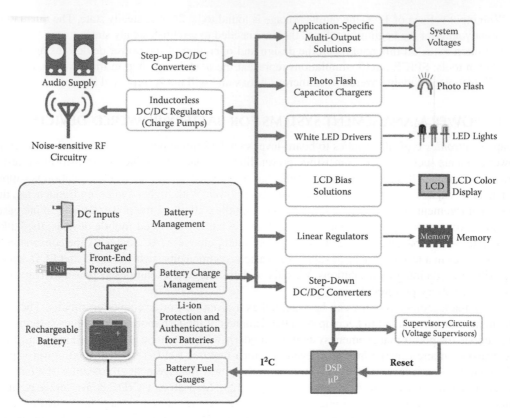

FIGURE 1.31 A more detailed power management architecture for a battery operated system (after Texas Instruments' Power Management Guide 3Q 2008).

1.12 SUMMARY

This chapter is intended to give a broader view to power device engineers since the main applications of power devices are power electronics and power management systems. In the next chapter, we will continue with an introduction to power management ICs where integrated power devices are typically used. Later chapters focus on the description of various types of power semiconductor devices such as diodes, power MOSFETs, and LIGBTs. In particular, we will provide detailed explanations on how TCAD tools can be used to analyze and design these devices.

2 Power Converters and Power Management ICs

Chapter 1 briefly reviewed the topologies of power converters. This chapter focuses on integrated power converters and power management ICs. Smart power ICs are a special class of integrated circuits that combine the decision-making circuits (the smart) and the power switches (the muscle) to deliver power to a load in an intelligent and efficient way. This chapter introduces the concept of power management and the corresponding smart power IC implementation via several examples. The fabrication technologies and semiconductor devices that are used to implement smart power ICs will be discussed in Chapters 7 and 8.

2.1 DYNAMIC VOLTAGE SCALING FOR VLSI POWER MANAGEMENT

Power management is the technique used to deliver power or to manipulate the power supplies efficiently without interfering with the normal operation of the load. An example of power management architecture for a battery-operated system is already introduced in Figure 1.31. This power management architecture can be further subdivided into different DC/DC converters, voltage regulators, and output drivers to accommodate the specific requirements of each load. A popular example of power management techniques is the dynamic voltage scaling (DVS) for optimizing the power consumption of the digital signal processor/microprocessor (DSP/µP) chip. In DVS, the supply voltage to a DSP/µP chip is reduced to a level that is just sufficient to achieve a target operating frequency. This technique relies on the fact that real-time computing tasks (such as frame decoding) often result in predictable workloads, and therefore the power management algorithm can dynamically select a target frequency to allow "just-in-time" completion of the tasks. A competing strategy for power reduction, known as dynamic threshold scaling (DVTS), uses external substrate biasing to dynamically control V_{TH} and reduce both dynamic and leakage power consumption [24] [25]. DVTS is more advantageous in nanotechnologies below 0.13 µm, where subthreshold leakage power dominates. The simulated power savings using DVS and DVTS for a ring oscillator in a 0.18 µm CMOS process are as shown in Figure 2.1a.

From basic CMOS logic gate analysis, it is well-known that the dynamic power dissipation $P_{dynamic}$ is proportional to

$$P_{dynamic} \propto \cdot C \cdot V_{DD}^2 \cdot f \tag{2.1}$$

where C is the load capacitance (usually the input gate capacitance of the next stage), V_{DD} is the supply voltage, and f is the clock frequency. The dynamic power dissipation in a generic digital circuit can be reduced by simply reducing f. This results in a linear reduction of the power consumption as shown by the "no scaling" line in Figure 2.1b. It is also interesting to note that in modern deep submicron and nanoscale CMOS technologies this linear curve does not intersect the origin as the frequency is scaled to DC. This indicates that a significant amount of static power is being dissipated due to leakage current. In DVS, both V_{DD}, and f are scaled to achieve a quadratic power saving during low demand cycles. DVTS provides additional power saving. However, the range of operation is limited due to the need to generate the extra substrate bias voltage.

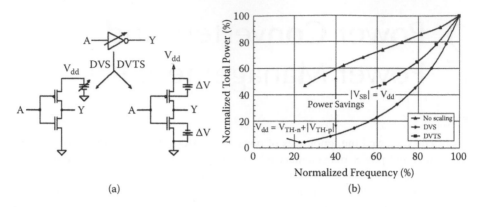

(a)   (b)

FIGURE 2.1   A simple CMOS inverter circuit showing the concept of (a) dynamic voltage scaling (DVS) and dynamic threshold voltage scaling (DVTS), and the corresponding normalized power saving (b) for a 0.18 μm CMOS ring oscillator using DVS and DVTS.

A large DSP/μP chip can be organized into multiple macro blocks. Figure 2.2 is a conceptual representation of DVS for VLSI power management system. Depending on the tasks on hand, each block can be programmed to operate at different speed (different f) with optimized power consumption using DVS. In this case, the macro blocks can be considered as separate voltage islands, powered by individual DC/DC converters. The output voltage of each converter is controlled by a power management circuit that receives commands from a task scheduler that is part of the operating system (OS). Early attempts in DVS development demonstrated substantial power saving at the DSP/μP chip level. However, the power saving for the entire system is often less significant because the power conversion efficiency of the DC/DC converter has peak efficiency only at a particular output current. A typical synchronized buck converter with all the major sources of power loss is as shown in Figure 2.3. This efficiency rolls off quickly at low output current due to gate drive and controller losses. It also rolls at high output current due to conduction loss from various parasitic resistances. As a result, good power saving can be achieved only for certain operating conditions. The design of integrated DC/DC converters that are suitable for DVS application will be described in the next section.

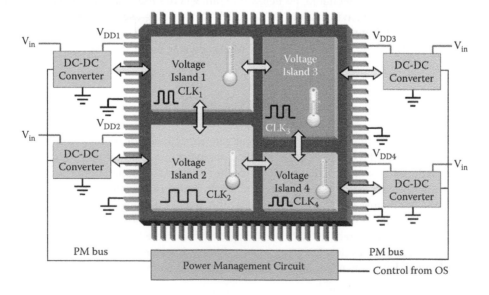

FIGURE 2.2   Conceptual representation of dynamic voltage scaling for a VLSI power management system.

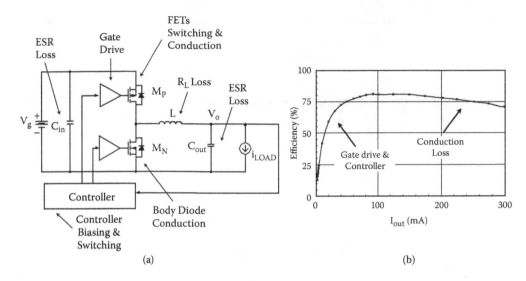

FIGURE 2.3 (a) A typical buck converter topology with all the major sources of loss identified. (b) The power conversion efficiency for a buck converter as a function of output current level.

2.2 INTEGRATED DC/DC CONVERTERS

To enjoy the full benefits of DVS power management, high-performance DC/DC converters with a flat power efficiency curve is required. At the same time, the form factor of the converters must be reduced to provide a high power density (power-to-volume ratio). Switched mode power supplies (SMPS) such as the DC/DC converters rely on an inductor as the energy storage element and a capacitor as a smoothing filter. The size of the inductor and capacitor is inversely proportional to the switching frequency of the DC/DC converter. As can be seen from Figure 2.4, the increase in switching frequency leads to a progressively smaller form factor. As the switching frequency approaches the multiple MHz range, it is also possible to combine the inductor and capacitor into the smart power IC package. Even though it is possible to implement integrated DC/DC converters with switching frequencies in the 10s to 100s MHz range, the high-frequency magnetic core losses prevented the use of ferrite core inductors. However, air core inductors have a large form factor, reducing the attractiveness of high switching frequency. In addition, switching loss also increases with frequency, causing a roll-off in the power conversion efficiency as shown in Figure 2.5. Currently, converters with switching frequencies in the multiple MHz range appear to be a good compromise between form factor and power efficiency.

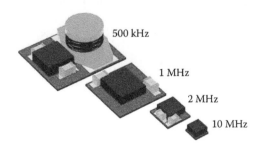

FIGURE 2.4 Benefits of miniaturization via high level integration using smart power IC fabrication technology, specialized packaging, and increased switching frequency.

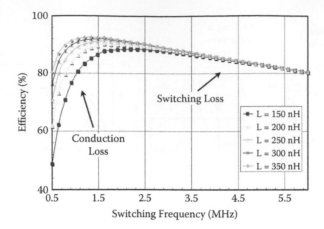

FIGURE 2.5 Power conversion efficiency for a typical DC/DC converter as a function of switching frequency and inductor size.

The characteristics of the power conversion efficiency in DC/DC converters is dependent to a large extent on the choice of the power transistors for the output stage. The typical output stage is a half-bridge configuration as shown in Figure 2.6. The totem pole design, which uses two N-channel power transistors for the high side and low side drivers or the push–pull design, which uses a P-channel power transistor for the high side driver and an N-channel power transistor for the low side driver, are two approaches to implement the half-bridge output stage. For integrated DC/DC converters, the choice on which configuration to use depends on the available power devices in the particular smart power IC fabrication technology. This topic will be covered in Chapters 4 and 8. Basically, the low-side (LS) power transistor is always an N-channel device due to the lower specific on-resistance (in mΩ·cm², resistance normalize to a unit area) and simple gate drive signal requirement. The high-side (HS) power transistor can be either an N- or P-channel device. Generally, the N-channel device is preferred due to its inherent low specific on-resistance. This is because the electron mobility is approximately three times higher than the hole mobility. As a result, for the same on-resistance, the N-channel device can be three times smaller than a P-channel device with similar rating. The smaller device area will translate directly to a lower cost. However, the trade-off in choosing an N-channel HS power transistor is the more complex gate driver requirement. A gate signal higher than the

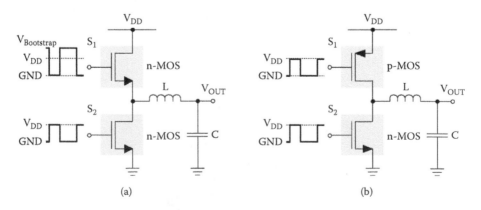

FIGURE 2.6 Output stage designs for a buck converter: (a) a totem pole output stage using N-channel power transistors for both the HS and LS drivers; (b) a push–pull output stage using complementary N- and P-channel power transistors.

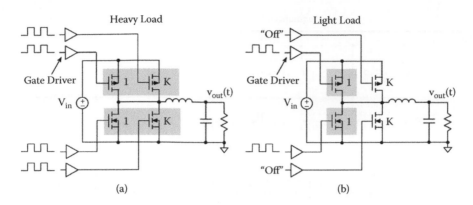

FIGURE 2.7 The "Dual-Gate W-Switched Power MOS" originally proposed by R. Williams et al. in ISPSD'97 [26]. For a heavy load, both large (K) and small (1) power transistors are activated (a); for a light load, only the small power transistor is used (b).

supply voltage is required to fully turn on the N-channel transistor. This usually involves a bootstrap circuit, consisting of an external or integrated diode and an external capacitor. However, the most critical decision is based on whether the smart power IC fabrication technology offers a floating source N-channel device (the source terminal has to swing between GND and V_{DD}).

The on-resistance, gate capacitance, and switching speed of the transistors are the critical parameters that determine the peak power conversion efficiency. In general, the size of the transistors is proportional to the nominal output current.

2.2.1 Segmented Output Stage

In a practical situation, the load current can change drastically as the system operates between full power and standby mode. In a DVS system, the DC/DC converter is also required to provide efficient power delivery at different output voltages and currents. The concept of using a combination of small and large transistors for the output stage was first introduced by Williams et al. [26] as shown in Figure 2.7. For a light load, it is more advantageous to use a small size transistor to reduce the gate drive loss. Since the output current is small, the high on-resistance from the small transistor will not lead to significant conduction loss. Therefore, good power conversion efficiency can be achieved. For a heavy load, there is no choice but to use the largest transistor available to reduce the conduction loss. Depending on the switching frequency, the gate drive and switching loss incurred by the large size transistor will be less dominant when compared with the conduction loss.

While this concept is straightforward, the implementation is rather difficult because it not easy to monitor the output current (especially for low output voltage, such as 1 V for most µP chips). In addition, switching the size of the power transistors on the fly may also lead to instability of the control loop. As a result, this remained as an interesting concept until the advent of the digital controller for DC/DC converters. With the development of the digital controller, various nonintrusive, sensorless sensing techniques have also been developed [27]. For example, the output current of a buck converter can be easily deduced if the input voltage, output voltage, PWM duty ratio, and the on-resistance of the power transistors are known [28]. In addition, the digital controller can also reconfigure itself on the fly to ensure stable operation on an as needed basis. These developments lead to the practical demonstration of a simple DC/DC converter with automatic selection of output transistor sizes according to the output current level. The architecture of a digitally controlled DC/DC converter with a segmented output stage is as shown in Figure 2.8. The block diagram is similar to a conventional digitally controlled DC/DC converter with an ADC to monitor the output voltage (V_{out}) and compare it with a reference voltage (V_{ref}). The error voltage $e[n]$ is fed to a digital

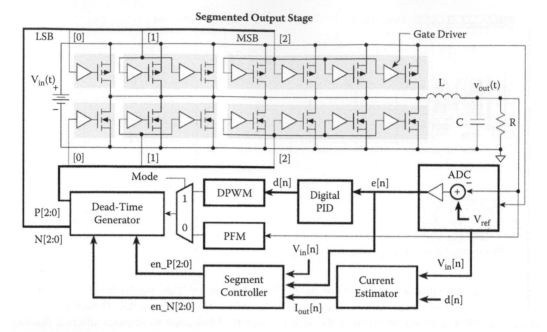

FIGURE 2.8 Architecture of the segmented output stage and the digital controller. Seven identical segments are paired to achieve three binary-weighted segments.

PID controller, which in turn generates the duty command *d[n]*. The PWM waveform is then generated via the DPWM module. After the appropriate dead times are incorporated, the gate drive signals are sent to the power transistors via the gate driver. If PFM (pulse frequency modulation) operation is desired, a mode selector can be used to choose PWM or PFM. The major difference is the segmented output stage. In this design, the sizes of the power transistors (both high side and low side) are actually the same as in a conventional design with the same output current rating. The only difference is in terms of organization. In this case, the HS and LS power transistors are organized into seven equal segments and basically occupy the same area as one big transistor. The segments are connected in binary weighted 1, 2, and 4 segments such that a three-bit control signal can activate seven different transistor sizes (from size 1 to 7). The only additional control circuits introduced are the current estimator and segment controller blocks. The output current is estimated based on

$$I_{out} = \frac{DV_{out} - V_{out}}{r} \qquad (2.2)$$

where r is the converter's series resistance, $r = DR_{on,P} + (1 - D)R_{on,N} + R_L \approx R_{on} + R_L$, with R_L being the inductor's series resistance, and D is the steady-state duty ratio. In a digital controlled DC/DC converter, all these parameters are known. As a result, the current estimator can calculate the output current level and decide on the most appropriate size of the HS and LS power transistors.

The die micrograph of the integrated DC/DC converter with segmented output stage is as shown in Figure 2.9. This test chip includes extra circuitries and look-up tables. The actual die size for a 5 V to 1 V, 1 A output converter can be 2 mm × 1.25 mm for this 0.18 μm CMOS implementation. The diagonal metal lines in the power transistors are the results of the use of a hybrid waffle layout. This technique allows an optimized trade-off between the metal interconnect resistance and the intrinic on-resistance of the power MOSFET [29]. In addition, an IC package showing a built-in energy storage inductor to save a circuit board footprint is also included in Figure 2.10. The measured power conversion

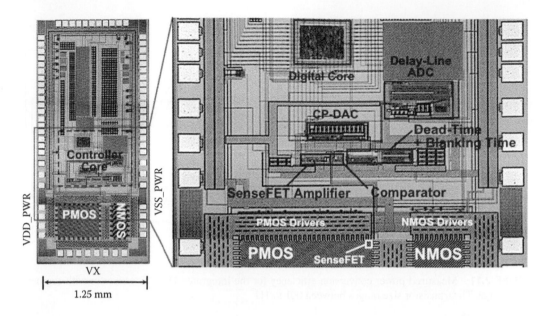

FIGURE 2.9 Die micrograph of an integrated DC/DC converter with segmented output stage. The test chip contains extra test circuits and look-up table. The actual die size for a 5 V to 1 V, 1 A output converter can be 2 mm × 1.25 mm for this 0.18 μm CMOS implementation.

efficiency of the DC/DC converter running at 4 MHz with V_{in} = 2.7 V, V_{out} = 1.8 V for different output transistor sizes are as plotted in Figure 2.11. The benefit of having the ability to switch to different power transistor sizes according to output current level is immediately apparent. For example, it is more advantageous to use large power transistors (size 111 = 7) for high output current and a small size transistor (size 001 = 1) for low output current. At very low current level, the power conversion

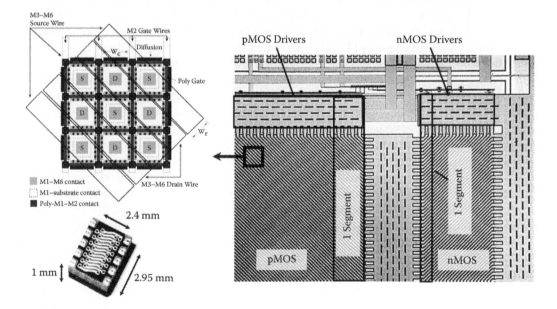

FIGURE 2.10 Zoomed-in view of the segmented output stage. The diagonal metal lines are the result of the hybrid waffle layout for the optimized trade-off between metal interconnect resistance and the intrinsic resistance of the power MOSFETs. Also shown in this figure is the IC package with a built-in inductor.

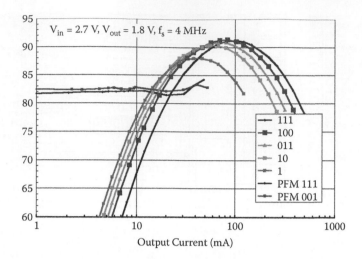

FIGURE 2.11 Measured power conversion efficiency for the integrated DC/DC converter with segmented output stage. The transistor size ranges between 001 to 111.

efficiency can be further improved by changing to a PFM operation. This type of power conversion efficiency optimzation is essential for a true DVS power management system for VLSI applications.

2.2.2 TRANSIENT SUPPRESSION WITH AN AUXILIARY STAGE

In addition to providing efficient power conversion, point-of-load (POL) DC/DC converters driving modern microprocessors need to provide low output voltage, high output current, and good dynamic performance during load transients. Digitally controlled DC/DC converters have shown more flexibility over their analog counterparts with the introduction of intelligent control techniques such as the auto-tuning system that can tolerate passive component variations [30], the segmented output stage that can dynamically adjust the size of the output transistors according to load conditions to maintain high-power conversion efficiency [31], and the one-step dead-time correction that can optimize the turn-on and turnoff dead time for power transistors on the fly [32]. They also have the ability to switch seamlessly between linear and nonlinear operation modes and achieve near-optimal load transient performance [33].

In this design example, a fully integrated DC/DC converter utilizes digitally controlled dual output stages to achieve fast load transient recovery. As shown in Figure 2.12, it consists of a main converter output stage connected in parallel with an auxiliary output stage. The main output stage is responsible for steady-state operation and is designed to achieve high conversion efficiency using large inductor and power transistors with low on-resistance. The auxiliary stage is responsible for transient suppression and is active only when a load transient occurs.

The need for the auxiliary output stage can be illustrated by the theoretical current waveforms during a heavy-to-light load transient as shown in Figure 2.13. In general, the large inductor in the main output stage is too slow to react to a sudden change in output current. As a result, the output voltage will exhibit an undesired overshoot or droop until the PID controller recovers. This change in output voltage may cause the digital circuits in the POL to change states inadvertently. Therefore, it is important to make use of a smaller size inductor in the auxiliary stage to suppress this disturbance within a short time. The proposed control method involves three control parameters: the main stage recovery time (T_R), turn-on time (t_{on}), and turnoff time (t_{off}), which represent the on-time for transistors LS_{main}, LS_{aux}, and HS_{aux}, respectively. In the case of heavy-to-light load transients, during T_R, the inductor current in the main output stage (I_{LM}) ramps down with a slope of k_1 until it reaches the targeted load current. In the meantime, the auxiliary stage is activated to help sink current with a high slew rate: during t_{on} the auxiliary stage inductor draws current (I_{LA}) from the output filter capacitor and then ramps

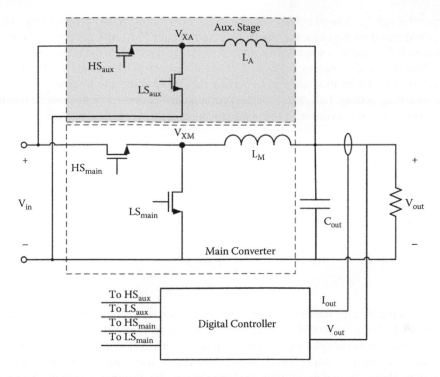

FIGURE 2.12 A digitally controlled POL DC/DC converter with an auxiliary output stage for fast load transient recovery.

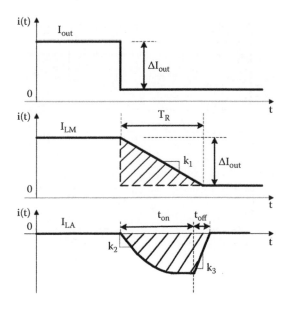

FIGURE 2.13 Theoretical waveforms of load current (I_{out}), main stage inductor current (I_{LM}), and auxiliary stage inductor current (I_{LA}) under heavy-to-light load transient taking into account the auxiliary switches' on-resistances.

back to zero through t_{off}. The switching commands for both the main and the auxiliary output stage are determined based on the capacitor charge balance principle [34], such that the amount of charge released through the auxiliary stage equals the excess charge injected by the main stage. By equating the areas of the two shaded regions in Figure 2.13, the three control parameters can be obtained from solving (2.3)–(2.5). This method aims at recovering the output voltage to its steady-state value with one set of switching actions. Full details on the operation and controller design can be found in [35]. The focus in this section is to describe the implementation of this type of converter.

$$T_R = \frac{\Delta I_{out}}{k_1} \tag{2.3}$$

$$\int_0^{t_{on}} I_{LA}(t)dt + \frac{I_{LA}(t_{on})^2}{2k_3} = \frac{\Delta I_{out}}{k_1^2} \tag{2.4}$$

$$t_{off} = \frac{|I_{LA}(t_{on})|}{k_3} \tag{2.5}$$

If this dual output stage transient suppression converter were to be built using discrete components, the PCB footprint would be effectively doubled, making this design twice as costly as well. For example, even though one can choose power transistors with different ratings, the size of the packages would essentially be the same. As a result, there will be no real area, size, and weight savings. However, in a monolithic implementation, the designer is free to choose different transistor sizes and the complete converter can be integrated onto one silicon die inside one package. The die micrograph of the DC/DC converter with a main output stage and an auxliary output stage for transient suppression is as shown in Figure 2.14. This converter is implemented using TSMC's 0.25 μm CMOS technology with 12 V option. The specifications and test conditions for this converter are summarized in Table 2.1. The main output stage transistors are designed to achieve a peak efficiency of around 90%. The auxiliary stage transistors are sized following the

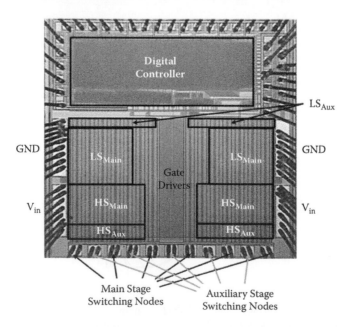

FIGURE 2.14   Micrograph of the integrated DC/DC converter (Die size is 2.35 mm × 2.35 mm).

TABLE 2.1

Summary of Specifications and Test Conditions

Parameters	Values
Die area	2.35 mm × 2.35 mm
PMOS_$R_{on(HS)}$	200 mΩ
NMOS_$R_{on(LS)}$	20 mΩ
PMOS_$I_{max(DC)}$	2.34 A
NMOS_$I_{min(DC)}$	3.24 A
V_{in}	6 V
V_{out}	1 V
L_M (main stage)	2.2 µH
L_A (aux stage)	820 nH
C_{out}	200 µF
Switching Frequency	390 kHz
ΔI_{out}	Switching from 2.25 A to 0.25 A

criteria discussed in [36]. From the die micrograph, it can be observed that the auxiliary output stage imposes less than 20% area overhead. Transient performance of the prototype converter was measured under the test condition specified in Table 2.1. Transient output voltage waveforms for a conventional single-stage PID controller and the proposed dual-output-stage controller are compared in Figures 2.15a and b. For a 2.25 A to 0.25 A load transient, a reduction in recovery

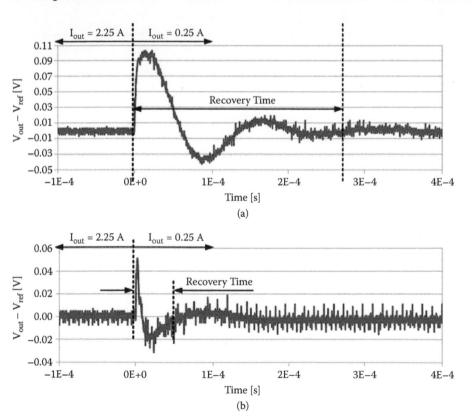

FIGURE 2.15 Dynamic response under a 2.25 A to 0.25 A load transient using (a) conventional single-stage PID controller and (b) dual output stages.

time from 280 μs to 50 μs and in overshoot from 105 mV to 51 mV is observed. Since the auxiliary stage is disabled in steady state, the two control methods have a similar efficiency profile.

2.3 SUMMARY

In this chapter, the DVS power management technique for just-in-time processing has been introduced. Examples of two power management ICs have also been given. The segmented output stage can be used to effectively extend the power conversion efficiency over a wide range of output current levels. This is highly suitable for driving DVS systems. The second example involves a transient suppression technique using a dual output stage with a large inductor for high efficiency and a small inductor for fast transient. In both cases, clever arrangement of the power transistors and the use of control methods can greatly improve the functionality and performance of PMICs.

The next chapter discusses the semiconductor industry and the status of smart power IC technology.

3 Semiconductor Industry and More than Moore

As mentioned in the Preface, this book adopts a top-down approach. In Chapter 1, we discussed the power electronics system, which is the system-level application of power devices (mostly discrete). Power management ICs are presented in Chapter 2. Power management ICs are IC-level application of power devices (mostly integrated). Now we step further down the ladder to explore the foundation of the power management IC: the semiconductor industry.

3.1 SEMICONDUCTOR INDUSTRY

On July 1, 1948, the *New York Times* made an announcement about a new invention, "a device called a transistor, which has several applications in radio where a vacuum tube ordinarily is employed" [37]; the inventors did not yet know how much their invention would change the world. The semiconductor industry is now one of the most dynamic industries in the world and starting with the very first transistor at Bell Labs, it has now grown into a $US 280 billion industry. The total impact of the semiconductor industry can be even greater: when electronics systems and services are included, it could be over $US 5 trillion, representing close to 10% of world gross domestic product (GDP) [38]. Figure 3.1 gives some key applications of semiconductor technology [39].

3.2 HISTORY OF THE SEMICONDUCTOR INDUSTRY

3.2.1 A BRIEF TIMELINE

A brief historical timeline of the semiconductor industry is listed in Table 3.1 [40]:

3.2.2 THE TRAITOROUS EIGHT

One of the most notable people in the history of semiconductors is Nobel Prize winner William Shockley, the co-inventor of the first transistor. After this great invention, Shockley left Bell Labs to form a company called Shockley Semiconductor in Palo Alto, California, and hired some of the brightest scientists and engineers available. However, due to some managerial conflicts, a group of these (the famous "Traitorous Eight" [41]) left Shockley Semiconductor. Those eight men later made their own tremendous contributions to the semiconductor industry and are widely credited with the creation of the present-day "Silicon Valley." Short biographies of these eight men are listed in Table 3.2.

3.2.3 HISTORICAL ROAD MAP OF THE SEMICONDUCTOR INDUSTRY

During the last decades, semiconductor chip performance advances have been mostly fueled by shrinking feature sizes and the use of larger wafers. Feature size is generally defined as the smallest gate length achievable by a certain lithography process. Supply voltage and junction depth need to shrink at the same time with the gate length to avoid the so-called short channel effect.

Table 3.3 illustrates the feature sizes and gate oxide thickness shrinking with time [42]. Please note that beyond 45 nm, materials with high dielectric constants (high-K) are used to reduce the

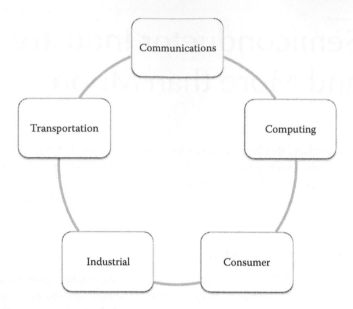

FIGURE 3.1   The semiconductor technology applications.

leakage current that comes from ever thinner gate oxide. With high-K materials, thicker physical thickness materials with equivalent oxide thickness (EOT) of 1.2 nm can be used to reduce gate leakage current while achieving the same performance at 32 nm node. At the time of writing, the most advanced feature size that is commercially available is 22 nm, first introduced by Intel with FinFET (Tri-gate) technology on its latest Ivy-bridge processors [43], which packed 1.5 billion transistors in its third generation "Core" processors [44].

The use of larger wafers brings no direct performance gain but lowers the average cost per chip for manufacturers since more chips can be packed onto the same wafer even if the feature size stays the same. From the 4 inch (100 mm) wafers used in 1975 to the 12 inches in use today, the wafers have gone through some major changes. If current trends continue, 18 inches (450 mm) will become the standard in a few years' time. Figure 3.2 illustrates the evolution of wafer sizes.

TABLE 3.1

Historical Timeline of the Semiconductor Industry

Time	Historical Event
1948	First transistor was invented at AT&T Bell Labs
1952	First Single Crystal Germanium introduced
1954	First Single Crystal Silicon introduced
1958	First Integrated Circuit invented by Jack Kilby of Texas Instruments
1961	First Integrated Circuit product by Robert Noyce of Fairchild
1971	Microprocessor was invented by Intel
1975	First commercial microprocessor unit (8086 and 6800)
1977	Apple and Radio Shack computers
1981	IBM personal computer introduced

TABLE 3.2
Brief Biographies of the Famous "Traitorous Eight" [41]

Name	Brief Bios
Julius Blank	Cofounder of Fairchild Semiconductor. He later founded Xicor, which was acquired by Intersil in 2004.
Victor Grinich	Cofounder of Fairchild Semiconductor. He was later with University of California at Berkeley and Stanford University.
Jean Hoerni	Cofounder of Fairchild Semiconductor. He was the inventor of the planar process, which allowed transistors to be created out of silicon rather than germanium. He later founded Intersil in 1967.
Eugene Kleiner	Cofounder of Fairchild Semiconductor. He was the cofounder of Kleiner Perkins, the Silicon Valley venture capital firm that later became Kleiner Perkins Caufield & Byers.
Jay Last	Cofounder of Fairchild Semiconductor. He later cofounded Amelco, which was acquired by Teledyne.
Gordon Moore	Cofounder of Fairchild Semiconductor. He cofounded Intel Corporation in 1968. He is the author of the famous "Moore's Law."
Robert Noyce	Cofounder of Fairchild Semiconductor. Nicknamed "the Mayor of Silicon Valley," he was the cofounder of Intel with Gordon Moore, credited (along with Jack Kilby) with the invention of the integrated circuit.
Sheldon Roberts	Cofounder of Fairchild Semiconductor. He later cofounded Amelco, which was acquired by Teledyne.

TABLE 3.3
Historical Road Map of Feature Sizes and Gate Oxide Thickness

Year	Feature Sizes	Gate Oxide Thickness
1975	20 μm	250 nm
1980	10 μm	150 nm
1985	5 μm	70 nm
1990	1 μm	15 nm
1995	0.35 μm	8 nm
2000	0.18 μm	3 nm
2005	65 nm	1.4 nm
2010	32 nm	1.2 nm (EOT)

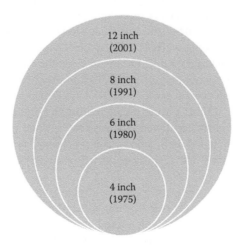

FIGURE 3.2 The evolution of wafer sizes.

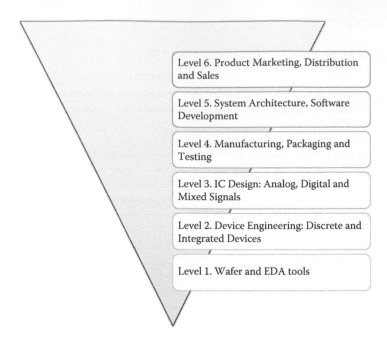

Level 6. Product Marketing, Distribution and Sales

Level 5. System Architecture, Software Development

Level 4. Manufacturing, Packaging and Testing

Level 3. IC Design: Analog, Digital and Mixed Signals

Level 2. Device Engineering: Discrete and Integrated Devices

Level 1. Wafer and EDA tools

FIGURE 3.3 The inverted pyramid illustration of the semiconductor industry.

3.3 FOOD CHAIN PYRAMID OF THE SEMICONDUCTOR INDUSTRY

The semiconductor industry is structured similar to an inverted food chain pyramid [39] (Figure 3.3), where the upper levels of the pyramid require the output of the lower levels to function properly while the lower levels need guidance of the upper levels. The pyramid is inverted because in terms of industry size the upper levels employ more people than the lower levels.

3.3.1 LEVEL 1: WAFER AND EDA TOOLS

The lowest level of the pyramid is the wafer suppliers and electronic design automation (EDA) tools suppliers. Please note that EDA in general supports all levels of the pyramid, from fundamental physics solvers (TCAD for device engineering) and IC design tools such as SPICE to system-level tools (thermal, mechanical, fluidic, and electromagnetic field).

The wafer business is the backbone of the electronics industry. Contrary to the dramatic reduction of cost per transistor, the cost per wafer is steadily increasing due to the increased cost of tools and stricter requirements on material quality. However, since the wafers themselves are also growing larger, the real cost per area is not that great. As an example, according to cost analysis from IC knowledge LLC [45], a 200 mm (8-inch) wafer is $1203.17 ($ 3.83 per cm^2) compared with $1936.11 ($2.74 per cm^2) for a 300 mm (12 inch) wafer. Since scaling of the feature size also produces more dies for the same area, lower costs per die can be achieved in spite of increasing wafer costs. For power IC applications, 6-inch and 8-inch bulk silicon wafers are typically used. Recently, Texas Instruments opened the industry's first 12-inch wafer fab for analog and power management applications in Texas [46].

To further reduce leakage current and increase digital gate density, sometimes Silicon on Insulator (SOI) wafers are used instead of bulk silicon. Companies such as Soitec can supply SOI wafers for power and analog applications with a rated breakdown voltage of up to 600 V [47]. SOI provides good isolation between power devices and analog/digital devices on the same die and may become the de facto standard with further feature size scaling. The biggest problems with SOI wafers at the moment are higher costs (usually double the price of bulk wafers) and reduced thermal handling capability.

For power devices applications, compound materials such as SiC and GaN are also drawing attention due to their superior FOM (Figure of Merits). However, compound substrates such as SiC and GaN substrates are more expensive and only available in much smaller sizes compared with silicon. It has been reported that 6-inch substrates are becoming available for GaN [48] and 6-inch SiC substrates are under mass production [49].

The EDA industry is gaining a lot of momentum in recent years due to advances in digital circuit design and high-level abstraction for synthesis and routing. The EDA industry reached $1,619.9 million total revenue in the third quarter of 2012, up 4.9 percent from the third quarter of 2011, according to the *EE Times* [50]. As of 2013, the top three EDA companies in the world are Synopsys, Cadence, and Mentor Graphics [51].

For analog and power IC design, computer-aided design (CAD) tools are widely adopted. These tools use compact device models like BSIM3, BSIM4 [52], and PSP [53] implemented by hardware description language such as Verilog-A to create netlists for simulation. Compact models are mostly semi-empirical in that not all the model parameters are physical. This is especially true for power devices such as lateral double-diffused DMOS (LDMOS), which uses subcircuit models. LDMOS models are made available from multiple research institutes such as HiSIM (Hiroshima-University STARC IGFET Model) [54] and companies like NXP [55] and Freescale [56]. Generally speaking, compact models are technology dependent, which means the parameters need to be extracted for different process technologies; software tools such as IC-CAP from Agilent are usually used for parameter extraction from measurement [57], and the models are subsequently provided to designers through a Process Design Kit (PDK).

Technology computer-aided design (TCAD) is a special category within EDA. Rather than being used to design ICs, these tools operate at a lower level of abstraction and are used by semiconductor device engineers to design process flows and various kinds of semiconductor devices. More detailed treatment of TCAD tools will be given in Chapters 5 and 6.

3.3.2 LEVEL 2: DEVICE ENGINEERING

The second level is device engineering, which consists of both process integration and device design. Device engineers are "customers" of the wafer suppliers and EDA (TCAD) vendors. They will define and select the most appropriate starting wafers with specific sheet resistance and substrate dopant type. TCAD tools are essential for device engineers to save time and cost. It helps device engineers understand the device physics and help design process recipes and structures for various devices. TCAD may even help them to predict the electrical, thermal, and optical properties under different operating conditions without actual fabrication (i.e., virtual prototyping) provided that proper calibration has been done [39]. Device design engineers need to work closely with process integration engineers to carry out test vehicles for new technology certification. In general, device engineers must plan ahead and choose the right technology flow for specific applications. For example, if the company is selling the power IC chips for wireless communication customers, the device breakdown voltage rating may not need to exceed 20 V. However, if the customer is making hybrid vehicles, then a 600 V LDMOS rating is necessary for insulated gate bipolar transistor (IGBT) drivers.

Different voltage ratings will also demand different technology flows. For example, power ICs are suitable for low to medium power applications; even high-voltage (> 600 V) and low-current applications are feasible in most cases. On the other hand, applications that require high current levels demand large silicon space and render it noneconomical to use an integrated solution so discrete devices are chosen instead. It has been estimated that monolithic integration of the power stage will be too expensive if the power stage area approaches 30% of the total IC [58]. We will have a very detailed discussion of silicon integrated power devices and TCAD simulation in Chapters 7, 8, and 9. Chapter 10 is dedicated to a brief introduction of GaN devices.

3.3.3 Level 3: IC Design

The level above device engineering is called the circuit level, which uses the processes and devices designed by the device/process engineers to build functional ICs. The IC designers are "customers" of device engineers; they cooperate with each other on the circuit and technology development. Nowadays, a digital circuit designer does not have to know much about device and process technology, while an analog and power IC designer has to be quite familiar with device and process information.

Analog and power ICs rely largely on device properties and process conditions to achieve their performance goals. Details such as MOSFET layout orientation, mismatch, well-proximity effect, and optical proximity correction (OPC) often affect the IC performance. Analog layout is therefore very time-consuming, and companies should either have a team of dedicated layout engineers or outsource and assign the job to contract workers. ESD engineers should also be counted in this group; they provide specially designed ESD device structures to protect the main circuitry from catastrophic electrostatic discharge. Engineers from ESD groups work closely with device engineers to design the process steps; it is also possible to have one or more dedicated layers specifically for ESD purposes even though this should be avoided to save on cost.

3.3.4 Level 4: Manufacturing, Packaging, and Testing

The next level is IC manufacturing, packaging, and testing. In the past, companies were almost always vertically integrated with a semiconductor company owning all the facilities necessary to manufacture their products from device design to packaging and even to final consumer products. "Real men have fabs" was a famous statement by AMD founder Jerry Sanders; now, even AMD has become fabless. Still, many top companies like Texas Instrument, Freescale, NXP, ST microelectronics, ON semiconductor, and Infineon continue to manufacture power and analog ICs using their existing fabrication facilities. These companies are called integrated device manufacturers (IDMs).

Except for Intel and a few others, most IDMs these days use their own facilities to fabricate analog and power ICs only. This is because analog and power ICs are not scaling as fast as their digital counterparts: since the power stage occupies a large portion of the ICs, it becomes uneconomical to scale down the feature size. The industry is also still using 8-inch wafers (and smaller) compared with the 12-inch wafers that are widely used by digital ICs. There are certain advantages for IDMs to reuse their relatively old fabs. Besides the obvious advantage of reusing otherwise obsolete equipment, unique or application-specific development of analog and power IC technology gives IDMs a competitive advantage since the process technology becomes part of the intellectual know-how for analog and power IC vendors. Process technology can be a product differentiator for IDMs: analog and power IC designers need to work closely with device engineers and process integration engineers to solve problems and boost chip performance. Semiconductor foundries like TSMC, UMC, Global Foundries, and XFab are getting more and more interested in analog and power ICs. These "pure players" provide certified process technologies for certain breakdown voltages aiming at fabless analog and power IC companies and even some IDMs.

Packaging and testing are often done offshore, since they are traditionally considered less technology intensive. This is not true, however, for today's more and more complicated packaging technologies. For power ICs, engineers are trying to find ways to integrate the inductors and even transformers on the same chip, which often requires advanced packaging. Testing is also not as straightforward as it would appear at first glance: setting test conditions that allow "bad dies" to be distinguished from "good dies" without further damaging the good dies requires a lot of skill. This is especially true for applications like automobile airbag deployment circuitry, where consumer safety is a primary concern.

3.3.5 LEVEL 5: SYSTEMS AND SOFTWARE

These functional IC packages are the building blocks for system engineers. For power management applications, system engineers here refer to power electronics engineers. Power electronics engineers are the "customers" of power controller ICs and power discrete devices. For various reasons, power electronics systems are extremely cost-sensitive. Modern power electronics systems require high power densities and high efficiency. Unfortunately, high power density requires high switching frequencies to reduce the size of magnetic components, and this also brings about more switching power loss and thermal dissipation. A trade-off has to be made even if advanced technologies such as soft switching methods are applied. Some of the world's top power electronics and power supply manufactures includes Delta Electronics, Lite-On Technology, and AcBel Polytech. Power electronics systems are themselves building blocks for the final electronics systems, such as desktop PCs. Many power electronics systems now have digital control, with A/D and D/A converters and DSPs.

Software development is essential for many electronics systems: it is often the software that determines the system popularity for an end product. Software is making machines "smarter" and serves as the interface between human beings and machines. Typical examples include tablets and smartphones, which integrate the essential functions of a PC plus MEMS-based sensors and actuators. Software systems thus control all the hardware and make them smart enough to accomplish complicated functions.

3.3.6 LEVEL 6: MARKETING AND SALES

The topmost level, product marketing, is often neglected by engineers but is actually very important. Innovation is important, but it is useless if it is not directed toward establishing defensible market positions and comparative advantages [59]. Innovations should be combined with in-depth market research for any new technology development to lead to a more successful future.

Engineers too often brag about the sophistication of the design under the hood and how advanced the technology is, but the real problem is whether or not the consumers will buy it. There are countless examples of start-up companies that failed because of no customer engagement until it is too late [60]. Understanding what customers need and implementing a proper technology to satisfy these needs is often the key to success. Advanced wafer technologies, such as SOI wafers, may provide a lower parasitic capacitance and resistance to latch-up but still do not guarantee that the chip performance and price are better than that of bulk wafers. New compound materials, such as SiC, GaAs, and GaN, do not magically cause the public to accept them. In fact, even in the traditional GaAs radio frequency (RF) power amplifier (PA) market, the growing trend is to use CMOS technology instead of the compound material to provide companies such as Qualcomm a competitive edge [61]. The SiC material for high-power devices has long been anticipated to replace silicon like GaAs did in the RF field, but SiC has continued to hold a relatively small market share compared with silicon. This is partly due to lack of reliable transistor technology (no oxide growth on SiC) as well as formidable device cost: the bill of material accounts for 75% of total device cost for SiC compared with less than 10% of that for silicon-based devices [62].

A good marketing example involves CPU makers, which put their logos on the outside of PCs. Consumers are aware of what is inside and are willing to pay a higher price tag for better performance. On the other hand, power management system providers seldom market their products to end users, and consumers automatically assume that all power supplies are the same. This is a pity because a consumer who is willing to pay the price for a state-of-the-art CPU or GPU to get maximum performance is rarely motivated to opt for an advanced power management system that provides a more efficient, more stable, and cleaner (with power factor correction) power supplies. Consumers assume all power management systems are the same as long as the specs are similar, so original equipment manufacturers (OEMs) have to compete on price to the extent that little profit margin is retained.

That is partly why power management systems are almost always under cost constraints and why initially costly new technologies are difficult to introduce. One suggestion for power management systems providers is to try to differentiate themselves by making a sustained marketing effort to educate end users about the advantages of using newer and better technologies. For example, higher efficiencies lead to reduced power consumption and environmental benefits, while weaker electromagnetic field leakage may appeal to health-conscious consumers.

3.4 SEMICONDUCTOR COMPANIES

Over the years, semiconductor companies have developed into several categories, due to market demand and the shift to more specialized entities. For historical reasons, semiconductor companies are divided into three categories: IDMs; fabless; and pure players (foundries) (Table 3.4). In the past, all semiconductor companies were IDMs: vertical-integrated companies that do both manufacturing and design. Some companies like Sony and Samsung even go beyond this level and do consumer products design and manufacturing. IDM companies have certain advantages; they can often differentiate their products from their competitors'. A typical example is Intel: with its most advanced fabrication technology (now at the 22 nm node), it differentiates itself from its competitors without fabs of their own. Table 3.5 shows the top 10 semiconductor companies in the years 1987, 1997, 2007, and Q1 of 2013 [63] with data from Gartner [64], iSuppli [65], and IC insights [66]. From the most recent data, two fabless companies have entered top 10 lists, with Qualcomm in fourth place (third place if foundries are excluded from the list) and Broadcom in tenth place; this would have been unthinkable back in the 1980s and 1990s. A growing trend of fabless companies in the top 20 lists can also be observed as shown in Figure 3.4.

The semiconductor industry is highly dynamic so the rankings change every year: more fabless companies and foundries might replace IDMs in the top 10 list if the current trend continues. However, it is also possible that some IDMs may regain momentum: even though the fabless model has been thriving for more than 20 years now, some experts predict the collapse of the fabless model in future [67]. While it is hard to forecast future trends, another possibility is that some of the largest fabless companies may eventually acquire fabs or foundries and become IDMs to secure their manufacturing supply. IDMs like Intel may offer part of its advanced manufacturing capability to other companies, as it already did recently for a FPGA company called Achronix [68]. Foundries may also provide their own designs and more complicated IPs for certain customers in the near future. Needless to say, the boundaries between IDMs, fabless and foundries are becoming vaguer than they once were.

TABLE 3.4

Types of Semiconductor Companies

Semiconductor Company Types	Descriptions
IDMs	Have their own fabs to fully or partially support their product manufacturing. Nowadays even IDMs outsource the logic manufacturing capabilities from foundries.
Fabless	Have no manufacturing capabilities and must outsource all the manufacturing capabilities from foundries. Recently a new business model was created that allows companies to license their IPs rather than manufacturing them.
Foundries	Have no intention of designing their own ICs, even though they may also design some IP blocks.

TABLE 3.5
Global Semiconductor Companies Rankings [69]

Rank	1987 [63]	1997 [63]	2007 [63]	2013 Q1 [70]
1	NEC Semiconductors	Intel Corporation	Intel Corporation	Intel Corporation
2	Toshiba Semiconductors	NEC Semiconductors	Samsung Electronics	Samsung Electronics
3	Hitachi Semiconductors	Motorola Semiconductors	Texas Instruments	TSMC*
4	Motorola Semiconductors	Texas Instruments	Toshiba Semiconductor	Qualcomm**
5	Texas Instruments	Toshiba Semiconductor	STMicroelectronics	Toshiba Semiconductor
6	Fujitsu Semiconductors	Hitachi Semiconductors	Hynix	Texas Instruments
7	Philips Semiconductors	Samsung Semiconductors	Renesas Technology	SK Hynix
8	National Semiconductor	Philips Semiconductors	Sony	Micron
9	Mitsubishi Semiconductors	Fujitsu Semiconductors	Infineon Technologies	STMicroelectronics
10	Intel Corporation	SGS-Thomson	AMD	Broadcom**

Note: Companies with ** are fabless companies and with * are foundries.

3.5 MORE THAN MOORE

On April 19, 1965, *Electronics Magazine* published a paper by Gordon E. Moore, who later cofounded Intel. In that paper he observed that the number of transistors on a chip roughly doubled every two years [71]; this observation was later extrapolated into a prediction called Moore's law. Shown in Figure 3.5 is a plot of transistor count versus year from 1971 to 2012 [72] using Intel's CPUs along with a line that corresponds to exponential growth with transistors count doubling every two years. Amazingly, this trend has continued for nearly half a century and should continue for at least a few years into the future [73].

As technology advances, an increasing diversity of semiconductor applications is emerging. These applications are mostly in the form of analog signals, which interact with people and environment: analog/RF, passives, high voltage power, sensors/actuators (MEMS), and even biochips (Lab on Chip) are all changing our daily lives. For these applications, Moore's law does not govern development. Unlike advanced digital electronics, which move toward extreme miniaturization, this trend of diversification is more like a complement to Moore's law; this has been called "More

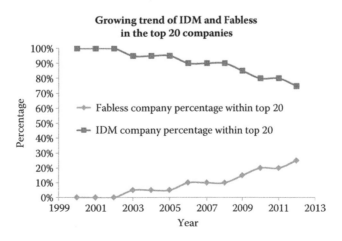

FIGURE 3.4 The growing trend of fabless and IDM within the top 20 companies.

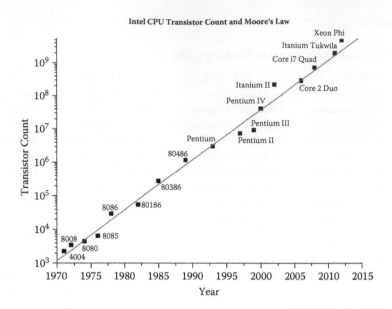

FIGURE 3.5   Transistor counts and Moore's law shown with Intel's CPUs.

than Moore" (Figure 3.6) by the world-renowned organization ITRS (International Technology Road Map for Semiconductors) [74]. As the ITRS mentioned in the 2011 edition of its executive summary [75], "Difficult Challenge for High Voltage MOS," several aspects of high-voltage (HV) devices and the associated base technology make it difficult and unlikely that the HV road map for the future will follow the lithographic shrink seen for CMOS because the HV designs cannot take advantage of the lithography capability to shrink the intrinsic HV device dimensions. Analog devices are usually large to improve the noise and mismatch. The digital content of a HV chip that could be shrunken is usually a small fraction of the chip area [75].

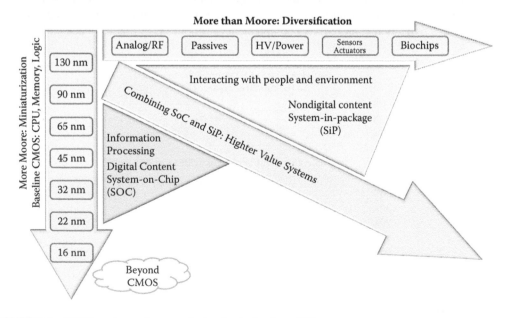

FIGURE 3.6   ITRS road map for nanoelectronics technology [74].

High-voltage devices and power ICs are used extensively for virtually every electronic gadget. The future of power management is to minimize the size, weight, and cost. System on chip (SOC) is possible for some applications but difficult for power management due to ubiquitous use of energy storage components such as inductors and even transformers. Unlike a radio frequency integrated circuit (RFIC), for relatively low frequency (< 5 MHz) switching it is hard to integrate these components on the silicon chip without increasing the cost. On-chip power supply is not commercially available at this time of writing [76]. On-chip inductors and transformers have been demonstrated recently by using wire bonding with Ferrite-epoxy glob coating [77]; however, it may take a long time for commercial availability.

3D integration and system in package (SIP) seem like a better fit, especially for current levels lower than a few amperes. It has been estimated that the global power system in package will grow from approximately $65 million in 2011 to $284 million in 2016, a compound annual growth rate (CAGR) of 34.4% [76]. The emerging SIP market is expected to encroach in the traditional market occupied by LDOs, switching mode power supplies, and controllers and power field-effect transformers (FETs).

Recently, 3D integration of coreless inductors and transformers in bulk of silicon chips has been reported [78] [79]. These passive devices are embedded in the substrate of silicon through the backside; due to the very thick metal used and a large number of turns available, inductors with a high inductance to resistance ratio were obtained. By incorporating a magnetic core, embedded inductors and transformers were also shown to have very high inductance density [80]. Similar advances in other integrated passive devices are needed for future development of power system-on-chip technologies.

4 Smart Power IC Technology

In the previous chapter, we provided a guided tour to the semiconductor industry. Now we are ready to delve deeper into an important subdivision of the semiconductor industry: the smart power IC industry.

4.1 SMART POWER IC TECHNOLOGY BASICS

Traditionally, power devices were available only in discrete/vertical form because they were expected to handle large current and power. With the ever-increasing demands from consumer electronics and mobile devices that require compact and cost-effective power management designs, laterally integrated solutions are becoming the mainstream. Since power devices can now be made as lateral devices where all terminals are on the surface and the current flows laterally in the silicon, a natural instinct is to integrate them with digital and analog devices by borrowing some process steps from CMOS technology. Indeed, LDMOS shares some process steps with CMOS with a few additional masks and process steps needed to create high-voltage capability. The integration of digital, analog, and power devices has created a new technology category: smart power, or BCD technology. Here, BCD technology stands for bipolar-CMOS-DMOS technology and can be regarded as an extension of the existing bipolar-CMOS (BiCMOS) technology.

The word *smart* is often applied to anything that has a brain: smartphones, smart grids, smart cards. The brains of a smart power technology come from the monolithically integrated digital and logic controllers; they instruct the muscles (power devices) to perform desired jobs. Analog components act like sensory organs; they link the digital chip to the outside analog world. Figure 4.1 illustrates the relationship between these functional parts in a smart power IC.

However, integrating different components onto a single chip is not an easy task. Without proper isolation, these components can suffer undesirable cross-talk that sometimes results in disastrous events such as latch-up in CMOS by substrate minority carrier injection from power devices [81].

A simple implementation of four integrated devices—NPN, n-LDMOS, NMOS, and PMOS—is illustrated in Figure 4.2 with two layers of metal and an N-type buried layer (NBL). A more detailed smart power process flow will be given in Chapter 7.

4.2 SMART POWER IC TECHNOLOGY: HISTORICAL PERSPECTIVE

The first smart power IC technology that combined CMOS, BJT, and DMOS devices was introduced in 1986 by SGS Thomson and was called multipower-BCD [82]. It was a 60 V, 4 µm process created by merging a conventional junction-isolated bipolar IC process with vertical DMOS technology. The total mask count was 12, similar to that of modern BJT technology. It was considered the first real BCD technology because unlike earlier technologies it packed multiple power DMOS on-chip that allowed efficiency to exceed that of BJT-only technology.

The second generation of BCD technology (BCD-2) was developed based on 2.5 µm CMOS technology to increase digital design complexity and approximately doubled the digital and power components density [83]. In this technology, LOCOS isolation was added with two metal layers.

The third generation of BCD technology (BCD-3) appeared in the early 1990s, based on 1.2 µm twin-well CMOS technology; nonvolatile memories (NVMs) such as EEPROM and EPROM were integrated for the first time. This generation of BCD technology was also regarded as the first system-oriented technology because of its capability to integrate an entire monolithic system solution. Three levels of metal were used in this technology: two of them were for high-density digital

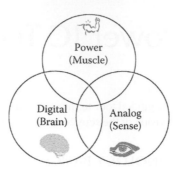

FIGURE 4.1   Smart power IC technology.

devices, and the top thick metal layer was reserved for power devices. The current density was doubled in power devices compared with the second generation, while digital component density almost tripled [83].

The fourth generation of BCD technology, developed at ST Microelectronics in the mid-1990s, was a 0.8 μm technology and was much more complex to satisfy the need for system-oriented technologies. Figure 4.3 illustrates the first three generations of BCD technology developed at ST Microelectronics, one of the pioneers in the smart power IC field [83].

Since its first appearance, BCD technology has garnered wide acceptance in the semiconductor industry. Many companies like ST, Freescale, Texas Instruments, TSMC, UMC, Global Foundries, and SMIC are actively engaged in developing new generations of smart power IC technology. The scaling trend is similar to the digital counterpart, but at a much slower pace due to the trade-off between digital gate density and power device silicon real estate. More devices and metal layers are integrated as technology moves forward. As an example, ST's BCD-8 uses 0.18 μm technology and is capable of supporting NVMs. Up to six layers of metal interconnects may be included in this technology [84].

A road map of BCD technologies developed by two companies (ST Microelectronics and Motorola/Freescale) is shown in Figure 4.4 and may be used to highlight technology trends in the semiconductor industry. Please note that the data are taken from various published articles and brochures [85], [84]. The timeline and feature sizes are also approximate values.

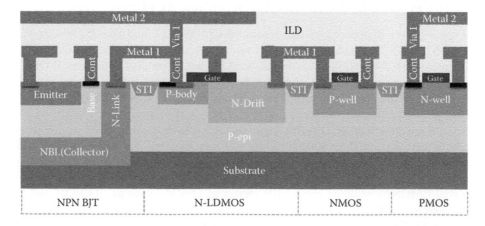

FIGURE 4.2   A simple schematic of BCD integration.

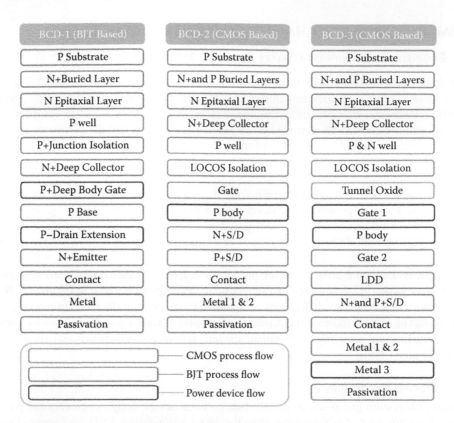

FIGURE 4.3 Process flowchart of first three generations of BCD technology.

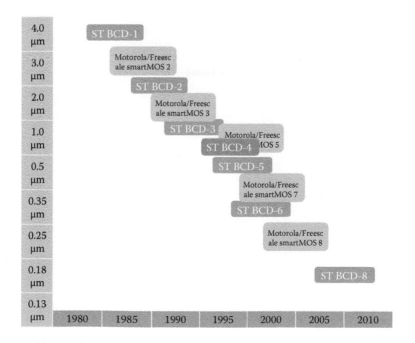

FIGURE 4.4 BCD technology road map of two companies from 1980 to 2010. Timeline and feature sizes are approximate values [85] [84].

4.3 SMART POWER IC TECHNOLOGY: INDUSTRIAL PERSPECTIVE

4.3.1 ENGINEERING GROUPS OF A SMART POWER IC TECHNOLOGY

Semiconductor process technology is an important know-how for IDMs and foundries; each company has its own process recipes for similar applications. This is like different chefs and recipes in different restaurants, with similar menus on the table. Process flows are different from company to company and are considered as trade secrets since it is an important differentiator just like one particular restaurant may add more salt and sugar for its pizza while another may opt for a lighter taste. The performance of analog and power IC depends heavily on the process technologies and manufacturing equipment.

Building a semiconductor process technology flow is a complicated task. Depending on past experience and technology advances, developing a new process technology generation (from planning stage to final certification) for analog and power IC usually takes a couple of years. Many groups of engineers with different technology backgrounds must work side by side to achieve this goal.

To give the reader some insight into the process, a simplified organization chart of engineering groups of smart power IC technology in a semiconductor company is given in Figure 4.5. Note that the exact names and functions differ from company to company, but the basic group functions are often similar. The arrows indicate the direction of support between groups and are for illustration purposes only. In a real company environment, the interactions between functional groups are much more complicated [39].

4.3.1.1 Process Integration

The process integration group acts as an interface between device design engineers and process/fab engineers. It spends a lot of its time supporting the device engineering, ESD, and compact modeling groups. Engineers in a process integration group will visit fabs from time to time and work with fab

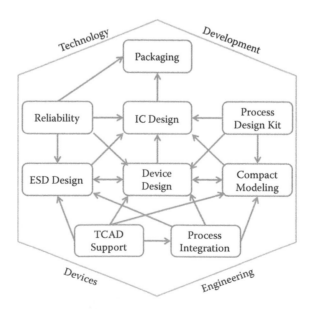

FIGURE 4.5   Organization chart of a typical smart power technology development group.

engineers from individual process (e.g., CVD, implantation, diffusion) and provide process integration solutions to device design engineers, ESD design engineers, and compact modeling engineers.

4.3.1.2 TCAD Support

The TCAD support group basically uses commercially available TCAD tools or internal proprietary simulation tools to simulate the process flow and perform design of experiments (DoE) analysis with different device structures and process conditions. It is critical for TCAD software vendors to provide DoE functionality to enable batch processing and accelerate the development cycle. For larger companies, TCAD simulation is performed with a dedicated group of engineers, whereas smaller companies generally do not distinguish TCAD engineers from device engineers. TCAD simulation plays an important role throughout the process technology development cycle as TCAD tools can help explain important physics that are hard to visualize or calculate manually; well-tuned simulators may even be able to predict the device performance before tape-out. The TCAD support group needs to support device design, compact modeling, ESD design, and process integration.

4.3.1.3 Compact Modeling

Compact modeling groups are very important for IC designers. Sometimes an inaccurate compact model can lead to a design failure. This is especially true for novice designers who rely too much on simulations rather than experience. Even though there are excellent models like BSIM4, PSP, and VBIC, the model parameters usually need to be determined before they can be used. For BCD technology, LDMOS also needs a more complicated model that is structure dependent. Unlike the fabless companies, in which the device models are provided by the foundries in the process design kit (PDK), IDMs, or foundries themselves have very different process flow and device structures for analog and power ICs. Even within the same company, an older 0.35 μm technology is by no means the same as the latest 0.13 μm technology in terms of compact models and their parameters. Compact modeling groups will work closely with device design engineers, since they are the ones who will provide device types, structures, and test wafers for parameters extraction. Compact modeling groups need to work closely with PDK groups to provide simulation tools for IC designers.

4.3.1.4 Device Design

The device design group is often a core group in the process technology development. Engineers in this group will work closely with IC design engineers to determine the next generation of process technology on the road map. A good practice should always look more than a few years ahead of time. However, since the scaling of power ICs is not keeping pace with their digital counterparts, an overly agreesive design without market and cost considerations is not a good practice for analog and power IC development. Besides device structure and process flow design, device design engineers will generally coordinate with almost all other groups to define the technology road map. The most important of these other groups is process integration; device designers need to get constant feedback from process integration engineers to run process splits and optimize the process recipe.

4.3.1.5 ESD Design

The ESD design group directly supports IC designers to meet their electrostatic discharge requirements as ESD protection devices have special structures that may be different from any other devices in use for the IC. Engineers in the ESD group utilize the layers from the process flow to build the ESD devices. Sometimes, one or more dedicated layers need to be created especially for ESD devices; however, this should be avoided whenever possible to save on cost. Different models are used in ESD tests to mimic different situations and ESD events, such as the human–body model (HBM), machine model (MM), and charged-device model (CDM).

4.3.1.6   Process Design Kit

The PDK group is a CAD support team. Engineers in this group provide all technology development engineers with files that contain standard cell libraries, design rule checking (DRC), device compact models, and parameterized cells (Pcells) for layout. These files are meant to be used in circuit-level CAD tools like Cadence by the device and IC design engineers.

4.3.1.7   IC Design Group

The IC design group is at the heart of any analog/power IC company and can be regarded as the customer of the whole technology development group. Analog and power IC designers need to interact with process development engineers; this is very different from digital design where IC designers rarely care about what a transistor looks like or how the device size or layout orientation can impact on the device performance.

4.3.1.8   Reliability Group

The reliability group deals with wafer-level and package-level reliability tests. For new technology to be certified as mature, accelerated wafer-level tests need to be carried out. The standard tests include time-dependent dielectric breakdown (TDDB) to qualify the gate oxide, hot carrier injection (HCI) to test the long-term physical parameter drift, and electromigration (EM) to exam the robustness of the metal interconnect, vias, and contacts. Package-level tests like high-temperature operating life (HTOL) and IDDQ (test supply current I_{dd} leakage in the quiescent state for CMOS) may be performed as well. The reliabilty group meets with engineers from other groups to make sure the technology being developed meets the standards for the intended applications, which is important for final technology certification.

4.3.1.9   Packaging

Packaging is becoming ever more important with novel packaging methods constantly being developed. System in package (SIP) is an emerging trend for integration of power management in a single package. Compared with system on chip (SOC), SIP provides more flexibility, faster time to market, lower research and development (R&D) costs, and thus lower production costs for many applications [86]. However, SIP also has disadvantages such as internal parasitic effects and the difficulty of assembling multiple chips.

4.3.2   Smart Power IC Technology Development Flow

Analog and power IC process technology development is a time-consuming project even if consumer electronics with its narrower time window is driving the process. A typical turnaround time for an 8-inch 0.25 μm power IC fabrication is at least a month or two; unless one can ensure as few test vehicles as possible, the total development time for a new generation of power IC process will be over two years. Looking a bit more than a few years ahead is always necessary as long as the estimated chip cost is under control. A typical process technology development flow is shown in Figure 4.6. In the following sections, each step will be explained in detail.

4.3.3   Planning Stage

The very first and most critical step of a new technology development is the planning stage: the entire project cannot be carried out smoothly without proper planning. Engineers should get advice and guidance from the business and strategy department to make sure they are targeting the right

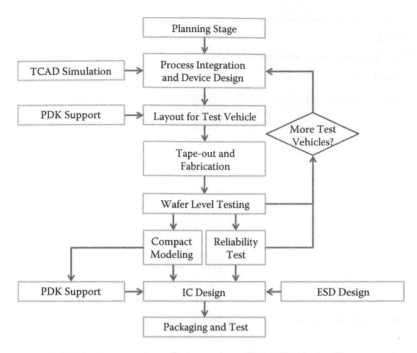

FIGURE 4.6 Simplified process flow of analog and power technology development.

applications. At this stage, engineers need to foresee future needs, usually a few years ahead, to make sure the new technology is neither too advanced (which may be costly) nor too conservative (which will make it obsolete even before release). Engineers need to understand the applications of the new technology since different applications will require different technologies. For example, standard automotive application generally requires higher voltage (> 80 V) than wireless communication application (~20 V). IGBT driver IC for hybrid cars usually requires HVIC with over 700 V of breakdown voltage capability.

The proper selection of technology node (i.e., the feature size) is critical. For power IC technologies, there is no equivalent to Moore's law [87] guideline of improvement found in advanced CMOS technologies. It is true that with smaller technology feature sizes a higher logic device density can be achieved and a larger number of gates can be packed onto the same die. However, the largest portion of the chip is often occupied by power devices that are much less influenced by scaling trends. There is always a trade-off between higher gate density and chip cost so a more advanced node does not necessarily mean a lower cost per chip. The choice of wafer is determined by several aspects. For example, a silicon-on-insulator (SOI) wafer may be good for better immunity of cross-talks but are very expensive for cost sensitive applications such as power management systems. Table 4.1 lists several considerations for the planning stage.

4.3.4 PROCESS INTEGRATION AND DEVICE DESIGN

The device design stage is the core stage that defines the road map of smart power IC technology. Since there are usually many device types in a smart power IC technology, engineers take ownership of devices, and most of the time advanced power devices such as LDMOS with high breakdown voltages are assigned to the most senior engineers as they require years of experience. Analog and digital MOSFETs have fewer considerations and are usually assigned to relatively

TABLE 4.1

The Planning Stage of a Simplified Analog and Power Technology Flow

Items Checklist	Possibilities
Target applications	Consumer electronics (e.g., RF, power)
	Industry applications
	Automotive applications
Technology nodes	0.35 μm
	0.25 μm
	0.18 μm
	etc…
Wafer technologies	Bulk silicon
	Silicon on insulator (SOI)
	Through-silicon via (TSV)
Isolation methods	LOCOS
	STI
	DTI
	PN junction isolation
Approval from business unit	Mask and stage count
	Cost

junior engineers. However, this is not absolute. Sometimes, the bottleneck in a design can stem from seemingly "easier" device types. For example, Zener diodes are easier to design than a power super junction LDMOS or a lateral insulated-gate bipolar transistor (LIGBT), but if proper care is not given to the defects and dislocations that are associated with high implant dose and energies fluctuations of leakage current caused by those defects can occur. Other considerations such as reducing the effects of orientation and mismatch in analog MOSFETs are also important requirements from IC designers.

Device engineers work closely with process integration engineers during the process development stage. Process integration engineers integrate the individual process stages (e.g., CVD, implantation, diffusion, etching) to create a full process flow. As with any other technologies, the best starting point for any new smart power IC technology is the previously successful generation: the technology node may move forward to smaller features sizes and wafers may become even bigger (e.g., TI's 12-inch wafer for analog and power), but the fundamentals always remain the same. Normally, a smaller feature size will bring the benefit of higher gate density of digital devices, but this is not necessarily a primary consideration for power devices.

TCAD tools are used extensively by engineers since device design is a rather complicated job for smart power ICs, which are very different from digital ICs. Many design considerations have to be balanced, and even highly skilled engineers need to run TCAD simulations from time to time to verify their intuitions. TCAD simulation can also provide some valuable information that is not available by measurements. For example, engineers need to detect the "hot spot" where the power devices breakdown. Measurement can provide data such as the final breakdown voltages and values of on-state resistance, but only TCAD can plot the distribution of electric field and impact ionization in a way that helps device engineers understand the critical locations and guides them in making adjustments to improve the device performance. A well-calibrated TCAD tool can even predict the device performance before real fabrication. DoE analysis is a very useful tool for engineers to find the most optimized combination of parameters that maximizes device performance. However, like

TABLE 4.2

Process Integration and Device Design Considerations

Type of devices	Digital MOSFETs (1.5 V); Analog MOSFETs (3.3 V); Power MOSFETs (> 12 V)
	Zener diode (5 V); HV diode (45 V); BJTs
	Capacitors (poly-poly capacitor, metal-insulator-metal capacitor)
	DRC test shapes, etc…
	Resistors (diffused, poly)
Possible improvement over last generation	Higher breakdown voltage; lower on-state resistance
	Better safe operating area (SOA); more robust ESD protection
	Better temperature coefficient and voltage coefficient for resistors
	Higher current gain for BJTs; better power handling capability for power devices
	Lower substrate leakage with suppressed minority carrier injection
	Higher capacitor density
	Higher digital gates density with more advanced technology node
	Reliability test shapes for TDDB, HCI, NBTI, and EM
	Metal interconnect schemes for large power LDMOS to reduce the metal debiasing effect
	Thick copper or gold last metal layer for devices to handle larger current flow
	Fuse/anti-fuse/NVM for trimming
TCAD simulation	Various device structures are simulated with TCAD process and devices simulators to predict and verify performance and to help designers find the best parameters for both process recipes and device structures
	Process integration steps like buried layer, well implants, and STI can be simulated to help reduce turnaround time
	Important performance parameters like threshold voltage and breakdown voltage can be simulated
	Calibrated data from previous technology are reused so that engineers can have more confidence with the simulations
	DoE to simulate and guide process splits

any other software tools, engineers should use their own experience and insights when designing a process recipe or device structures: blindly following simulation results can lead to off-target designs and frustration.

Table 4.2 is an outline of some important considerations during the device design and process development stage [39]. While this table provides a starting list, there are many other considerations in the real world. Some of these can be unexpected, such as performance variations from lot to lot and die to die that come from the fabrication equipment in spite of the best efforts of device and process engineers.

4.3.5 LAYOUT, TAPE-OUT, FABRICATION, AND TEST

After device and process design, engineers can now lay out the device test structures using layout tools such as Cadence Virtuoso. Often, engineers will place all types of devices including capacitors and resistors onto the same test wafer. Many design variations have to be considered for the devices, such as LDMOS with rectangle or racetrack shapes, analog MOSFETs with different orientations, digital MOSFETs with minimum feature sizes, and various capacitors. Before tape-out, a design rule checking (DRC) is normally performed to make sure that all the layouts obey the design rules. These design rules are made by collaboration between the various function groups, and many of them are derived from past experience and measurement data from experimental device structures. For a new generation of design, the DRC test structures should also be placed onto the test wafer. Obeying the design rules can help improve process yield by checking important parameters such as minimum width and spacing between devices. For readers that are not familiar with the word *tape-out*, it means to deliver the finished design masks to the fab engineers to make photomasks. The word is derived from the early days when the enlarged "artwork" for the photomask was "taped out" manually using black line tape and adhesive-backed die cut elements on sheets of polyester PET film [88] [89]. For most design engineers, the busiest months of the year are just before tape-out: since the mask-making process is very costly, they have to make sure everything they draw is correct according to the best of their knowledge. In addition to the costs involved, the turnaround time can be in months, so errors at this stage can introduce delays of several months for the next tape-out, which is unacceptable for time-sensitive products like cell phone chips.

After tape-out, engineers can relax a little bit while waiting for the silicon wafers to come back. In the meantime, the compact modeling group can ramp up to build some "paper models" based on their past experience and TCAD simulations. During the fabrication period, process integration engineers give feedback based on the current status of the process and may perform some tests by withdrawning some wafers during specific stages. TEMs or SEMs are taken and send back to device design engineers to verify the design or process if something unusual happens.

After the fabrication is finished, some of the wafers are held back by fab engineers for scribe grid process control (SGPC) tests and data gathering. The SGPC test structures are located at the unused scribe lines between dies, redundant locations where the wafer is cut into separate dies for packaging. These test structures are prepared by device engineers, and they are very important to monitor and control the process. Usually, automatic wafer-level tests are performed, and statistical data are collected and analyzed using software tools such as JMP from the SAS Institute [90]. After the SGPC tests, wafers are shipped back to the device engineers for further tests and characterization. Compact modeling engineers can use the measurement data to extract parameters for compact models used by IC designers.

Most of the time, one test vehicle is not enough, since new issues always pop up; multiple test vehicles (test mask sets) are often necessary. In that case, more tape-outs are carried out before finalizing the process technology and device designs.

4.3.6 RELIABILITY AND QUALIFICATION

Reliability engineers are responsible to make sure the devices designed and process technologies meet the contingent requirements for various applications. Note that reliability is not the same as quality. Reliability is something measured over the long term as a semiconductor device's performance gradually degrades over time depending on the environment and working conditions. Typical reliability tests include the TDDB, HCI, EM, and NBTI.

Gate dielectric materials can suffer a sudden breakdown after being subject to electrical stress over a prolonged period of time; depending on the gate oxide's quality, the exact time to breakdown varies from device to device. The mechanism for this breakdown is the formation of a conductive

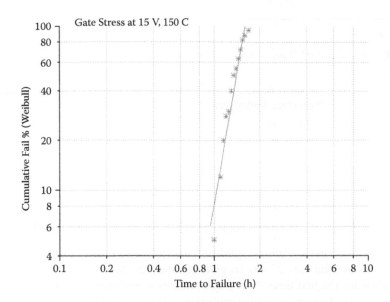

FIGURE 4.7   A typical cumulative fail reliability assessment using the Weibull distribution.

path within the thin gate oxide due to stress: many models [91] [92] [93] have been proposed by researchers in this field to analyze this phenomenon. For example, researchers from the University of California at Berkeley in [93] proposed an anode hole injection (AHI) model for reliability projection of oxide. Ultimately, the exact causes of oxide failure and the proper models to use to predict oxide reliability are still under debate. In industry, time to breakdown is projected with the help of accelerated measurements under high-voltage and high-temperature conditions. These tests often involve the extrapolation of the dielectric lifetime for test structure using scaling coefficients to account for area, temperature, and voltage. The Weibull distribution and percolation theory are used for the cumulative fail assessment [93]. Figure 4.7 shows a typical cumulative fail accelerated test result for TDDB.

HCI affects all MOSFETs. In this process, electrons and holes can gain sufficient kinetic energy to overcome a potential barrier necessary to break an interface state [94]. For digital and analog MOSFETs, hot carriers can tunnel through the thin gate oxide and contribute to gate leakage current. In LDMOS with STI in the drain-drift region, hot carriers injected and trapped in the STI corner can alter the device performance. For example, researchers have found the on-state resistance can be modified by hot carrier injection [95].

EM is the gradual movement of ions in a conductor due to the momentum transfer between conducting electrons and diffusing metal atoms [96]. Modern power IC technologies use copper instead of aluminum to reduce resistance from the metal interconnects. It also turns out that copper has better electromigration robustness than aluminum or Al/Cu alloy; however, while the advantage is significant it is not as great as originally anticipated [97]. Like TDDB, electromigration tests are routinely carried out at elevated temperatures to make sure it meets the desired specs.

Other tests like NBTI for PMOS, plasma-induced damage or antenna effect [98], and IDDQ are typically carried out.

4.3.7 SURVEY OF CURRENT SMART POWER TECHNOLOGY

A brief survey (as of the year 2013) of smart power technology foundry providers is shown in Table 4.3. Most foundries have multiple technology offerings for different power IC applications. For example, TSMC provides both lower voltage (60 V) and high voltage (800 V) services [99].

TABLE 4.3

A Brief Survey of Smart Power Technology Foundry Providers

	Smart Power IC Technology Nodes (μm)	Voltage Rating (V) (for PIC)	High-Voltage IC Technology Nodes (μm)	Voltage Rating (V) (for HVIC)
TSMC [99]	0.18 to 0.35	12 to 60	0.5 to 1.0	500 to 800
UMC [100]	0.18 to 0.60	12 to 120	0.5 to 0.6	700 to 800
Global Foundry [101]	0.13 to 0.18 (BCDlite)	5 to 60	-	
SMIC [102]	0.13 to 0.35	20 to 100	-	600

4.4 SMART POWER IC TECHNOLOGY: TECHNOLOGICAL PERSPECTIVE

Nowadays, advanced manufacturing has pushed the feature size from submicron to nanoscale. The latest Intel Ivy–bridge chip uses 22 nm Tri-gate technologies that brings 3D concepts to advanced CMOS fabrication for the first time. With improvements in extreme ultraviolet (EUV) lithography and other innovations, the semiconductor technology is aiming at 14 nm in a few years' time, so it will not be long before we reach the end of scaling where Moore's law is no longer valid. "More than Moore" is rapidly picking up momentum, with smart power IC technology being a typical application of this concept. As mentioned before, the feature size shrinking for smart power technology is several generations behind due to the trade-off between cost and performance so there is still lots of room for improvement. Today's smart power technology uses technology with a feature size of 0.13 μm and above. Some companies are still using feature sizes greater than 1 μm for high-voltage ICs since only a small number of digital gates are required.

A comparison of CMOS technology and smart power technology as of the year 2013 [39] is given in Table 4.4. A variety of device types like the passives (resistors and capacitors) are available for smart power technology, which makes system-on-chip applications possible. However, inductors and transformers are very difficult to integrate on the chip: they are either stand-alone on the PCB board or integrated in the package by advanced packaging technologies.

4.4.1 DEVICES FOR SMART POWER TECHNOLOGY

In advanced CMOS technology development, the ability to realize continued scaling without jeopardizing leakage and reliability is the key to success. Novel technologies such as high-k

TABLE 4.4

Comparison between CMOS Technology and Smart Power IC Technology as of 2013

	CMOS Technology	Smart Power Technology
Feature size	45 nm and below	0.13 μm and above
Wafer size	Mostly 12 inch	8 inch and below
Device types	NMOS and PMOS	LDMOS, BJTs, diodes, capacitors, resistors
Isolation	STI	STI, LOCOS, DTI, P-N Junction
Gate insulators	Oxide or high-K materials	Oxide
Gate materials	Metal	Polysilicon
Strain engineering	Available	Not available

TABLE 4.5

Typical Devices in a Smart Power Technology

Categories	Devices
Digital	NMOS and PMOS
Analog	NMOS and PMOS, BJT, Zener diode
Power	LDMOS, BJT, LIGBT, power diode
Capacitors	MOS capacitor, double-poly capacitor, metal-insulator-metal capacitor
Resistors	Diffused resistor, poly resistor
Trim	Fuse, anti-fuse, NVM

materials and metal gate stacks, 3D current flows with FinFET-like structures, stress and strain to increase carrier mobility, and phase-shifted masks and immersion lithography have driven silicon technology into the nanoregion. However, CMOS technology has only two basic device types (NMOS and PMOS), whereas a smart power technology may have over a dozen of device types. The considerations of smart power technology development are therefore quite different from those of CMOS. Leakage current is still a big consideration for smart power applications, but gate leakage is less of a concern due to the much thicker oxide layers used in this technology. In the context of smart power ICs, leakage refers to the substrate leakage current that should be suppressed to prevent cross-talk between devices. Since both power devices and digital devices coexist on the same chip, improving the isolation between devices is absolutely critical.

Further challenges for smart power ICs depend on the specific application. For example, an LDMOS may be used in high-side applications, where the substrate is still grounded while source and body are lifted to high potential; in that case, preventing the influence of the substrate depletion on the active devices is an important task. Another example is the electron-hole plasma generated during the on-state of LIGBT, which can generate cross-talk between high-voltage devices or between high-voltage and low-voltage devices. Special precautions are needed if one would like to take advantage of the lower on-resistance of LIGBT over LDMOS. Many other such examples exist and for every single case, multiple device technology parameters need to be optimized at the same time.

Table 4.5 gives list of devices used in smart power technology [39]. The requirements on these devices and the choice of technology depend on the intended application. Design considerations and optimizations will be discussed in the following sections.

4.4.2 DESIGN CONSIDERATIONS FOR SMART POWER IC TECHNOLOGY

4.4.2.1 Power MOSFETs

For power devices, breakdown voltage is one of the first design parameters to be considered; different applications require different breakdown values. For real-world applications, voltage spikes need to be considered so that a 12 V application may require a power device rating of 20 V and above. Additionally, a 20 V power LDMOS does not mean the breakdown voltage is exactly 20 V; quite often an extra 20% or 30% safety margin is added to account for device degradation and ensure the device meets specifications over its entire lifetime.

In power electronics, N-type LDMOS is often preferred over P-type LDMOS because electrons have higher mobility than holes in silicon; this means the nLDMOS will have a lower specific on-state resistance. For power device engineers, breakdown voltage and specific on-state resistance are almost always a trade-off, and meeting both goals at once is not easy. Over the past decades both RESURF (Reduced Surface Field) and Super Junction concepts were widely used, especially for high voltage ICs (> 100 V for RESURF, > 300 V for Super Junction). Both RESURF and Super Junction use the same charge coupling effect; RESURF is about two-dimensional optimization that can be applied multiple times (e.g., multiple-RESURF), whereas Super Junction is more about three-dimensional effects and usually consists of alternating N and P-regions in a direction perpendicular to the channel.

Gate oxide integrity is also important. For power devices, a high voltage is usually applied to the gate (> 5 V) to fully turn on the device so that low on-state resistance is achieved. However, high-voltage and high-temperature stresses on the gate oxide over a prolonged period can cause gradual oxide degradation and an eventual rupture. The previously mentioned TDDB test is an accelerated method of determining the oxide quality and its long-term reliability. Modern smart power technologies use different oxide thicknesses for different device categories (power, analog, and digital) so individual TDDB tests are necessary for every oxide thickness.

Other considerations of power devices include its energy capability, safe operating area (SOA), and thermal runaway (second breakdown) prevention. The suppression of the parasitic BJTs in power LDMOS is a very important design consideration; since these parasitic BJTs can prove catastrophic for power devices, a lower resistance path in the P-body region of LDMOS is often used to reduce this risk.

4.4.2.2 LIGBTs

For power switching applications with high current density, the on-resistance of an LDMOS will be too large unless the device size is very large; LIGBTs are usually used to avoid this problem [103] [104] [105] [106]. By using conductivity modulation in the structure, the on-resistance of an LIGBT can be many times lower than that of an LDMOS. For the same current and voltage rating, the chip size of an LIGBT can also be many times smaller than that of an LDMOS, but there is a trade-off between on-resistance and turnoff time: the longer turnoff time of an LIGBT is due to the existence of minority carriers (related to the conductivity modulation), which reduce the on-state resistance of the device significantly. Therefore, for low current density and fast switching applications, LDMOS should be used; for high current density and slow switching applications, LIGBTs are the better choice.

In terms of breakdown voltage design, both structures are similar except that an N buffered layer is used in LIGBT to prevent premature breakdown at the anode. There are two unique design parameters in LIGBT, which are not a major concern for LDMOS. The first one is the device latch-up due to the turn-on of the P+ anode/N-drift/P-body/N+ cathode parasitic thyristor. Once this parasitic thyristor latches on, gate turnoff of the device will not be possible. The turn-on mechanism of LIGBT latch-up is similar to the second breakdown mechanism for LDMOS. The difference is that LIGBT latch-up might cause only loss of gate control and not necessarily a catastrophic failure as in LDMOS. However, both devices require the P-body doping to be high enough to provide a low resistive path for the hole current. The second major concern for the LIGBT is the spreading of the minority carriers in the substrate. As LIGBTs are integrated with the control circuits in power IC applications, spreading of the minority carriers might cause cross-talk between power devices [107] or between power and low-voltage devices (e.g., CMOS and BJT [108]). Problems such as CMOS latch-up in neighboring circuits might be induced if proper precautions are not taken. Plasma confinement during system operation [109] and effective isolation between power devices and low-voltage devices were reported as effective methods for cross-talk prevention [110].

4.4.2.3   Analog Active Devices

Analog NMOS and PMOS devices mostly work in the saturation region, so important considerations for these devices include transconductance (G_m), output resistance (R_{out}), transistor matching, off-state current, and noise. Device orientation (channel orientation) may cause discrepancy for device parameters like threshold voltage (V_t). Even a small difference in V_t will cause a mismatch problem. For BJTs, one needs to consider the current gain, early voltage, matching, and noise.

Diode characteristics will depend on the kind of behavior desired. Power diodes have similar structures and voltage ratings as power LDMOS. Zener diodes, on the other hand, have heavily-doped P- and N-regions and are often meant to work under reverse bias to clamp and protect the power device gates from voltage spikes. The breakdown voltage of these devices is usually low. In fact, for Zener diodes with breakdown voltage up to 5.6 V, the Zener effect is the predominant effect and shows a marked negative temperature coefficient. Above 5.6 V, the avalanche effect becomes predominant and exhibits a positive temperature coefficient [111]. Strictly speaking, Zener diodes with breakdown voltage above 5.6 V are not pure Zener diodes.

4.4.2.4 Resistors

Depending on the application requirement, the integrated resistors can be diffused-well resistors or poly resistors. The sheet resistivity is determined by the process and by the doping type (N+ or P+). Temperature and voltage coefficients are two important considerations for resistors: in general, poly resistors have better accuracy and a lower voltage coefficient than diffused resistors, but diffuse resistors are a good choice if a large sheet resistance value is required.

4.4.2.5 Capacitors

There are several types of capacitors in a smart power IC technology: the most widely used are MOS gate capacitors (MOS-caps) and poly-insulator-poly (PIP or double-poly) capacitors. PN junction capacitors and metal-insulator-metal (MIM) capacitors are used as well. The voltage ratings for MOS caps are always lower than that of double-poly capacitors due to thin MOS gate oxide and thick poly to poly insulator, but MOS caps have higher capacitor density (Farads per unit area). The linearity, temperature, and voltage coefficients are also important design parameters.

4.4.2.6 Voltage and Frequency Trims

Fuse, anti-fuse, and or NVM are used as voltage and frequency trims in analog ICs. Since the accuracy of an analog IC is limited by the realities of the manufacturing process, trimming is often used to make adjustments to the IC after it is fabricated. This is typically done at wafer level on each individual die but may be performed on the die after packaging [112]. There are several ways to achieve anti-fuses in smart power IC, the technical details of which are beyond the scope of this book.

4.4.2.7 Logic/Digital NMOS and PMOS

For smart power IC technology, the most important aspect for logic design is the gate density. A 0.25 μm technology can pack more than 250 K gates for the digital part [113]. Other design considerations include the cost due to the added complexity of state-of-the-art CMOS process and noise isolation to protect the digital circuits from the analog and power devices.

4.4.2.8 System-Level Design and Fabrication Considerations

In summary, the following should be considered when doing system-level design:

1. Substrate injection isolation
2. Embedded parasitic BJTs
3. Noise immunity
4. Heat dissipation
5. High-voltage ESD

In practice, many other considerations related to physical effects of the fabrication must be considered. This includes the loading effect, which is caused by mask pattern density variations and the ion implantation proximity effect. The net result of these effects is a variation of device parameters and IC performance from die to die, from wafer to wafer, and from lot to lot. For economic

FIGURE 4.8   Schematic illustration of isolation methods.

reasons, device designs should therefore be sturdy and accommodate these variations while still meeting specifications.

4.4.3   ISOLATION METHODS

Isolation methods play an important role in determining how the integrated devices influence each other. Four isolation methods are commonly used: PN junction isolation; local oxidation of silicon (LOCOS); shallow trench isolation (STI); and deep trench isolation (DTI). Junction isolation relies on reverse biased PN junctions to isolate devices. It is easily realized by selective implantation and diffusion. The drawback of junction isolation is the high leakage current from the reverse-biased PN junctions, particularly at high temperatures. This approach also takes up a huge silicon area, so it is only popular for feature sizes larger than 0.5 μm: high-voltage ICs with breakdown voltage ratings exceeding 600 V is an example of this. LOCOS isolation also occupies a large silicon space due to the "bird's beak" extension and is typically applied to technology with a feature size larger than 0.35 μm.

Trench isolation methods such as STI and DTI are more advanced techniques that require much less silicon space and allow ICs to be much more compact; they are typically reserved for smaller feature sizes such as 0.25 μm and below. STI has been widely applied to advanced CMOS technologies. DTI is especially designed for smart power ICs to prevent interactions between devices. Sometimes DTI comes with SOI wafers to provide full isolation [114]. Unlike STI, a DTI trench is typically etched and filled with polysilicon and oxide liners. The quality of the trench sidewall (i.e., the silicon/trench oxide liner interface) is important for reliability and leakage issues [115]. Combining DTI with the buried oxide that comes with SOI wafers, the DTI + SOI method achieves the best isolation that current technology can offer. Figure 4.8 provides schematic illustrations of LOCOS, STI, and DTI isolation methods.

5 Introduction to TCAD Process Simulation

In previous chapters, we discussed power electronics systems and integrated circuits (ICs) the semiconductor industry in general, and the fundamentals of smart power IC technology by following a top-down approach. Starting from this chapter, we are getting into the lowest and most fundamental level of device engineering. Before we explain the integrated power devices that are used in today's power IC industry, we need to prepare the readers with some basic knowledge of technology computer-aided design (TCAD) simulation, since these will be used extensively from Chapters 7 to 10. TCAD simulations can be divided into two categories: process simulation and device simulation. This chapter discusses process simulation, and device simulation will be the topic of Chapter 6.

5.1 OVERVIEW

In the past decades, semiconductor manufacturing has gotten increasingly complex. The cost of development is rising sharply in spite of the ever-decreasing unit price of semiconductor chips. Driven by consumer electronics, electronic gadgets' lifetime and time to market are getting shorter and shorter. The traditional "trial and error" methodology is no longer feasible due to cost and time constraints. Computer-aided design (CAD) is universally accepted for design and development of ICs; for semiconductor process and device design, a special tool within this CAD category is called TCAD.

TCAD is widely used in industry and research institutes to simulate semiconductor manufacturing process technology; a different set of tools is also used to study the electrical, thermal, and optical properties of semiconductor devices. This chapter and Chapter 6 focus on general TCAD simulation applications. Since a detailed explanation of the physical models implemented in TCAD software can be found in most commercial software manuals, this chapter and Chapter 6 concentrate on the most useful and important aspects of TCAD simulation that an ordinary TCAD user will encounter for power devices. Advanced readers can refer to [39] for detailed explanations.

Process simulation is similar to virtual manufacturing of semiconductor devices. It starts with a wafer substrate and simulates most of the important process steps like ion implantation, deposition, oxide growth and diffusion, and etching. A modern process simulator can simulate both front-end of the line (FEOL) and back-end of the line (BEOL) processes with reasonable accuracy. This chapter gives the reader an introduction of process simulation based on one of the most popular industry standard process simulators: Suprem IV, which was originally developed by Stanford University.

A simplified process simulation flowchart is illustrated in Figure 5.1.

5.2 MESH SETUP AND INITIALIZATION

Process simulation always starts with mesh definition and initialization. Let us first take a look at a simple example:

```
line x location = 0.00000 spacing = 0.250000 tag = left
line x location = 5.00000 spacing = 0.250000 tag = right

line y location = 0.00000 spacing = 0.1 tag = top
line y location = 5.00000 spacing = 0.5 tag = bottom
```

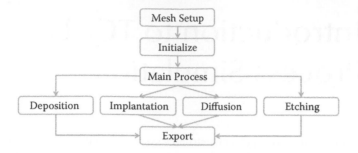

FIGURE 5.1 Simplified process simulation flowchart.

```
region silicon xlo = left xhi = right ylo = top yhi = bottom
boundary exposed xlo = left xhi = right ylo = top yhi = top
boundary backside xlo = left xhi = right ylo = bottom yhi = bottom

initialize boron conc = 1.0e15 orient = 100
structure outfile = sub.str
```

This example simulates a silicon substrate with a width of 5 μm (x direction, pointing to the right) and a thickness of 5 μm (y direction, pointing downward). The simulation result is shown in Figure 5.2.

Every process simulation starts with mesh definition, and a well-defined mesh can increase simulation accuracy and efficiency. A rule of thumb is denser mesh required at locations where a large gradient of doping concentration exists, where contacts will be defined or at particularly interesting locations like a MOSFET channel region. The "line" command specifies the location of a mesh line. Vertical and horizontal lines are used, and the software divides the device into a triangular mesh by splitting the rectangles defined by these lines. This means that at the very least two lines need to be defined to enclose the meshed area. We note that in the Stanford convention for Suprem, the positive Y direction is pointing downward into the substrate.

An additional parameter called "spacing" is used to automatically put mesh lines between the specified lines of the input file. While additional mesh lines can always be added manually, using spacing allows mesh lines to be added in a smooth manner. The spacing of the supplemental lines will vary geometrically from one end of the interval to the other [116] so that the specified

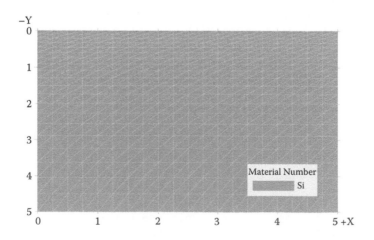

FIGURE 5.2 Process simulation result of a simple silicon substrate.

spacing value (in microns) is used at both ends. Additional lines are then introduced using a smoothing procedure so that adjacent lines intervals have a ratio no greater than 1.5.

Another parameter in the line command is "tag," which allows user to specify which line command has one of the four boundaries: left, right, top and bottom. These tags will be used in the "region" and "boundary" commands to define the initial simulation conditions.

The "region" command should follow "line" command to set the rectangular mesh boundary. The string value given in the "tag" can be used to set the parameters of "xlo," "xhi," "ylo," and "yhi," which are abbreviations of the low point of x, high point of x, low point of y, and high point of y. At least one material should be specified for one particular region. In this example, silicon is the only material within the mesh boundary.

The "boundary" command is used to specify what conditions to apply at each surface in a rectangular mesh. One of three boundary conditions can be chosen for a particular surface: reflecting, exposed, or backside. In this example, the first boundary command set the top surface of the substrate as exposed. Materials are deposited only on exposed surfaces. Impurity pre-deposition also happens at exposed surfaces, as does defect recombination and generation [116]. The second boundary command defines the bottom of the substrate as backside. Backside surfaces roughly correspond to a nitride-capped or oxide-capped backside; defect recombination and generation happen here. Reflecting surfaces correspond to the sides of the device and are also good for the backside if defects are not being simulated: this is the default setting for all surfaces.

With the mesh and boundaries defined, we can now initialize the process simulation. The initialize statement sets up the mesh and also initializes the background doping concentration and wafer surface orientation. In this example, the substrate has an initial boron doping concentration of 1E+15 cm^{-3} and wafer surface orientation is (100).

Finally, the "structure" command allows the user to save the file to a certain file format so that the structural information such as material and net doping can be visualized. The saved data can also be used as a restart point for future simulations.

5.3   ION IMPLANTATION

The most widely used doping technique in modern integrated circuits technology is ion implantation, in which ionized dopants atoms are accelerated by an electrostatic field that then strikes and penetrates the target material. The dose can be controlled by controlling the ion current and the profile depth is controlled by the implantation energy which can be adjusted through the field strength. Since dose and depth can be adjusted electrically, ion implantation is highly controllable and reproducible [117] [118]. The implanted ions cause implantation damage by displacing the wafer atoms (defects generations) or by creating clusters and precipitates: a subsequent thermal annealing is therefore necessary to heal as much of the implantation damage as possible. A rapid thermal anneal (RTA) at very high temperatures can be used to repair the damage with minimum dopant diffusion. Ion implantation into a single crystal causes another problem due to the anisotropy of the target: along the major crystal axes, the implanted impurities collide with fewer target atoms and can thus penetrate deeper into the crystal. This phenomenon is called channeling [4] and can be significantly reduced by slightly tilting the ion beam against the preferred orientation [116] [117] [119] [120]. For process simulation of the implant depth profile, analytical methods based on both collision theory and tabulated experimental results are used.

5.3.1   ANALYTICAL MODELS

Using the analytical model, for one-dimensional analysis the distribution of implanted dopants $C(y)$ is given by

$$C(y) = N_d f(y) \qquad (5.1)$$

TABLE 5.1

Comparison of Gaussian, Pearson IV, and Dual Pearson IV Models

Models	Description
Gaussian	The Gaussian model is simple with only two parameters (projected range R_p and standard deviation σ_p). However, this distribution is only a first-order approximation and can be substantially different from real SIMS data. As an example, both phosphorus and boron can travel deep into the substrate if channeling along the [121] direction is not minimized.
Pearson IV	Pearson IV is more complicated with four parameters (projected range R_p, standard deviation σ_p, skewness γ and the kurtosis or excess β) and is more accurate than the Gaussian model to predict a doping profile in many applications. However, it is not really suitable to predict the typical tail observed when channeling phenomenon are presents [116].
Dual Pearson IV	Dual Pearson IV is the most accurate analytical model for Suprem. It features a combination of Pearson IV distribution in both amorphous material and cystalline material so that the channeling tail can be properly modeled. The original Suprem IV only had the Pearson IV model, which was extended by commercial versions such as CSUPREM into the Dual Pearson IV [116] model.

---

where N_d is the implantation dose per unit area. $f(y)$ is the probability density (or frequency function) and can be either a Gaussian or Pearson IV distribution.

The 2D ion implantation profile is based on the convolution of the 1D lateral (x-coordinate) and 1D longitudinal (y-coordinate) profiles and can be formulated in general form as

$$C(x, y) = N_d C_1(y, t(x) * C_2(x)) \tag{5.2}$$

where $*$ is the convolution product, and $t(x)$ represents the thickness of the mask or coating when the implantation is done through a mask (e.g., as pad oxide); if no mask is present then $t(k) = 0$. C_1 is the depth-dependent longitudinal concentration, which is a Gaussian or Pearson IV distribution. C_2 is the lateral doping concentration and is usally represented with a Gaussian distribution.

A comparison of Gaussian, Pearson IV, and Dual Pearson IV models is shown in Table 5.1.

5.3.2 MULTIPLE-LAYER IMPLANTATION

Quite often, ion implantation has to go through multiple layers. For example, a screen oxide is frequently used in manufacturing for ion implantation; to reach the silicon bulk region, charged ions have to go through the oxide layer first. Simulators such as Suprem IV have to caculate the ion distribution profile in all the layers. A simulation example is shown in Figure 5.3 with three implant layers: from top to bottom; silicon nitride (Si_3N_4); oxide and silicon. All three layers have a uniform thickness of 0.1 μm. Phosphorus is implanted from the top surface, and we see the discontinuity of the profile at the material interfaces, which is caused by different stopping-power coefficients in those regions [121].

5.3.3 MONTE CARLO SIMULATION

The Monte Carlo method is based on statistical description of interatomic scattering events. A sophisticated Monte Carlo simulator would detail the scattering events using interatomic potential fitted to quantum mechanical calculations.

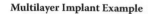

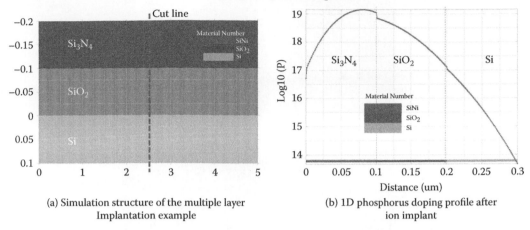

(a) Simulation structure of the multiple layer
Implantation example

(b) 1D phosphorus doping profile after
ion implant

FIGURE 5.3 Multiple-layer implant example: (a) Material structure; (b) Phosphorus implant profile in three layers.

Monte Carlo simulation uses repeated sampling of random collision events to compute the results. This can provide very accurate simulation results for ion implantation, but the drawback of this approach is that it is very time-consuming. An example is made using a complimentary 2D Monte Carlo simulator tool from Axcelis [122] for a phosphorus implant in silicon. The total number of ions used in this simulation is 1 million, as illustrated in Figure 5.4. Monte Carlo simulation results can be saved in the implant table and used by process simulators to achieve highly accurate ion implantation simulations.

As a simple example, a command in Suprem IV for ion implantation is

```
implant phosphorus dose = 1e13 energy = 30
```

This statement will instruct the simulator to implant phosphorus with a dose of 1E+13 cm^{-2} and implant energy of 30 keV.

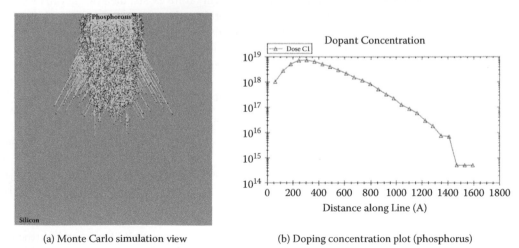

(a) Monte Carlo simulation view

(b) Doping concentration plot (phosphorus)

FIGURE 5.4 Monte Carlo simulation example: (a) simulation view; (b) doping concentration plot along the y direction at the center point of the x-axis.

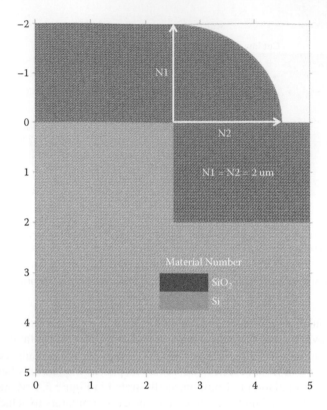

FIGURE 5.5   Illustration of deposition simulation with curvature.

5.4   DEPOSITION

In most process simulators, deposition is purely geometrical. The algorithm works by finding exposed surfaces and adding a new mesh layer on top of the existing mesh to mimic the physical progress of deposition. For flat surfaces, this task is easy, and the new layer just inherits the original x mesh while the y mesh is defined by user input. When a curvature exists, then two normal vectors (N1 and N2) are drawn from the original mesh line and projected for a given deposition thickness; a series of mesh points following the arc connecting the tips of N1 and N2 is then drawn to create a smooth deposit surface line [39]. Shown in Figure 5.5 is a deposition simulation example with curvature; here, a 2 µm thick oxide is deposited at a stepped silicon surface.

Please note that epitaxial growth in manufacturing is often simulated as a deposition; *in situ* doping of this deposited layer is also possible with Suprem IV. As a simple example, a command in Suprem IV for deposition is

```
deposit oxide thick = 0.01
```

This statement will deposit an oxide layer with a thickness of 0.01 µm.

5.5   OXIDATION

Oxidation is the process of converting silicon on the wafer surface into oxide. This chemical reaction already starts at room temperature, and the resulting very thin oxide film is called native oxide. The oxidation rate can be accelerated by exposing the wafer to oxygen or water vapor at high temperature. The growth of the thermally grown oxide depends on temperature, pressure, crystal orientation, doping level, the impurity contamination of the silicon substrate, and stress due to nonplanar wafer structures [117] [123]. The grown oxide is commonly used as an insulator between single devices to act as a gate oxide in MOS structures and to serve as a mask against dopant implantation [117]. There are two techniques to get the silicon dioxide (SiO$_2$) from pure silicon [116]: dry oxidation and wet oxidation.

5.5.1 DRY OXIDATION

The silicon wafer is settled to a pure oxygen atmosphere at a temperature above 900°C. At the surface of the wafer a chemical reaction

$$Si + O_2 \Rightarrow SiO_2 \tag{5.3}$$

takes place. Because of a very low oxidation velocity (< 100 nm/h), this method is often used to generate high-quality thin oxide layers [117].

A simple command for dry oxidation is

```
diffuse time = 1 temp = 1000 dryo2
```

This statement will ask the simulator to grow a dry oxide at 1000°C for 1 minute.

5.5.2 WET OXIDATION

The silicon wafer is settled to a water vapor atmosphere and the reaction equation

$$Si + 2H_2O \Rightarrow SiO_2 + 2H_2 \tag{5.4}$$

This process has a rather high oxidation velocity (>20 nm/min) and is therefore often used to get thick oxide layers for insulation and passivation techniques.

A simple command for wet oxidation is

```
diffuse time = 1 temp = 1000 weto2
```

This statement will ask the simulator to grow a wet oxide at 1000°C for 1 minute.

5.5.3 OXIDATION MODELS

From a modeling point of view, process simulators provide both analytical and numerical models of the oxidation. Numerical models treat the oxide as a viscous incompressible fluid [124] [125] [126] [127] [128]; this involves solving the simplified Navier-Stokes equations inside the growing oxide [124] [125]. Interstitials injection phenomenon (a.k.a. oxidation enhanced phenomenon) from oxide to the bulk of silicon [129], [130], segregation of dopants (e.g., segregation of boron to oxide) phenomenon and stress effects are also included in the Suprem numerical models.

5.5.3.1 Moving Boundaries

The main problem encountered in the simulation of the oxidation is the movement of the oxide/ silicon interface. This means that for each time step we solve the equations of the numerical model, regenerate the mesh, and then recalculate the interface boundaries. This requires an efficient mesh generator and efficient data management for this huge and frequently updated data [131]. From a numerical point of view, finite element discretization is preferred for oxidation simulation since the grid quality can be relaxed compared with the Voronnoi finite box technique. New process technologies involve oxidation of nonrectangular structures [132] [117] so a three-dimensional oxidation simulator is needed to model arbitrary structures and include the three-dimensional effects [126].

5.5.3.2 2D Analytical Models

The analytical models in the Suprem process simulator are based on the Deal Grove model [133] [134]. The basic idea of the model is the assumption of a steady state between three fluxes:

1. *Oxidant's flux* (F_1)

F_1 is the flux of the oxidant (O_2, H_2O,...etc.) from the bulk of gas to the gas/oxide interface.

2. *Diffusion flux* (F_2)

F_2 is the diffusion flux across the oxide given by Fick's law.

3. *Reaction flux* (F_3)

F_3 is the reaction flux corresponding to the oxidation reaction at the oxide/silicon interface.

At steady state, these fluxes are identical and can be expressed as

$$F_1 = F_2 = F_3 = F \tag{5.5}$$

The flux of oxidant reaching the oxide/silicon interface is described in 1D by the differential equation developed by Deal Grove:

$$N_1 \frac{dx_{ox}}{dt} = F \tag{5.6}$$

where N_1 is the number of oxidant molecules incorporated into a unit volume of oxide and (dx_{ox})/dt represents the oxide growth rate. The solution to the 1D-oxide rate equation yields the final oxide thickness x_{ox}. It has been found that initially the oxide thickness (x_{ox}) versus time (t) curve is similar to linear growth rate, which then becomes parobolic. This is seen in Figure 5.6, which is a typical representation of oxide growth [135].

The analytical models can be used only for simple planar structures. Most of the time, numerical models are used in TCAD simulation.

5.5.3.3 Numerical Models

From a mathematical point of view, the numerical oxidation models can be described by a coupled system of partial differential equations. The diffusion equation describes the diffusion of the

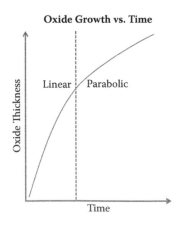

FIGURE 5.6 Oxide growth curve.

TABLE 5.2
Comparison of Numerical Models for Oxidation

Models	Planar or Nonplanar	Stress Effect	Accuracy	Speed
Vertical	Only for Planar	No stress effect	Low	Fast
Compress	Planar and Nonplanar	No stress effect	Medium	Medium
Viscous	Planar and Nonplanar	With stress effect	High	Slow

oxidant in the existing silicon dioxide. The chemical reaction equations describe the transformation of silicon to silicon dioxide. Finally, the displacement equations are about the displacement and expansion of the oxide layer and neighboring materials.

At high temperatures, the oxide layer is modeled as an elastic [126], viscoelastic [136] [137] [126] [128], or viscous compressible or incompressible flow [124] [125] [126] [127] [128]. Due to unknown motion of the interface between silicon and silicon dioxide this leads to a free boundary problem [126] [127] [128].

Three numerical models are provided by Suprem: vertical, compress, and viscous. The vertical model indicates that the oxidant should be solved for and that the growth is entirely vertical. The compress model treats the oxide as a compressible liquid. The viscous model treats the oxide as an incompressible viscous liquid. Real oxide is believed to be incompressible; however, the compressible model is a lot faster, so there is a trade-off between accuracy and speed [116]. These models are compared in Table 5.2.

5.5.3.4 LOCOS Growth Example

Local oxidation of silicon (LOCOS) is a typical process step in power integrated circuits. Here a simulation example is given so that readers can get familiar with the oxidation simulation. First an oxide layer of about 40 nm is thermally grown and followed by a deposition of silicon nitride (Si_3N_4) of about 80 nm thick as an oxidation mask. Silicon nitride is then selectively etched, leaving a window for LOCOS growth. A wet oxide thermal process at 1000°C for 90 minutes is subsequently carried out and the nitride is etched away at the end, leaving the bird's beak-shaped LOCOS. Interested readers can refer to [135] for more detailed process steps of LOCOS growth. Figure 5.7 illustrates the LOCOS growth simulation with enlarged view of the "bird's beak" after wet oxidation (before nitride is removed).

5.6 ETCHING

Like deposition, etching in the process simulation is purely geometrical, and no chemical reaction is included. This is relatively easy if the required etching shape simply has straight edges, be they purely vertical or at an angle. However, if an arbitrary shape is required, we need to specify some parameters for the Suprem process simulator.

Reactive ion etching (RIE) is a mainstream etching method in modern IC technologies: put simply, the unmasked silicon surface is subjected to high-energy ion bombardment that knocks out the material being exposed [39]. Ideally, no side wall should be created by RIE, but in reality there is always a sidewall effect due to the isotropic etching rate. In this case, using a geometrical etching method is sometimes necessary; the etching rate (μm/s) and etching time (s) in both the vertical and horizontal directions must be provided. Figure 5.8 is an example of geometrical etching creating a curved surface.

A simple example of etching command is something like this:

```
etch silicon thick = 0.2 dry
```

This will ask the simulator to perform a dry etching of silicon with an etching depth of 0.2 μm.

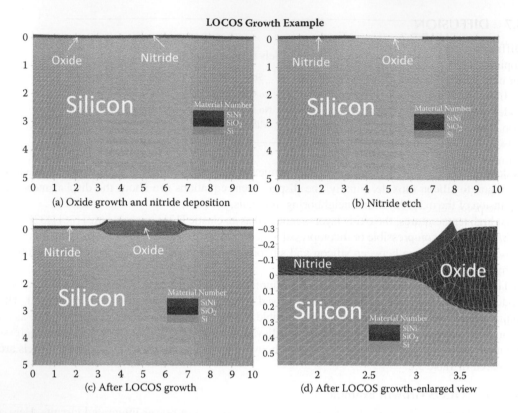

FIGURE 5.7 LOCOS growth example.

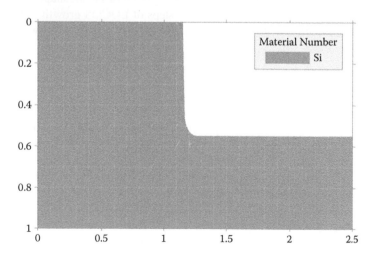

FIGURE 5.8 A geometrical etching example.

5.7 DIFFUSION

Diffusion is one of the main challenges in designing a process flow, since the placement of the active doping regions need to be accurately controlled. Proper selection of diffusion conditions is crucial for obtaining desired device electrical characteristics [135].

In terms of process simulation, diffusion is by far the most time-consuming step if the number of nodes is large. The latest models of impurity diffusion are often based on the concept of pair diffusion. These models account for interactions not only between impurities and space charge but also between impurities and lattice point defects such as interstitials and vacancies [39]. Impurity atoms cannot diffuse on their own and need neighboring point defects as diffusion vehicles so it is common to denote a diffusing dopant involved in a diffusion mechanism as a dopant–defect pair. When the binding energy between an impurity and a neighboring defect is not too weak, the impurity and the the point defect will move as a dopant–defect pair until they get separated by recombination or other effects. Two types of defects exist: vacancies and interstitials. Thus, we label a dopant A paired with a vacancy V as an AV pair and a dopant A paired with an interstitial I as an AI pair. Diffusivities of dopants always represent the diffusivities of the dopant–defect pairs.

The recombination between the paired dopants and other species (as unpaired defects) is not included in the equations in Suprem. The chemical reaction rate equations of these types of recombination are $AV+I \Leftrightarrow A$ in case of the Frank–Turnbull recombination and $AI+V \Leftrightarrow A$ in case of a dissolution process [117].

5.7.1 DIFFUSION MECHANISMS

Suprem includes three basic diffusion mechanisms for dopant diffusion models [138] [139] [140] [117].

5.7.1.1 Direct Diffusion Mechanism

In direct diffusion, impurities with small ionic radii can travel directly from one interstitial site to another. Group I and Group III elements in particular tend to diffuse directly and are therefore fast diffusers.

5.7.1.2 Vacancy Mechanism

In this diffusion mechanism, a substitutional dopant exchanges its position with an adjacent or neighboring vacancy. The vacancy should move at least to a third neighbor site away from the dopant to prevent oscillations. For more details about this regime, see [139].

5.7.1.3 Interstitials Mechanism

This mechanism is also called a kick-out mechanism. In the case of silicon, the dopant at a substitutional site is approached by a silicon interstitial and kicked out to reside at an interstitial position while the original self-interstitial has disappeared by occupying the regular lattice site. This interstitial dopant can now move toward an adjacent lattice site to regenerate a self-silicon interstitial by the same kick-out mechanism.

The most commonly used P-type dopant (boron) tends to diffuse via interstitials. The N-type dopant phosphorus tends to diffuse via interstitials at low doping and via a dual mechanism at high doping concentrations, with a strong bias toward vacancy diffusion. Arsenic also uses interstitials and vacancies for its diffusion. However, its interstitial mechanism is limited due to the relatively large ionic radii [117], and large impurities such as antimony diffuse only via vacancies. The rate equations for the basic atomistic diffusion mechanisms are

$$A+V \Leftrightarrow AV \tag{5.7}$$

$$A+I \Leftrightarrow AI \tag{5.8}$$

Unlike impurities, point defects can travel on their own, so their diffusion models are based on two concepts: paired and unpaired diffusion. Paired diffusion is already implicitly included in the impurity diffusion models, so an extra equation only for the unpaired defect diffusion needs to be

TABLE 5.3

Chemical Reactions of Pair Diffusion Models with Interstitials (I) and Vacancies (V)

Silicon	Boron	Phosphorus	Arsenic	Antimony
$I+V \Leftrightarrow 0$	$B+I \Leftrightarrow BI$	$P+I \Leftrightarrow PI$	$As+I \Leftrightarrow AsI$	$Sb+V \Leftrightarrow SbV$
	$BI+V \Leftrightarrow B$	$P+V \Leftrightarrow PV$	$As+V \Leftrightarrow AsV$	$SbV+I \Leftrightarrow Sb$
		$PI+V \Leftrightarrow P$	$AsI+V \Leftrightarrow As$	
		$PV+I \Leftrightarrow P$	$AsV+I \Leftrightarrow As$	

added as the diffusivity of paired and unpaired point defects can be very different. The diffusion of defects is also limited or stopped by the recombination effects. The recombination reaction between interstitials and vacancies is also included in the Suprem models. Table 5.3 lists the chemical reactions of pair diffusion models with interstitials (I) and vacancies (V).

A simple command for diffusion works like this:

```
diffuse time = 1 temperature = 1050 nitrogen
```

This will allow the simulator to perform a diffusion process simulation at a furnace temperature of 1050°C for 1 minute within the nitrogen enviroment.

5.7.2 Diffusion Models

There are four different diffusion models in Suprem. The accuracy of each model depends mainly on how the point defects are modeled and simulated.

5.7.2.1 Fermi Diffusion Model

This model is the fastest in CPU time. In this simple model, the free point defects equations are not solved since they are assumed to be in thermodynamic equilibrium. In this model, only the equations representing the diffusion of dopant–defect pairs are solved, and their diffusivity coefficients are supposed to be time-independent. This model therefore cannot be used to simulate transient-enhanced, oxidation-enhanced, or retarded diffusion in which the free point defects are not in thermodynamic equilibrium.

5.7.2.2 Two-Dimensional Diffusion Model

In this model, the free point defects can evolve over time, and the equations representing their diffusion are solved. The diffusivity coefficients of the dopant–defects pair diffusion equations are enhanced to include the transient effects of the free point defects so the diffusion of the dopants is strongly affected by the diffusion of the free point defects. In this model, the diffusion of free point defects is not affected by the diffusion of dopants: mathematically speaking, the diffusion equations of free point defects are decoupled from the diffusion equations of dopants, and there is only one-way coupling between impurities and defects. This model can be used to include the oxidation-enhanced or retarded-diffusion effects and costs more CPU time than the Fermi model.

5.7.2.3 Fully Coupled Diffusion Model

In this model, the diffusion of the free point defects is affected by the diffusion of the dopants. This coupling adds the dopant–defect pair diffusion fluxes to the diffusion fluxes of the free point defects. This model is an extension of the two-dimensional model, which in turn is an extension of the Fermi model. The fully coupled model is physically more accurate, and it was used in different process simulations to predict different physical effects observed in the processing. However, even this model still suffers from not being able to take into account the explicitly paired and unpaired dopants point defects diffusion.

TABLE 5.4

Oxide Enhanced (or Retarded) Diffusion of Several Impurities

	Boron (B)	Phosphorus (P)	Arsenic	Antimony (Sb)
Oxide-enhanced diffusion	X	X	X	
Oxide-retarded diffusion				X

5.7.2.4 Steady-State Diffusion Model

This model is a special case of the two-dimensional model in which the point defects are assumed to be at steady state.

5.7.2.5 Oxide Enhanced (Retarded) Diffusion

It has been found that some dopants appear to have their diffusion coefficients enhanced or reduced when the surface of the silicon is oxidized. When oxide grows, there is a 30% expansion that occurs to form the SiO_2 structure and interstitials are injected to relieve the compressive stress. Impurities such as boron and phosphorus have a tendency to diffuse through interstitials so an enhanced diffusion rate is observed. On the contrary, antimony seems to diffuse with vacancies. Since the injected interstitials reduces the population of vacancies in the silicon bulk (through recombination [135]), the diffusion of antimony is retarded during oxidation. Table 5.4 shows the oxide-enhanced (or retarded) diffusion of several impurities.

Figure 5.9 is a simulation result of buried n+ (phosphorus implant) layer diffusion with oxidation. We see a noticeable diffusion enhancement of phosphorus at the location where oxidation took place.

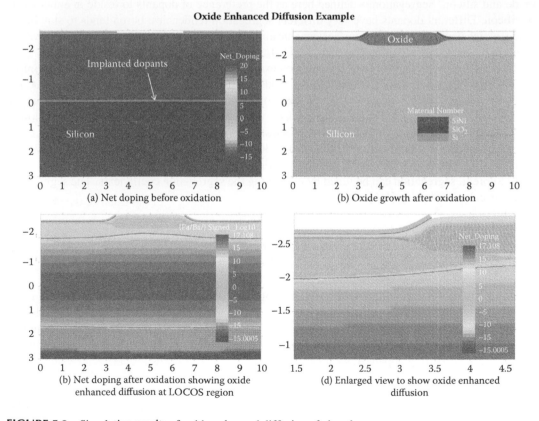

FIGURE 5.9 Simulation results of oxide enhanced diffusion of phosphorus.

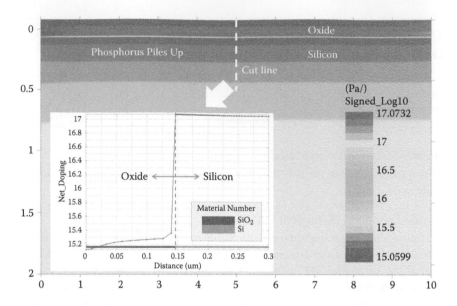

FIGURE 5.10 Phosphorus segregation example.

5.8 SEGREGATION

A problem device design engineers often faced with is dopant segregation at the interface between oxide and silicon. Segregation is defined here as the preference of dopants to reside in either oxide or silicon. Different dopants have very different segregation tendencies: boron tends to stay in the oxide, which depletes the silicon surface; arsenic and phosphorus prefer staying in the silicon, which result in a pile up at the oxide/silicon surface [121].

To clarify the segregation effect, a simulation example starts with silicon substrate with an initial phosphorus concentration of 1E+17 cm^{-3}. A dry oxidation step is then done at a temperature of 1000°C for 300 minutes, which yields an oxide thickness of about 800 Å. From Figure 5.10 we see that the phosphorus dopants are piled at the silicon surface, while the doping level in the oxide is much lower.

A comparison is made with a boron-doped substrate with a doping concentration of 1E+17 cm^{-3}. After dry oxidation (temperature of 1000°C and 300 minutes), the boron is depleted in the silicon surface while piling up in the oxide, as shown in Figure 5.11.

In Suprem, segregation can be tuned by modifying the segregation coefficients. The segregation model is calculated using [116]:

$$\vec{J}_{seg} \cdot \vec{n} = T_r \left(C_1 - \frac{C_2}{M_{12}} \right) \qquad (5.9)$$

where $\vec{J}_{seg} \cdot \vec{n}$ is the normal flux of dopants flowing from material 1 to material 2 across the interface. C_1 and C_2 are the concentrations in materials 1 and 2, respectively. T_r is the transport velocity, which takes into account the rate of dopants crossing the interface. M_{12} is the segregation term taking into account the variation of the threshold solubility of dopants in the two neighboring materials [39]. For each diffusing dopant A, we have

$$M_{12} = \frac{TS_1(A)}{TS_2(A)} \qquad (5.10)$$

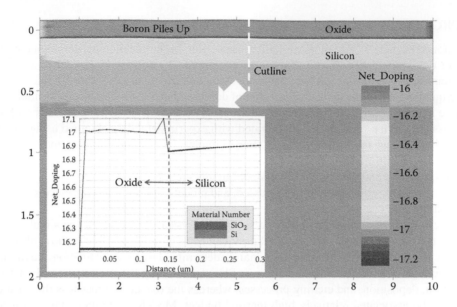

FIGURE 5.11 Boron segregation example.

where $TS_1(A)$ is the threshold solubility of the dopant A in material 1. $TS_2(A)$ is the threshold solubility of the dopant A in material 2, while M_{12} and T_r are calculated using a standard Arrhenius relationship. The process simulator allows the user to set these parameters using the impurity name as a command with parameters such as Seg.0, Seg.E, Trn.0, and Trn.E. The meanings of these parameters are listed in Table 5.5 [39].

The Suprem default segregation coefficient parameters for various dopants (at (100) orientation) are listed in Table 5.6. Note that these segregation coefficients are for silicon/oxide interface only.

TABLE 5.5

The Segregation Coefficients Parameters and Meanings

Parameters	Meaning
Seg.0	Pre-exponential constant for segregation
Seg.E	Activation energy for segregation
Trn.0	Pre-exponential constant for transport
Trn.E	Activation energy for transport

TABLE 5.6

The Suprem Default Segregation Coefficient Parameters for Various Dopants

Dopants	Seg.0	Seg.E	Trn.0	Trn.0
Boron	1126	0.91	1.66e-7	0
Phosphorus	30	0	1.66e-7	0
Arsenic	30	0	1.66e-7	0
Antimony	30	0	1.66e-7	0

5.9   PROCESS SIMULATOR MODELS CALIBRATION

As mentioned earlier, TCAD plays an indispensable role in the technology development of semi-conductor devices and ICs. TCAD enables the experimental results to be simulated and analyzed before running any experimental lots; the results obtained can be used to improve the DoE (design of experiment) analysis and reduce the number of development lot, which reduces the total development time and cost [141] [142]. However, the accuracy of TCAD simulation depends on how well the process and device simulators are calibrated with experimental data obtained using a particular set of equipment and fabrication facilities. For example, if the simulators are not calibrated and are used with default model parameters only, an unacceptable discrepancy between the simulated results and experimental data might be obtained. In this section, a general approach to calibrate the models for a process simulator will be discussed, and models calibration for a device simulator will be given in the next chapter.

A process simulator has to be calibrated such that physical properties of the simulated device, especially the doping profiles and film thickness, can be as close as possible to those of the experimental device [143]. The doping profiles are mainly determined by ion implantation and thermal diffusion processes. The thickness of the different films for building the device is mainly determined by oxidation, deposition, and etching processes, whereas the oxidation process is the most critical one due to the associated relatively high thermal budget. Models calibration starts with running shoot-loop experiments with various process conditions. For examples, in addition to the species of impurities, dose and energy are varied in those shoot-loop experiments for the ion implantation, and temperature and duration are varied in those shoot-loop experiments for the thermal diffusion and oxidation processes. The range of the process conditions of the short-loop experiments should at least cover the range of the process conditions of the desired process flow. The number of short-loop experiments needed depends on the required accuracy of the calibration. It can be over 20 [144] or even over 100 [145] for very accurate calibration. The experimental doping profiles can be extracted by using secondary ion mass spectroscopy (SIMS) or spreading resistance profiling (SRP), and the film thickness can be extracted by using scanning electron microscope (SEM) or transmission electron microscope (TEM). The experimental data are compared with the simulated results generated by the process simulator. If the discrepancy between the simulated results and experimental data is unacceptable, the default models and model parameters used in the simulator have to be adjusted until the discrepancy becomes acceptable.

Calibration of a popular process simulator, TSUPREM-4 [146], is given in the following section as an example for illustration. The default analytical model for the ion implantation process is the Pearson distribution. Four model parameters (range, standard deviation, skewness, and kurtosis) with respect to different impurities, dose, and energy of the implantation can be adjusted to calibrate the simulated doping profiles with the experimental ones extracted by SIMS or SRP. The damages to silicon (e.g., point defects) produced during the implantation, which are critical to transient-enhanced diffusion (TED), can also be modeled in the process simulator. Furthermore, the diffusivities of the various impurities in the process simulator can be adjusted for the doping profiles calibration. Modeling of thermal diffusion of point defects and segregation of impurities at material interfaces is also allowed for a more precise calibration of doping profiles. Moreover, the oxidation rate in the process simulator can be adjusted so that the simulated oxide thickness matches the experimental oxide thickness extracted by SEM or TEM.

5.10   INTRODUCTION TO 3D TCAD PROCESS SIMULATION

With ever more complicated structures, three-dimensional effects are getting a lot of attention from power semiconductor device designers. 3D TCAD simulation is becoming more and more important, since a successful 3D simulation can lead to better understanding of 3D effects in modern device designs. In the past, 3D TCAD was very difficult to realize because of constraints on computing

Some 3D Process Simulation Examples

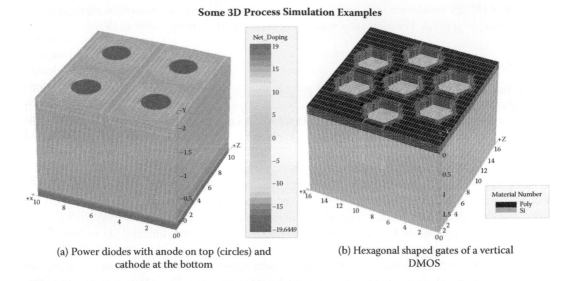

(a) Power diodes with anode on top (circles) and cathode at the bottom

(b) Hexagonal shaped gates of a vertical DMOS

FIGURE 5.12 Some 3D TCAD process simulation examples.

power. With the dramatic improvement of CPU speed and multicore architecture, 3D simulation is becoming a daily practice for device designers.

3D TCAD mesh generation can be done in two different ways. The traditional method used in mechanical and fluid simulations relies on pyramid-like mesh elements and is popular because of relatively mature mesh generation algorithms. However, it comes at the price of huge total mesh count when mesh density changes rapidly.

Another method uses stacked planes with prism-shaped mesh elements; this results in less overall mesh and much better convergence. One main advantage of plane stacking is that an optimized 2D mesh can be directly imported to a 3D simulation. Interested readers can refer to [39] for more information. Figure 5.12a is a 3D TCAD process simulation of a diode, with anode (circle) on top and cathode at the bottom. Figure 5.12b is a simulation mesh structure of vertical DMOS with multiple hexagonal-shaped poly gates [147] [148].

A three-dimensional circular shaped LOCOS oxidation is shown in Figure 5.13. The finished 3D LOCOS with a nitride layer removed looks like a "saucer." The process parameters are the same

Some 3 D Process Simulation Examples

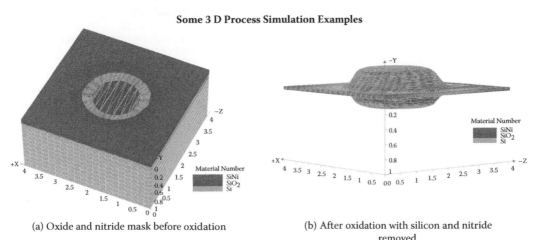

(a) Oxide and nitride mask before oxidation

(b) After oxidation with silicon and nitride removed

FIGURE 5.13 3D oxidation of a circular-shaped LOCOS.

as the 2D LOCOS simulation shown in Figure 5.7, except a 3D process simulation method called "bent planes" is applied. For more information about bent planes, please refer to [39].

5.11   GPU SIMULATION

The computational requirements for the 3D process and device simulations are very cumbersome due to the large mesh size and memory requirements. Depending on the complexity and computer configuration, a complicated 3D simulation can takes days or even weeks. In recent years, parallel processing such as the CUDA platform from NVidia is becoming more and more popular among scientists and engineers. CUDA enables dramatic increases in computing performance by harnessing the power of the graphics processing unit (GPU) [149].

Modern computers primarily divide the tasks they are assigned to do between two different processing units: the CPU, which is the "brain" of the computer and handles all general-purpose tasks; and the GPU, which handles specialized instructions related to the video display. As shown in Figure 5.14a, a simplified architecture view shows the CPU has a control unit and one or more arithmetic logic units (ALUs), which extract instructions from memory, decode, and then execute them [150]. The GPU, on the other hand, is designed to rapidly manipulate and alter memory to accelerate the creation of images in a video display's output buffer [151]. The GPU is thus designed in a very different way from the CPU: as illustrated in Figure 5.14b, the number of processing cores is greatly increased, which gives it the ability to run a large number of highly specialized jobs in parallel.

This means that the GPU excels at tasks that can be well parallelized such as 3D shaders, which can accommodate the display of millions of pixels in a video game. The trade-off is that the individual cores are slower than those of a CPU, so the GPU fares poorly on tasks that cannot easily be done in parallel. This limitation in the speed gains due to parallelism is known as Amdahl's law [153].

The key to a successful GPU-accelerated simulation is thus to separate the tasks in the program that need to be done in a serial manner from those than can be run in parallel. The serial portions of the code run on the high-speed CPU, while parallel portions run on the GPU [154]. For 3D device simulation, the most time-consuming part of the code is the sparse matrix linear solver at the heart of the nonlinear Newton method, so specialized algorithms that maximize parallelism are needed.

In the case of a 3D process simulation, the most time-consuming aspect of the simulation is the 3D diffusion. Benchmark tests have been carried out to determine how the GPU + CPU simulation

Simplified Architecture Representation of a CPU and GPU

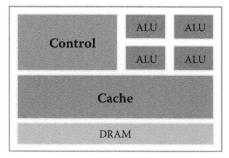
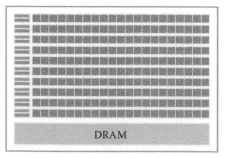

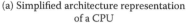

(a) Simplified architecture representation of a CPU

(b) Simplified architecture representation of a GPU

FIGURE 5.14   Simplified architecture comparison between a CPU and a GPU [152].

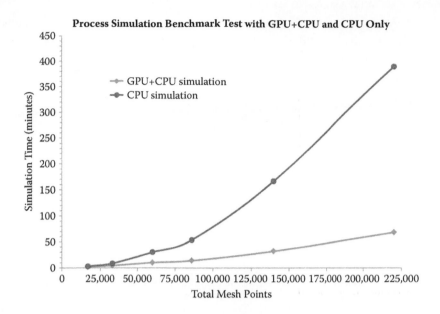

FIGURE 5.15   Benchmark comparison between GPU+CPU and CPU only simulation.

can be compared with traditional CPU-only simulation. As shown in Figure 5.15, 3D dopant diffusion simulations with different total mesh points are compared. At small mesh sizes (< 50,000), the GPU + CPU simulation time is close to CPU simulation time. However, as the mesh size becomes larger, the simulation time difference widens dramatically. For example, with 220,000 mesh points, CPU-only simulation takes about 400 minutes to finish compared with only 68 minutes for CPU + CPU simulation. The computer that performs this benchmark test has a quad-core Intel i7-3770 CPU, with NVidia GeForce GTX 690 (3072 cores, 4 G GDDR5) and 32 G system memory.

6 Introduction to TCAD Device Simulation

Chapter 5 introduced the TCAD process simulation. This chapter focuses on TCAD device simulation.

6.1 OVERVIEW

Device simulation investigates the electrical, thermal, and optical properties of semiconductor devices. This section highlights some of the most important aspects for power device simulation, such as impact ionization and mobility. Experienced readers should refer to [39] for a more advanced treatment of simulation equations and explanations. Figure 6.1 shows a simplified flow of TCAD device simulation.

6.2 BASICS ABOUT DEVICE SIMULATION

6.2.1 DRIFT-DIFFUSION MODEL

The basic semiconductor model is the drift-diffusion model. Even with more advanced models now available, drift-diffusion remains the workhorse of the industry since it can be used to explain many of their important characteristics such as the ability to amplify electrical signals. The most important parts of drift-diffusion theory are the Poisson and current continuity equations. Two optional equations (hydrodynamic model) are sometimes used to describe the carrier energy/temperature distribution; this should not be confused with the lattice temperature, as these equations describe how the carrier distribution deviates from the Fermi–Dirac distribution. For each mesh point, these equations are solved self-consistently for the electrostatic potential, the electron (hole) concentrations and the electron (hole) energies. The basic relationship between these equations is shown in Figure 6.2.

These differential equations are discretized as described in the next section. The resulting set of equations is coupled nonlinearly, and consequently no method exists to solve these equations in one direct step. Instead, solutions must be obtained by a nonlinear iteration method, starting from some initial guess.

6.2.2 DISCRETIZATION

To solve the device equations on a computer, they must be discretized on a simulation grid. That is, the continuous functions of the partial differential equations (PDE) are represented by vectors of function values at the nodes and the differential operators are replaced by suitable difference operators. This means that the simulator solves for 3N unknowns (potential and electron–hole concentrations), where N is the number of grid points; additional variables for the electron–hole energy are used if the hydrodynamic model is turned on.

The key to discretizing the differential operators on a general triangular grid is the box method [155]. Each equation is integrated over a small polygon enclosing each node, yielding 3N nonlinear algebraic equations for the unknown potential and carrier concentrations. The integration equates the flux into the polygon with the sources and sinks inside it so that conservation of current and electric flux is built into the solution. Compared with the finite element method, which performs

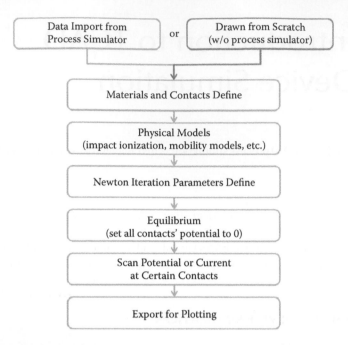

FIGURE 6.1   A simplified flow of TCAD device simulation.

integrals on a triangle-by-triangle basis, the box method produces an elegant method of handling general surfaces and boundary conditions: the integral is simply replaced by a summation of the integrand evaluated at the mesh node, multiplied by its surrounding area.

6.2.3   NEWTON'S METHOD

In this section, we will discuss the numerical solution of drift-diffusion equations. In Newton's method, all of the coupling between variables is taken into account, and these variables are allowed

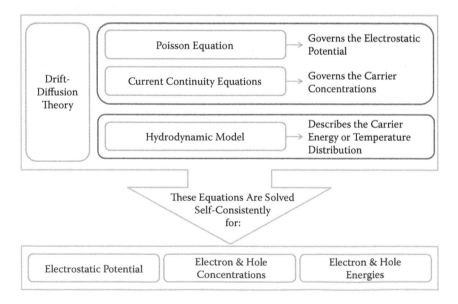

FIGURE 6.2   Drift-diffusion theory.

to change at every iteration. Due to this, the Newton algorithm is very stable, and the solution time is nearly independent of bias conditions. The basic algorithm is a generalization of the Newton–Raphson method used to find the root of a single equation. The Poisson equation and current continuity equations can be written as

$$F_v^j\left(V^{j1}, E_{fn}^{j1}, E_{fp}^{j1}\right) = 0 \tag{6.1}$$

$$F_n^j\left(V^{j1}, E_{fn}^{j1}, E_{fp}^{j1}\right) = 0 \tag{6.2}$$

$$F_p^j\left(V^{j1}, E_{fn}^{j1}, E_{fp}^{j1}\right) = 0 \tag{6.3}$$

where j runs from 1 to N, and $j1$ includes j itself plus its surrounding mesh points. These equations represent a total number of 3N equations, which are sufficient to solve for 3N variables: $\left(V^1, E_{fn}^1, E_{fp}^1, V^2, E_{fn}^2, E_{fp}^2, \dots V^N, E_{fn}^N, E_{fp}^N\right)$.

Once the equations are discretized in the previous form, standard Newton techniques can be used to solve for them. These involve evaluating the Jacobian matrix to linearize the equations, followed by a linear solver (involving lower and upper triangular matrix [LU] factorization of the matrix), and finally nonlinear iterations to get the final solution. Since the Jacobian matrix is sparse, sparse matrix techniques are used to improve the computation speed.

A major speed-up of Newton iterations is accomplished with the Newton–Richardson method, whereby the Jacobian matrix is refactored only when necessary. When it is not necessary to factorize, the iterative method using the previous factorization is employed. The iterative method is extremely fast provided the previous factorization is reasonable. Frequently, the Jacobian needs to be factorized only once or twice per bias point using the Newton–Richardson method, as opposed to the 20 to 30 times required in the conventional Newton method. The decision to refactor is made on the basis of the rate of decrease of the maximum error of the equation residuals and variable differences per nonlinear iteration. This mechanism has been automated within most simulators and no user intervention is needed [156].

6.2.4 INITIAL GUESS AND ADAPTIVE BIAS STEPPING

Several types of initial guess solutions are used in the TCAD simulator. The first is the simple charge neutral assumption used to obtain the first (equilibrium) bias point. This is the starting point of any device simulation.

Any later solution with applied bias needs an initial guess of some type, obtained by modifying one or more previous solutions. When only one previous solution is available, the solution currently loaded is used as the initial guess, modified by setting the applied bias at the contact points. When two previous solutions with two different biases are available, it is possible to obtain a better initial guess by extrapolating the two previous solutions.

All of these steps are automated within the program. A schematic of the biasing strategy is given in Figure 6.3.

As in any Newton method, convergence strongly depends on the choice of the initial guess solution. In principle, the Newton nonlinear iteration should always converge as long as the initial guess is close enough to the solution; the closer the initial guess is to the solution, the fewer nonlinear iterations are required to reach convergence. The program takes this fact into account and implements an adaptive method to control the bias step after the successful convergence of the previous solution.

Assuming that the previous solution takes K_1 number of iterations for a bias step of ΔV_1 to reach convergence and the user believes (by experience) that the optimal number of iterations

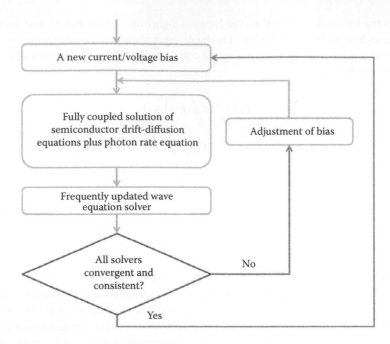

FIGURE 6.3   Flow diagram of how the simulator performs in a typical device simulation.

should be K_0, the current bias step ΔV_2 is adjusted to $\Delta V_2 = \Delta V_1 K_0 / K_1$. In this way an optimal step can be determined such that the nonlinear Newton iteration is controlled to converge within K_0 iterations.

One common cause of convergence problems in a Newton method is oscillations in the solution. This usually occurs when the initial guess is poor and the solver overshoots the solution. To prevent this, a damping value can be implemented, which prevents the solver to change the solution too much between successive iterations. A small damping value will damp the oscillations effectively but may cause slower convergence. A large value allows faster convergence when the initial guess is poor but carries a larger risk of oscillations.

6.2.5 CONVERGENCE ISSUES

Quite often, convergence issues plague TCAD users. The choice of voltage or current bias affects the convergence and stability of the Newton solver. To guarantee convergence, small changes in the applied bias should always result in small changes in the overall solution.

One particular situation where current bias is not appropriate is when the total amount of current flowing in the device is very small. Under these conditions, the actual current amount may fluctuate due to lack of numerical precision, making it difficult to use current bias. This situation can be detected by observing the net current over all the electrodes: if the sum is not zero, then Kirchhoff's current law is violated, and the current is too low to use as a control variable.

If we take a forward-biased diode as an example, the solver can enter a nonconvergent state if the applied (anode) voltage bias is much higher than the turn-on voltage. Since the conductivity increases exponentially with bias in a typical diode, seemingly small changes in voltage can result in very large changes of the solution.

This leads us to a simple general rule:

1. Use voltage bias for devices with high resistance.
2. Use current bias for devices with low resistance.

For example, a typical diode under forward bias has low resistance past its turn-on point but high resistance at lower bias or under reverse bias conditions. With these two extremes in mind, the following general strategy is recommended when setting up a simulation under forward bias:

1. Solve for equilibrium solution.
2. Apply voltage until 80–90% of the built-in bias value is reached. Some software tools also allow the possibility of terminating the voltage increase once certain current conditions have been met.
3. Verify that Kirchhoff's current law is satisfied at this bias point.
4. Apply current bias until desired value is reached.

6.2.6 BOUNDARY CONDITIONS

The boundary conditions for the electrical part include Ohmic contacts, Schottky contacts, Neumann (reflective) boundaries, lumped elements, and current controlled contacts. For the hot carrier equations, the boundary condition is the contact carrier temperature.

6.2.6.1 Ohmic Contact

Ohmic contacts are implemented as simple Dirichlet boundary conditions where the surface potential and electron and hole quasi-Fermi levels (Vs, E_{fn}^s, E_{fn}^s) are fixed.

The minority and majority carrier quasi-Fermi potentials are equal and set to the applied bias of the electrode:

$$\phi_n^s = \phi_n^s = -E_{fn}^s = -E_{fp}^s = V_{applied} \tag{6.4}$$

The potential V_s at the boundary is fixed at a value consistent with zero space charge. With the solution of V_s and Equation (6.4), n_s and p_s can be calculated.

It should be pointed out that the purpose of creating an Ohmic contact in a simulation is to have a boundary condition that does not disturb the area of simulation but provides a path for current flow. This is in contrast to the Schottky contact that can be either injecting or depleting carriers, depending on the polarity of the bias.

In most device simulations, the ideal Ohmic contact with a charge neutral assumption is sufficient as a good carrier injector if the contact area is highly doped or if the carrier injection requirement is not too high. If the vicinity of the contact has low doping or if a high level of carrier injection is required, the ideal Ohmic boundary described above ceases to be a valid model for a realistic Ohmic metal contact.

6.2.6.2 Schottky Contacts

Schottky contacts to the semiconductor are defined by the barrier height of the electrode metal and a surface recombination velocity. The surface potential at a Schottky contact is defined by

$$V_s = \chi - \chi_{ref} - \phi_b + V_{applied} \tag{6.5}$$

where χ and χ_{ref} are the electron affinities of the semiconductor and a reference material, respectively, and ϕ_b is the Schottky barrier height.

Note that this applies only for n-contacts. For p-contacts, this equation needs to be modified by the semiconductor bandgap to calculate the proper hole barrier. The sum of the barrier heights for electrons and holes is expected to be equal to the bandgap [138].

In general, the quasi-Fermi levels E_{fn}^s and E_{fp}^s are no longer equal to $V_{applied}$ and are defined by a current boundary condition at the surface instead:

$$J_{sn} = \gamma_n \bar{v}_n^{thm} (n_s - n_{eq})$$

(6.6)

$$J_{sp} = \gamma_p \bar{v}_p^{thm} (p_s - p_{eq})$$

(6.7)

where J_{sn} and J_{sp} are the electron and hole currents at the contact, n_s and p_s are the actual surface electron and hole concentrations, and n_{eq} and p_{eq} are the equilibrium electron and hole concentrations if infinite surface recombination velocities are assumed (i.e., $\phi_n = \phi_p = V_{applied}$). The average thermal recombination velocities are given by

$$\bar{v}_n^{thm} = \sqrt{\frac{kT}{2m_n \pi}}$$

(6.8)

$$\bar{v}_p^{thm} = \sqrt{\frac{kT}{2m_p \pi}}$$

(6.9)

The constants γ_n and γ_p provide a mechanism to include any correction (e.g., due to tunneling) to the standard theory.

6.2.6.3   Neumann Boundaries

Along the outer noncontacted edges of simulated devices, homogeneous reflecting Neumann boundary conditions are imposed so that current flows out of the device only through the contacts. Additionally, the normal electric field component goes to zero in the absence of surface charge along such edges. In a similar fashion, current is not permitted to flow from the semiconductor into an insulating region; further, at the interface between two different materials, the difference between the normal components of the respective electric displacements must be equal to any surface charge present along the interface.

6.2.6.4   Lumped Elements

Lumped elements (resistors, capacitors, and inductors) have been created to reduce the number of grid points to discretize some device structures, thereby saving CPU time. Lumped resistance might be useful in a simulation of a semiconductor device structure with a substrate contact located far away from the active region. If the whole structure were to be simulated, a tremendous number of grid points, probably more than half, would be wasted to account for a purely resistive region of the device. Further expansion of lumped elements leads to mixed-mode (TCAD + SPICE) simulation.

6.2.7   Transient Simulation

The method used to discretize the differential equations in time is the backward Euler method. Its basic approximation is the following equation:

$$\frac{\partial S}{\partial t} = \frac{S(t) - S(t - \Delta t)}{\Delta t}$$

(6.10)

This method has the advantages of being highly stable and recovers the steady-state solution in the limit $\Delta t \rightarrow \infty$.

Once all the equations are discretized in terms of solution of the previous time step, one can treat the solution of the present time step using the same method as for the steady-state solution (i.e., one can linearize the discretized equations and use Newton's method to solve them).

6.2.8 MESH ISSUES

Correct allocation of the grid is a crucial issue in TCAD device simulation. The number of nodes in the grid (N) has a direct influence on the simulation time, where the number of arithmetic operations necessary to achieve a solution is proportional to N^p where p usually varies between 1.5 and 2.

Because the different parts of a device have very different electrical and optical behavior, it is usually necessary to allocate a fine grid to some regions and a coarse grid to others. Whenever possible, it is desirable not to allow the fine grid in some regions to spill into regions where it is unnecessary to maintain a reasonable simulation time.

The first step in mesh generation is to specify the device boundaries and the region boundaries for each material. Typically, triangles and trapezoids are used as the basic building blocks to describe an arbitrary device. Furthermore, the edge of each polygon can be bent to the desired shape; this curve will later be approximated by a piecewise linear function when drawing the mesh triangles.

The basic operation in the mesh generator after the boundaries have been defined is to draw lines parallel to the edges of the polygons. In this way we obtain smaller trapezoids that can be bisected into two triangles, the basic elements of the finite element method.

Automatic mesh generation and refinement can be useful when desired locations need denser or looser mesh for accuracy or simulation speed. The mesh is refined according to the variation of specified physical quantities. For example, in a case where the potential variation between two adjacent nodes is greater than a value specified by the user, the program will automatically allocate additional mesh points between the two nodes.

6.3 PHYSICAL MODELS

This section gives the reader an overview of device simulation models with a special focus on avalanche breakdown and impact ionization, carrier mobility, and thermal effects.

6.3.1 CARRIER STATISTICS

Carrier statistics refers to the relation between carrier concentration and quasi-Fermi levels. Most device simulators offer the option of using Boltzmann or Fermi–Dirac statistics. The former is more numerically convergent, while the latter is more accurate for highly doped regions. For a more theoretical description, please refer to Appendix A.

6.3.2 INCOMPLETE IONIZATION OF IMPURITIES

TCAD simulation programs can accurately account for the incomplete ionization of shallow impurities in semiconductors. The occupancy coefficients f_D and f_A are used to describe the degree of ionization. It is assumed that the shallow impurities are in equilibrium with the local carriers and therefore the occupancy of the shallow impurities can be described by

$$f_D = \frac{1}{1 + g_d^{-1} \exp\left[\frac{E_D - E_{fn}}{kT}\right]} \tag{6.11}$$

$$f_A = \frac{1}{1 + g_a \exp\left[\frac{E_A - E_{fp}}{kT}\right]} \tag{6.12}$$

where the subscripts D and A are used to denote shallow donors and acceptors, respectively. The degeneracy levels are automatically set to $g_d = 2$ and $g_a = 4$.

6.3.3 HEAVY DOPING EFFECT

The Mott transition occurs at high doping concentrations, when the distance between impurities/dopants becomes comparable to the Bohr radius. It has been shown [157] that the effective Bohr radius a_B^* and critical impurity concentration $N_{crit}^{1/3}$ are determined by

$$a_B^* = \frac{0.53 \, \epsilon_r}{m^*/m_0} \, \text{Å}$$

(6.13)

$$N_{crit}^{1/3} = \frac{1}{4a_B^*} \left(\frac{\pi}{3} \right)^{1/3}$$

(6.14)

where ϵ_r is the relative dielectric constant, and m^* is the carrier effective mass. The ionization energy of the impurity as a function of the concentration can then be obtained by

$$E_D = E_{D0} \left[1 - \left(\frac{N_D}{N_{crit}} \right)^{1/3} \right]$$

(6.15)

A similar expression exists for acceptors. Note that this relationship is derived for shallow acceptors; deep-level acceptors in some materials are more complex and can sometimes form impurity bands rather than discrete levels.

6.3.4 SRH AND AUGER RECOMBINATION

Shockley–Read–Hall (SRH) recombination is used to simulate carrier recombination due to deep-level traps for both steady and dynamic (transient) conditions; this term is linear in carrier density and characterized by a lifetime coefficient.

Bimolecular (spontaneous emission) recombination involves an electron and hole recombining to emit a photon, while Auger recombination involves two electrons and a hole (or two holes and an electron). These two processes are normally significant only in nonequilibrium conditions when the carrier density is very high or for light-emitting devices. As this book concentrates on power devices, we will skip further discussion of these terms: interested readers can refer to any device physics book for details.

6.3.5 AVALANCHE BREAKDOWN AND IMPACT IONIZATION

Breakdown voltage is of great concern for power semiconductor devices. There are several types of device breakdown: avalanche breakdown, Zener breakdown, and second breakdown (permanent damage due to thermal runaway). Zener breakdown is more about the breakdown at highly doped PN junction; the breakdown voltage is usually less than 6 V, and the I-V curve shows a "softer" current increment than avalanche breakdown. In this section, avalanche breakdown will be studied in detail.

6.3.5.1 Avalanche Breakdown

Power devices are designed to support high voltages in the depletion region formed by PN junctions, metal-semiconductor contact (Schottky barrier), or metal-oxide-semiconductor (MOS) interfaces [158]. Electrons and holes traveling in the depletion region come from the quasi-neutral regions by diffusion or are generated in the space charge region and are swept away by the electric field (E-field) that is applied at the terminals. As the E-field increases along with the applied voltage, a critical field value is reached; the carriers then gain enough kinetic energy that they create electron–hole pairs when colliding with atoms in the lattice. On a band diagram view, the impact excites an electron from the valence band, which jumps to the conduction band and leaves behind a free hole in the valence band. This phenomenon is called *impact ionization*. Since the newly created electrons and holes can themselves create more electron–hole pairs in the high E-field, a multiplicative process is launched much like a snowball that grows bigger as it progresses along the path: the phase "avalanche breakdown" is used to describe this effect. Breakdown means the applied voltage can no longer be supported by the depletion region and current increases rapidly. Junctions that are not designed to support such large amounts of current can fail rapidly under these conditions, so the voltage at avalanche breakdown sets the voltage limits under which the device can operate.

6.3.5.2 Impact Ionization Coefficients

Impact ionization coefficients for electrons and holes are defined as the number of electron–hole pairs created by an electron or a hole traversing 1 cm through the depletion layer along the direction of the electric field [158]. The general requirements for impact ionization to happen include a field strength that is larger than a threshold value (critical field) and a space charge region width greater than the mean free path between two ionizing impacts so that charge multiplication can occur. The ionization coefficient α is the reciprocal of the mean free path. There are several models for impact ionization coefficients of silicon material:

- Chynoweth's law
- Baraff model
- Fulop's approximation
- Okuto–Crowell model
- Lackner model
- Mean free path model

6.3.5.3 Chynoweth's Law

Chynoweth's law [159] is a widely used impact ionization model expressed as

$$\alpha(E) = a \cdot e^{\left(\frac{b}{E}\right)} \tag{6.16}$$

The default parameters for this equation vary in different sources. A summary table is listed in Table 6.1 [160].

TABLE 6.1
List of Impact Ionization from Various Sources

Sources	a (Electrons)	b (Electrons)	a (Holes)	b (Holes)
Lee et al. [161]	3.8E+6 /cm	1.77E+6 V/cm	9.9E+6 /cm	2.98E+6 V/cm
Ogawa [162]	0.75E+6 /cm	1.39E+6 V/cm	4.65E+6 /cm	2.30E+6 V/cm
Van Overstraeten and De Man [163]	0.703E+6 /cm	1.231E+6 V/cm	1.582E+6 /cm	2.036E+6 V/cm

6.3.5.4 Baraff Model

The Baraff impact ionization model is more complicated in form and is defined as

$$\alpha = \frac{e^{g(r,x)}}{\lambda} \tag{6.17}$$

$$g(r,x) = (11.5r^2 - 1.17r + 3.9 \times 10^{-4})x^2 + (46r^2 - 11.9r + 1.75 \times 10^{-2})x - 757r^2 + 75.5r - 1.92 \tag{6.18}$$

where

$$r = \frac{<E_p>}{E_I}, x = \frac{E_I}{qE\lambda}, E_I = f_e E_g, <E_p> = E_p \tanh\left(\frac{E_p}{2kT}\right), \lambda = \lambda_0 \tanh\left(\frac{E_p}{2kT}\right) \tag{6.19}$$

$<E_p>$ is the optical phonon energy, and is the optical phonon scattering mean free path. E_I is the ionization energy, f_e is a constant often chosen to be 1.5 [164]. As an example, the default parameters for TCAD simulation in the APSYS software are shown in Table 6.2 [156].

6.3.5.5 Fulop's Approximation

Fulop's approximation is a very simple power law based approximation for the impact ionization coefficient and is expressed as [158]

$$\alpha_F = 1.8 \times 10^{-35} E^7 \tag{6.20}$$

This approximation is good for analytical or hand calculation of breakdown voltage in simple cases. However, Fulop's approximation is not accurate enough for TCAD numerical simulation.

6.3.5.6 Okuto–Crowell Model

The Okuto–Crowell is an empirical model developed by Okuto and Crowell [165].

$$\alpha(F) = a \cdot [1 + c(T - T_0)] \cdot E^\gamma \cdot e^{-\left\{\frac{[b[1+d(T-T_0)]]}{F}\right\}^\delta} \tag{6.21}$$

The default values for the parameters are listed in Table 6.3.

6.3.5.7 Lackner Model

The Lackner model is derived from the original Chynoweth law. By introducing a temperature dependent-factor, the Lackner model is expressed as

---

TABLE 6.2
Parameters for the Baraff Impact Ionization Coefficient

Parameters	Values
Electron phonon energy (E_p)	0.063
Hole phonon energy (E_p)	0.063
Electron optical phonon mean free path (λ_0) at 0 Kelvin	76
Hole optical phonon mean free path (λ_0) at 0 Kelvin	58
f_e factor above for electrons	1.5
f_e factor above for holes	1.5

TABLE 6.3
Parameters for the Okuto–Crowell Model

Parameters	For Electrons	For Holes
a	0.426 1/V	0.243 1/V
b	4.81×10^5 V/cm	6.53×10^5 V/cm
c	3.05×10^{-4} 1/K	5.35×10^{-4} 1/K
d	6.86×10^{-4} 1/K	5.67×10^{-4} 1/K
γ	1	1
δ	2	2

$$\alpha_n = \frac{\gamma a_n}{Z} e^{-\frac{\gamma b_n}{E}}, \alpha_p = \frac{\gamma a_p}{Z} e^{-\frac{\gamma b_p}{E}} \tag{6.22}$$

where α_n and α_p are the impact ionization coefficients for electrons and holes, respectively.

$$Z = 1 + \frac{\gamma b_n}{E} e^{-\frac{\gamma b_n}{E}} + \frac{\gamma b_p}{E} e^{-\frac{\gamma b_p}{E}} \tag{6.23}$$

and

$$\gamma = \frac{\tanh\left(\dfrac{h\omega_{op}}{2kT_0}\right)}{\tanh\left(\dfrac{h\omega_{op}}{2kT}\right)} \tag{6.24}$$

The values of the parameters are listed in Table 6.4.

6.3.5.8 Mean Free Path Model
The mean free path model defines the critical electric field using the following equation:

$$E_c = \frac{E_g}{\lambda} \tag{6.25}$$

where λ is the mean free path of electrons and holes, as in the Baraff model:

$$\lambda = \lambda_o \tanh\left(\frac{E_p}{2kT}\right) \tag{6.26}$$

TABLE 6.4
Parameter List for the Lackner Model

Parameters	Electrons	Holes
a	1.316×10^6 cm^{-1}	1.818×10^6 cm^{-1}
b	1.474×10^6 V/cm	2.036×10^6 V/cm
$h\omega_{op}$	0.063 eV	0.063 eV

TABLE 6.5
Parameter List for the Mean Free Path Model

Parameters	Values
Electron phonon energy (E_p)	0.063
Hole phonon energy (E_p)	0.063
0 temperature electron optical phonon mean free path (λ_0)	76
0 temperature hole optical phonon mean free path (λ_0)	58

E_p is the optical phonon energy. Physically, electrons and holes that accelerate in the depletion region over a length of λ will gain an energy of $\lambda \cdot E$. If this energy is greater than bandgap energy E_g, impact ionization will take place. The mean free path model is similar to Baraff's model except that Baraff model is more accurate and fits the whole experimental range. Unfortunately, being more complicated means the Baraff model has issues with numerical convergence in TCAD simulations. The mean free path model addresses this issue with a simplified equation. The default parameter table is similar to the Baraff model, as seen in Table 6.5.

6.3.5.9 Multiplication Factor and Ionization Integral

Avalanche breakdown is defined by the impact ionization rate becoming infinite. To allow calculation of the total number of electron–hole pairs generated by a single electron–hole pair at a distance x from the PN junction in the space charge region, a new parameter called multiplication coefficient is derived [158]:

$$M(x) = \frac{exp\left[\int_0^x (\alpha_n - \alpha_p)dx\right]}{1 - \int_0^w \alpha_p exp\left[\int_0^x (\alpha_n - \alpha_p)dx\right]dx} \tag{6.27}$$

where w is the depletion width. The avalanche breakdown condition is met when the total number of generated electron–hole pairs or $M(x)$, goes to infinity. This means

$$\int_0^w \alpha_p exp\left[\int_0^x (\alpha_n - \alpha_p)dx\right]dx = 1 \tag{6.28}$$

This expression is known as an ionization integral. If we assume that electrons and holes have the same impact ionization coefficient, or $\alpha = \alpha_n = \alpha_p$, then the ionization integral can be rewritten as

$$\int_0^w \alpha dx = 1 \tag{6.29}$$

This ionization integral is valid for diodes and MOSFETs where current flowing through the depletion region is not amplified [158]. For BJTs and LIGBTs, current flowing through the depletion region is amplified by a current gain so it is necessary to solve the multiplication coefficient rather than the ionization integral.

Most device simulators can export or plot the impact ionization coefficient α. However, the ionization integral is not easy to evaluate since the path of ionization current may not be a straight line in a 2D/3D simulation. One simplification is to estimate the ionization integral (2D) by

$$\frac{\int_0^D \int_0^w \alpha dx dy}{D} = 1 \tag{6.30}$$

where D is the dimension lateral to the direction of E-field (x direction). Many device simulators (including APSYS) provide such a 2D/3D integral of α.

Based on the method of ionization integral as an estimate of breakdown voltage, we can develop a convergence technique as follows. Since impact ionization is a leading cause of divergence in TCAD simulation, we can evaluate the impact ionization term but not use it in the drift-diffusion solver. That is, the impact ionization calculation is treated as a postprocessing step: once the solver has converged, α will be calculated and the ionization integral will be used to determine the breakdown voltage.

6.3.5.10 Critical Electric Field

The critical field for silicon at which the impact ionization takes place is doping concentration dependent. Several models have been proposed for the critical value. For example:

From [158]:

$$E_c = 4010 N_D^{1/8} \tag{6.31}$$

From [166]:

$$E_c = \frac{4 \times 10^5}{1 - \left(\frac{1}{3}\right)\log_{10}\left(\frac{N}{10^{16}\,cm^{-3}}\right)} \tag{6.32}$$

A comparison chart is plotted in Figure 6.4. Both models yield similar critical field values until the doping concentration reached 1E+17 cm^{-3}, beyond which a large discrepancy occurs.

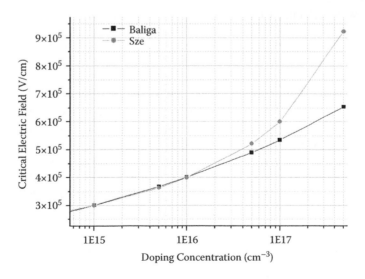

FIGURE 6.4 Critical electric field model comparison.

6.3.5.11 Analytical Breakdown Voltage

The method of using ionization integral with impact ionization coefficients to solve the breakdown voltage is most accurate and yet quite often results in a nonclosed form. Numerical iteration is necessary for most impact ionization coefficients models. Sometimes, engineers need a quick and simple estimate of breakdown voltage. The critical electric field is used to directly derive the breakdown voltage with calculated depletion width. From Poisson's equation the breakdown voltage can be expressed with a critical field as [160].

$$BV = \frac{\varepsilon_s E_c^2}{2qN} - V_{bi} \tag{6.33}$$

Here, V_{bi} is the built-in potential that can be safely neglected for most cases since it is much smaller than the breakdown voltage. N is the average doping concentration of the N- and P-regions and is calculated as

$$N = \frac{N_a N_d}{N_a + N_d} \tag{6.34}$$

Analytical closed-form breakdown voltage calculation is convenient for rough estimates or hand calculations. Several proposed approximations are available.

From Reference [167], assume $E_c = 5 \times 10^5 \, \text{V/cm}$ the breakdown voltage is expressed as

$$BV \approx 15 \times \left(\frac{10^{17}}{N} \right)^{0.6} \tag{6.35}$$

From reference [158], the analytical breakdown voltage is calculated by the ionization integral of power law (Fulop's law) derived impact ionization coefficient

$$BV = 5.34 \times 10^{13} N_D^{-3/4} \tag{6.36}$$

with a depletion width calculated as

$$W = 2.67 \times 10^{10} N_D^{-7/8} \tag{6.37}$$

6.3.5.12 Benchmark Comparison of Avalanche Breakdown Models

To give the reader a better view of different models for avalanche breakdown, both TCAD numerical simulations and analytical models are used to make a comparison. A P+/N junction is used. The highly doped P+ region has a constant doping concentration of 1E+20 cm^{-3}, with abrupt parallel junction. The doping concentration in the N-region varies from 5E+14 cm^{-3} to 5E+17 cm^{-3} as seen in Table 6.6. Figure 6.5 plots the resulting curves. Note that TCAD simulated breakdown voltages are higher than analytical results due to mesh roughness and more realistic depletion profiles.

6.3.6 Carrier Mobility

The carrier mobility μ_n and μ_p account for the scattering mechanism in electrical transport. One of the main effects on the mobility is the local electrical field [138].

TABLE 6.6

Benchmark Comparison of Breakdown Models Using the P+/N Parallel Plane

N-region Doping Conc. (cm^{-3})	Baliga (V) [158]	Hu (V) [167]	Chynoweth (V)	Baraff (V)	Lackner (V)	MeanfreePath (V)
5.00E+14	505	360	494	486	521	519
1.00E+15	300	238	316	293	334	304
5.00E+15	90	91	109	89	113	101
1.00E+16	53	60	72	55	73	65
5.00E+16	16	23	31	20	31	27
1.00E+17	9.5	15	23	14	23	19

6.3.6.1 Models Overview

Common analytical formulas for field-dependent mobility [156] are defined in this section:

- Constant mobility model
- Two-piece mobility model
- Canali or beta model
- Transferred electron model
- Poole–Frenkel field enhanced mobility model
- Impurity dependence of the low field mobility model
- Intel's local field models
- Lombardi model

6.3.6.2 Constant Mobility

The simplest mobility model uses constant mobility μ_{0n} and μ_{0p} for electrons and holes, respectively, throughout each material region in the device. This model is very simple but inaccurate.

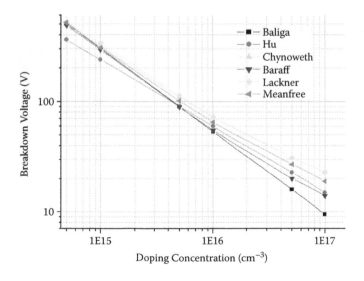

FIGURE 6.5 Benchmark comparison of breakdown voltage calculated or simulated by different models.

6.3.6.3 Two-Piece Mobility Model

Since carrier mobility is strongly dependent on an electric field, we can combine a low field model plus a high field model to approximately simulate the field dependent effect. For electrons:

$$\mu_n = \mu_{0n}, \quad for \ E < E_{0n} \tag{6.38}$$

$$\mu_n = \frac{v_{sn}}{E}, \quad for \ E \geq E_{0n} \tag{6.39}$$

$$v_{sn} = \mu_{0n} E_{0n} \tag{6.40}$$

where E_{0n} is a threshold field beyond which the electron velocity saturates to a constant. Similar expressions can be defined for holes:

$$\mu_p = \mu_{0p}, \quad for \ E < E_{0p} \tag{6.41}$$

$$\mu_p = \frac{v_{sp}}{E}, \quad for \ E \geq E_{0p} \tag{6.42}$$

$$v_{sp} = \mu_{0p} E_{0p} \tag{6.43}$$

6.3.6.4 Canali or Beta Model

The Canali or beta model [168] is commonly used in the literature and has the following form for electrons and holes, respectively:

$$\mu_n = \frac{\mu_{0n}}{\left(1 + \left(\frac{\mu_{0n} E}{v_{sn}}\right)^{\beta_n}\right)^{\frac{1}{\beta_n}}} \tag{6.44}$$

$$\mu_p = \frac{\mu_{0p}}{\left(1 + \left(\frac{\mu_{0p} E}{v_{sp}}\right)^{\beta_p}\right)^{\frac{1}{\beta_p}}} \tag{6.45}$$

6.3.6.5 Transferred Electron Model

The transferred electron model [168] is used in many III-V compound semiconductors, which exhibit negative differential resistance due to the transition of carriers into band valleys with lower mobility [138]:

$$\mu_n = \frac{\mu_{0n} + \frac{v_{sn}}{E_{0n}} \left(\frac{E}{E_{0n}}\right)^3}{1 + \left(\frac{E}{E_{0n}}\right)^4} \tag{6.46}$$

6.3.6.6 Poole–Frenkel Field Enhanced Mobility Model

Highly localized carriers with small mobility can be excited by an external field to make hopping movements, commonly used for organic semiconductors:

$$\mu = \mu_0 \exp\left[\left(\frac{E}{E_{cr}}\right)^{px}\right] \tag{6.47}$$

6.3.6.7 Impurity Dependence of the Low Field Mobility Model

Besides the field dependence of the mobility, another important effect is the impurity dependence of the low field mobility [138].

$$\mu_{0n} = \mu_{1n} + \frac{\mu_{2n} - \mu_{1n}}{1 + \left(\dfrac{N_D + N_A + \sum\limits_j N_{tj}}{N_{rn}}\right)^{\alpha_n}} \tag{6.48}$$

$$\mu_{0p} = \mu_{1p} + \frac{\mu_{2p} - \mu_{1p}}{1 + \left(\dfrac{N_D + N_A + \sum\limits_j N_{tj}}{N_{rp}}\right)^{\alpha_p}} \tag{6.49}$$

where the various parameters are fitting parameters from experimental data. Some of these values may also be temperature dependent.

6.3.6.8 Intel's Local Field Models

In many applications, the mobility is anisotropic and also depends on the vertical field. A number of vertical field dependent models (e.g., the Lombardi model) are implemented in TCAD device simulators. The anisotropic models are useful for vertical field dependent mobility in MOSFET channel.

The Intel1 and Intel2 models are based on PISCES 2ET manual of Stanford University [169] as follows. The Intel1 model defines a factor r_{perp} to reduce the mobility under the channel:

$$r_{perp} = \left(1 + \frac{E_\perp}{E_{crit}}\right)^{-\beta} \tag{6.50}$$

where $E_\perp$ is the field perpendicular to the SiO$_2$/Si interface. E_{crit} is a parameter with value of 4.2E+4 V/cm for electrons and 3E+4 V/cm for holes in silicon. By default, $\beta = 0.5$.

The model Intel2 uses the following formula for the reduction factor:

$$r_{perp} = \left(1 + \left(\frac{E_\perp}{E_{univ}}\right)^\alpha\right)^{-1} \tag{6.51}$$

where $\alpha = 1.02$ for electrons, and $\alpha = 0.95$ for holes. E_{univ} has a value of 5.71E+5 V/cm for electrons and 2.57E+5 V/cm for holes. Both of these models are for low-field mobility only.

6.3.6.9 Lombardi Model

A more physics-based model to account for the effect of the transverse field is developed based on extensive experimental data by Lombardi [169] [170]. Basically, this model is to summarize the contributions to mobility from different sources as follows:

$$\frac{1}{\mu} = \frac{1}{\mu_{ac}} + \frac{1}{\mu_{srf}} + \frac{1}{\mu_0} \tag{6.52}$$

TABLE 6.7
Parameter Values for μ_{ac} and μ_{srf}

	B	α	β	δ
Electrons	4.75E+7	1.74E+5	0.125	5.82E+14
Holes	9.93E+7	8.84E+5	0.0317	2.05E+14

where μ_{ac} is the mobility due to the acoustic phonon scattering, depending on both transverse field and lattice temperature. μ_{srf} is about surface scattering and is a function of the transverse field. μ_0 is the mobility due to the longitudinal field or hot-carrier effect. μ_{ac} and μ_{srf} are expressed as

$$\mu_{ac} = \frac{B}{E_\perp} + \frac{\alpha N^\beta}{T_L E_\perp^{\frac{1}{3}}} \tag{6.53}$$

$$\mu_{srf} = \frac{\delta}{E_\perp^2} \tag{6.54}$$

The values of the parameters used in Equations (6.53) and (6.54) are listed in Table 6.7.

6.3.6.10 Comparison of Different Diffusion Models

A comparison between different diffusion models is made using a simple silicon long channel N-type MOSFET structure. As shown in Figure 6.6 two gate voltages ($V_g = 5$ V and $V_g = 1.5$ V) are applied, and drain voltage is scanned from 0 to 10 V for both gate voltages. Four mobility models (Canali or beta model, Intel1, Intel2, and Lombardi) are compared. It has been found that for both gate voltages the Canali model yields the highest drain current, which means it predicts the highest electron mobility. This is somewhat expected because the Canali model does not take vertical field-induced mobility reduction into account while other three models are all anisotropic models. The Lombardi model and Intel2 model are relatively close, but the results depend on the gate voltage. Default model parameter values are used for all simulation cases.

6.3.6.11 GaN Mobility Model (MTE Model)

A modified transferred-electron mobility model designed for GaN devices is proposed by [168]. It has been found that GaN has a very specific drift velocity dependence on the electric field. A qualitative comparison (not drawn to scale) is shown in Figure 6.7. Different from silicon (where the Canali model can be applied), GaN has a negative differential mobility region similar to that of GaAs. However, GaN also has a kink at low field, which makes direct use of transferred-electron mobility model insufficient. A modified transferred-electron (MTE) mobility model is thus required to model GaN mobility. Based on the transferred-electron model, a more universal model should be like [168].

$$v_{MTE}(E) = \frac{F(E) + v_{sat}\left(\dfrac{E}{E_{MT}}\right)^{\beta_T}}{1 + \left(\dfrac{E}{E_{MT}}\right)^{\beta_T}} \tag{6.55}$$

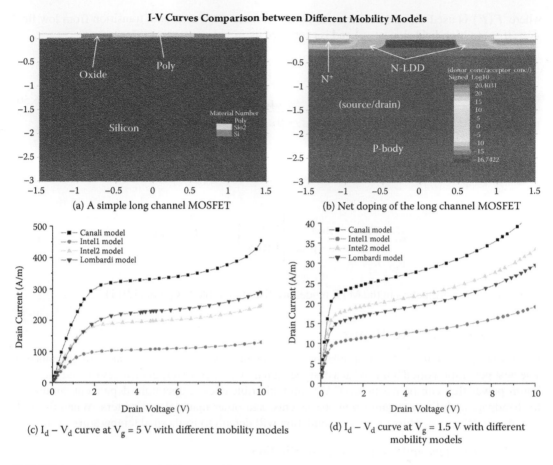

I-V Curves Comparison between Different Mobility Models

(a) A simple long channel MOSFET

(b) Net doping of the long channel MOSFET

(c) I_d – V_d curve at V_g = 5 V with different mobility models

(d) I_d – V_d curve at V_g = 1.5 V with different mobility models

FIGURE 6.6   Comparison of different mobility models using a long channel MOSFET.

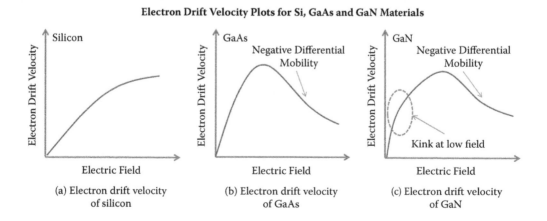

Electron Drift Velocity Plots for Si, GaAs and GaN Materials

(a) Electron drift velocity of silicon

(b) Electron drift velocity of GaAs

(c) Electron drift velocity of GaN

FIGURE 6.7   Qualitative comparison of electron drift velocity in silicon, GaAs, and GaN materials.

where $F(E)$ is used to mimic the low field kink effect to allow a smooth transition from low field slop (μ_{low}) to high field slop (μ_{high}) and is expressed as

$$F(E) = \frac{\mu_{low}E}{\left[1 + \left(\dfrac{\mu_{low}E}{(\mu_{low} - \mu_{high})E_k + \mu_{high}E}\right)^{\beta_k}\right]^{\frac{1}{\beta_k}}} \tag{6.56}$$

E_k is defined as

$$E_k = \frac{v_{sat}}{\mu_{low}} \cdot \frac{\mu_{high}}{\mu_{low} - \mu_{high}} \cdot \left[\left(\frac{\mu_{low}}{\mu_{high}}\right)^{\frac{\beta_c}{1+\beta_c}} - 1\right]^{\frac{1}{\beta_c}+1} + \delta E_k \tag{6.57}$$

Some of the parameters in the MTE model are listed in Table 6.8 [168] [171] [172].

6.3.7 THERMAL AND SELF-HEATING

For power semiconductor device simulations, thermal and self-heating effects are a major concern. The heat generation often limits what a designer can do with a passively cooled device and the maximum power that can be reached. TCAD simulation includes temperature-dependent simulation for bandgap, mobility, recombination coefficients, and other material parameters. When the self-heating models are not turned on, the simulation will default to isothermal temperature profiles.

6.3.7.1 Heat Flow and Temperature Distribution

In terms of simulation, one needs to figure out the temperature distribution from all possible heat sources. This often means considering a much larger simulation area than the small region near the PN junction as we must think about how the heat flows through the whole substrate as well as from any wire bonds. We must also examine how the heating affects the device performance, which means gathering as much data as possible on the temperature dependence of important material parameters. This is not a trivial task because virtually all variables and material parameters are temperature dependent. It is not immediately obvious which ones are most important, but in general bandgap and mobility should be at the top of this list.

TABLE 6.8

Some Parameters for the MTE Model

Parameters	Ref [171]	Ref [172]
μ_{high} cm²/V s	100	133.3
v_{sat} 1E+7 cm/s	1.91	1.44
E_{MT} kV/cm	257	195
β_c	1.7	3.2
β_T	5.7	3.9
β_k	4	4
δE_k kV/cm	0	5

We are concerned with the generation and flow of lattice heating power in a semiconductor, so we first consider the heat flux and then the heat source. We introduce the thermal conductivity such that the heating power flux (in Watt/m^2) is given by

$$J_h = -\kappa \nabla T \tag{6.58}$$

Conservation of energy requires that the temperature distribution satisfy the following basic thermal equation:

$$C_p \rho \frac{\partial T}{\partial t} = -\nabla \cdot J_h + H \tag{6.59}$$

or

$$C_p \rho \frac{\partial T}{\partial t} = \nabla \cdot \kappa \nabla T + H \tag{6.60}$$

where C_p is the specific heat, ρ is the density of the material, and H is the heat source.

The goal of TCAD thermal simulation is to provide a thermal modeling environment so that all possible sources of temperature dependence can be taken into account. Many different sources of heat are included in TCAD tools including joule heating from the current flow and recombination heat from the carriers interacting with defects. Consult the software's manual for more details on the models that are included.

6.3.8 BANDGAP NARROWING EFFECT

It is observed experimentally that the shrinkage of bandgap occurs when impurity concentration is high (e.g., $n_{imp} > 10^{17}$ cm^{-3}). This is called the bandgap narrowing effect, which is ascribed to the emerging of the impurity band formed by the overlapped impurity states. The bandgap narrowing model proposed by Slotboom is given by

$$\Delta E_g = E_{ref} \left\{ \ln \frac{n_{imp}}{n_{ref}} + \sqrt{ln^2 \frac{n_{imp}}{n_{ref}} + 0.5} \right\} \tag{6.61}$$

where E_{ref}, n_{ref}, and n_{imp} represent energy parameter, density parameter, and impurity concentration, respectively. For silicon, E_{ref} and n_{ref} were obtained by fitting the experimental result, which yields

$$E_{ref} = 0.009 \ V \tag{6.62}$$

$$n_{ref} = 10^{17} \ cm^{-3} \tag{6.63}$$

It should be noted that the bandgap narrowing effect is the doping-dependent effect, while the many-body effect, which can also reduce bandgap, is carrier dependent and thus depends on the applied bias.

6.4　AC ANALYSIS

6.4.1　INTRODUCTION

Most power semiconductor devices are driven under some sort of AC condition, and many of the device parameters (e.g., capacitance) are measured at a given frequency. Therefore, it is beneficial to have some basic understanding of how TCAD software performs AC analysis. We shall dedicate this section to the theory and application of AC analysis.

AC small signal analysis for semiconductor equations was first developed by Laux [173] in 1985. A modified version of this technique, including the deep level trap model, was later developed by Li and Dutton [174]. We shall follow their formulation in this section.

The basic problem of AC analysis is stated as follows: for a device at a given DC (or quasi-DC for slow transient cases) bias, how would the device behave if one or more device contacts were subject to an AC voltage signal of the form $\Delta\xi e^{i\omega t}$, where the modulation amplitude $\Delta\xi$ is much smaller than the DC bias.

6.4.2　BASIC FORMULAS

We rewrite the basic semiconductor equations including the trap dynamic model as follows:

$$-\nabla\cdot\left(\frac{\epsilon_0\epsilon_{dc}}{q}\nabla V\right) = -n + p + N_D(1-f_D) - N_A f_A + \sum_j N_{tj}(\delta_j - f_{tj}), \tag{6.64}$$

$$\nabla\cdot J_n - \sum_j R_n^{tj} - R_{sp} - R_{st} - R_{au} + G_{opt}(t) = \frac{\partial n}{\partial t} + N_D\frac{\partial f_D}{\partial t}, \tag{6.65}$$

$$\nabla\cdot J_p + \sum_j R_p^{tj} + R_{sp} + R_{st} + R_{au} - G_{opt}(t) = -\frac{\partial p}{\partial t} + N_A\frac{\partial f_A}{\partial t}, \tag{6.66}$$

$$R_n^{tj} - R_p^{tj} = N_{tj}\frac{\partial f_{tj}}{\partial t} \tag{6.67}$$

where

$$R_n^{tj} = c_{nj}nN_{tj}\left(1-f_{tj}\right) - c_{nj}n_{1j}N_{tj}f_{tj}, \tag{6.68}$$

$$R_p^{tj} = c_{pj}pN_{tj}f_{tj} - c_{pj}p_{1j}N_{tj}\left(1-f_{tj}\right). \tag{6.69}$$

The basic variables involved in the AC analysis are V, the potential; n and p, the electron and hole concentrations, respectively; and f_{tj}, the trap occupancy of the jth trap. To simplify our derivation, we use the following notation to represent Equations (6.64) to (6.67):

$$F_1(V, U_n, U_p, f_{tj}) = 0 \tag{6.70}$$

$$F_2(V, U_n, U_p, f_{tj}) = \left(\frac{\partial n}{\partial U_n} + N_D\frac{\partial f_D}{\partial U_n}\right)\frac{\partial U_n}{\partial t} \tag{6.71}$$

$$F_3(V, U_n, U_p, f_{tj}) = \left(-\frac{\partial p}{\partial U_p} + N_A \frac{\partial f_A}{\partial U_p} \right) \frac{\partial U_p}{\partial t} \tag{6.72}$$

$$H_j(U_n, U_p, f_{tj}) = N_{tj} \frac{\partial f_{tj}}{\partial t} \tag{6.73}$$

where we have changed the variables n and p into U_n and U_p, the normalized quasi-Fermi levels, respectively. For the electron or hole transport equations (F_2 or F_3) the first term on the right-hand side (RHS) corresponds to $\partial n/\partial t$ or $\partial p/\partial t$, while the second term is the response due to the trap filling and emptying.

In general, the AC input signal (voltage) and solution variables are expressed as

$$\xi = \xi_0 + \Delta \xi e^{i\omega t} \tag{6.74}$$

where ξ_0 is the DC solution, $\Delta \xi$ is the complex AC solution, and $\omega = 2\pi f$ is the modulation frequency. ξ can represent V, U_n, U_p, or f_{tj} as complex solutions. By substituting ξ back into the trap equation and doing a Taylor series expansion to first order around the DC solution, Equation (6.67) becomes

$$\frac{\partial H_j}{\partial U_n} \Delta U_n + \frac{\partial H_j}{\partial U_p} \Delta U_p + \frac{\partial H_j}{\partial f_{tj}} \Delta f_{tj} = N_{tj} i\omega \Delta f_{tj} \tag{6.75}$$

and solving for Δf_{tj} gives

$$\Delta f_{tj} = \left(\frac{\partial H_j}{\partial U_n} \Delta U_n + \frac{\partial H_j}{\partial U_p} \Delta U_p \right) \left(N_{tj} i\omega - \frac{\partial H_j}{\partial f_{tj}} \right)^{-1} \tag{6.76}$$

$$\Delta f_{tj} = \frac{-\frac{\partial H_j}{\partial U_n} \left(\frac{\partial H_j}{\partial f_{tj}} + i\omega N_{tj} \right) \Delta U_n - \frac{\partial H_j}{\partial U_p} \left(\frac{\partial H_j}{\partial f_{tj}} + i\omega N_{tj} \right) \Delta U_p}{(N_{tj}\omega)^2 + \left(\frac{\partial H_j}{\partial f_{tj}} \right)^2} \tag{6.77}$$

which can be rewritten for simplicity as:

$$\Delta f_{tj} = F_n(\omega) \Delta U_n + F_p(\omega) \Delta U_p \tag{6.78}$$

Similarly expanding the Poisson equation, electron and hole continuity equations as a first-order Taylor Series, we get

$$\frac{\partial F_1}{\partial V} \Delta V + \frac{\partial F_1}{\partial U_n} \Delta U_n + \frac{\partial F_1}{\partial U_p} \Delta U_p + \frac{\partial F_1}{\partial f_{tj}} \left(F_n \Delta U_n + F_p \Delta U_p \right) = 0 \tag{6.79}$$

$$\frac{\partial F_2}{\partial V} \Delta V + \frac{\partial F_2}{\partial U_n} \Delta U_n + \frac{\partial F_2}{\partial U_p} \Delta U_p + \frac{\partial F_2}{\partial f_{tj}} \left(F_n \Delta U_n + F_p \Delta U_p \right) = \left(\frac{\partial n}{\partial U_n} + N_D \frac{\partial f_D}{\partial U_n} \right) i\omega \Delta U_n \tag{6.80}$$

$$\frac{\partial F_3}{\partial V} \Delta V + \frac{\partial F_3}{\partial U_n} \Delta U_n + \frac{\partial F_3}{\partial U_p} \Delta U_p + \frac{\partial F_3}{\partial f_{tj}} \left(F_n \Delta U_n + F_p \Delta U_p \right) = \left(-\frac{\partial p}{\partial U_p} + N_A \frac{\partial f_A}{\partial U_p} \right) i\omega \Delta U_p \tag{6.81}$$

We rewrite these three equations in a 3 by 3 matrix form as

$$(J+D)\xi = 0 \tag{6.82}$$

For each mesh point in the semiconductor the J matrix or Jacobian contains the first three terms in each equation, while D contains the remaining terms concerning the traps and RHS of the equation. Using this notation with $\xi = (\Delta V, \Delta U_n, \Delta U_p)$ the D matrix at each point is of the form

$$D = \begin{pmatrix} 0 & \dfrac{\partial F_1}{\partial f_{tj}} F_n(\omega) & \dfrac{\partial F_1}{\partial f_{tj}} F_p(\omega) \\ 0 & -\left(\dfrac{\partial n}{\partial U_n} + N_D \dfrac{\partial f_D}{\partial U_n}\right) i\omega + \dfrac{\partial F_2}{\partial f_{tj}} F_n(\omega) & \dfrac{\partial F_2}{\partial f_{tj}} F_p(\omega) \\ 0 & \dfrac{\partial F_3}{\partial f_{tj}} F_n(\omega) & -\left(-\dfrac{\partial n}{\partial U_n} + N_A \dfrac{\partial f_A}{\partial U_p}\right) i\omega + \dfrac{\partial F_3}{\partial f_{tj}} F_p(\omega) \end{pmatrix} \tag{6.83}$$

Consider the boundary points and the internal mesh points separately so that

$$J \rightarrow j + J_B \tag{6.84}$$

and

$$D \rightarrow D + D_B \tag{6.85}$$

The matrix equation can then be rewritten in a separated form for internal and boundary mesh points as follows:

$$(J+D)\xi + (J_B + D_B)\xi_B = 0 \tag{6.86}$$

AC Dirichlet boundary conditions imply that ΔU_n and $\Delta U_p = 0$ so that the AC voltage input at the contact has the boundary solution

$$\xi_B = \begin{pmatrix} \Delta V_B \\ 0 \\ 0 \end{pmatrix} \tag{6.87}$$

where ΔV_B is the applied AC voltage of one volt. Since the first column in the D matrix must be zero, the matrix equation becomes

$$(J+D)\xi = -J_B \cdot \xi_B \equiv B \tag{6.88}$$

To create real-valued equations, the D and ξ complex matrices are written with the real and imaginary components separated:

$$D = D_r + iD_i \tag{6.89}$$

$$\xi = \xi_r + i\xi_i \tag{6.90}$$

The matrix is thus rewritten as

$$(J + D_r + iD_i)(\xi_r + i\xi_i) = B \tag{6.91}$$

Equating real and imaginary components provides the two equations

$$(J + D_r)\xi_r - D_i\xi_i = B \tag{6.92}$$

$$D_i\xi_r + (J + D_r)\xi_i = 0 \tag{6.93}$$

which can be rewritten as the expanded matrix

$$\begin{pmatrix} J + D_r & -D_i \\ D_i & J + D_r \end{pmatrix} \begin{pmatrix} \xi_r \\ \xi_i \end{pmatrix} = B \tag{6.94}$$

6.4.3 AC ANALYSIS IN TCAD

Our task has been reduced to solving the expanded matrix equation detailed in the previous subsection. The AC solution matrix $\begin{pmatrix} \xi_r \\ \xi_i \end{pmatrix}$ can be easily solved using a sparse matrix inverter to invert the expanded Jacobian and multiply by the matrix on the RHS. The inversion can be slow because the expanded Jacobian $J + D$ is twice the order of original J matrix obtained from the DC solution.

In most TCAD software (e.g., APSYS from Crosslight Software), the AC analysis is implemented as a postprocessor; that is, the analysis is done after the DC solution is obtained from the main solver input file (with the .sol extension). This is done in the following steps:

- The AC analysis matrix is constructed from Jacobian matrix elements used for DC simulation. Or in other words, the DC matrix is saved for later AC analysis.
- After the DC simulation, the AC analysis is activated by an AC analysis command defining which contact to apply the AC voltage and at what frequency range. For C-V simulation, one would fix the frequency at 1 megahertz and would define the DC bias range.
- Use various related TCAD commands to plot the AC conductance or AC capacitance. If necessary, various two-port parameters, such as Y-parameters or S-parameters, can be plotted.

As stated earlier, the basic problem of AC analysis is to study how the device would respond if a contact is subject to an AC voltage signal.

The basic responses are the complex AC currents at all the electrodes and the AC current distribution within the device. Once we have the AC current, the following important quantities can be extracted:

- The conductance matrix elements (such as transconductance) are computed from the real part of the electrode AC current with unit AC voltage applied at one of the electrodes.
- The capacitance matrix elements are computed from the imaginary part of the electrode AC current, divided by ω, with unit AC voltage applied at one of the electrodes.
- The current gain can be computed from the ratio of the AC current magnitudes between two electrodes, with a small AC voltage applied at one of the electrodes.

6.5 TRAP MODEL IN TCAD SIMULATION

As in most wide-gap compound materials, undoped GaN exhibits semi-insulating behavior due to the existence of deep-level traps. It is also believed that gate leakage, reliability, or degradation behaviors of GaN high electron mobility transistors (HEMT) are associated with surface traps. For proper understanding and analysis of GaN power devices, it is necessary to have some basic knowledge of how traps are modeled in TCAD simulation.

6.5.1 Trap-Charge States

A carrier trap can be completely described by trap density, charge state (donor or acceptor), energy level, electron, and hole capture coefficients.

Referring to the schematic of Figure 6.8, a trap has the charge property of being a donor or an acceptor. In the extreme case of energy level of a trap being close to the band edges, traps are no different than common dopants such as boron or arsenic in silicon. To be more specific, the space charge caused by the trap is either $N_t(1-f_t)$ for a donor trap or $-N_t f_t$ for an acceptor trap. Here, N_t and f_t are trap density and trap occupancy, respectively. In TCAD, these charge terms appear in the Poisson's equation along with the other charge terms: shallow dopants; free electrons/holes; and fixed bulk/surface charges.

As far as charge states are concerned, traps are similar to the usual intentional dopants such as boron or arsenic in silicon technology. They can be used to control the Fermi levels, and donors and acceptors can compensate each other. The formation of semi-insulating compound materials is caused by this compensation mechanism: the existence of both deep donors and deep acceptors pins the Fermi level near the middle of the bandgap, and this causes the high resistance of undoped substrate compound materials.

In silicon technology, traps are less important because their densities are low compared with intentional dopants. Since the charge contribution of deep-level traps in silicon is small, their charge states are usually ignored and a single minority lifetime is used to characterize traps in silicon TCAD simulation.

6.5.2 Trap Dynamics

The charge attributes of traps are relatively easy to understand and control in TCAD simulation. The dynamics of traps is far more complicated, and this section is dedicated to the description

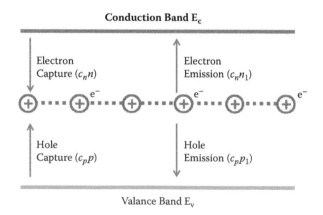

FIGURE 6.8 Schematic view of donor traps.

and understanding of trap dynamics. Dynamics refers here to the time-dependent behavior such as trapping, detrapping rates, and AC small signal analysis. The AC analysis is especially important since the existence of a large amount of deep-level traps strongly affects the AC capacitance and conductance of a GaN-based power device.

Our approach is to assume some simplified situation where traps are dominant in the drift-diffusion equation. We then derive capacitance and conductance of a semiconductor region and analyze their behavior. Finally, we set up a TCAD simulation to compute AC capacitance and conductance to confirm the simplified trap dynamic theory.

We introduce the trap dynamic equation (or sometime called trap rate equation) as follows:

$$N_t \frac{df_t}{dt} = R_n - R_p \tag{6.95}$$

$$R_n = c_n n N_t \left(1 - f_t\right) - c_n n_1 N_t f_t \tag{6.96}$$

$$R_p = c_p p N_t f_t - c_p p_1 N_t \left(1 - f_t\right) \tag{6.97}$$

where N_t is the bulk trap density; R_n and R_p are the bulk recombination rates for electrons and holes, respectively; c_n is the capture coefficient, which can be further expressed in terms of thermal velocity and capture cross section, $c_n = v_{th,n} \sigma_n$; and n_1 is the electron density when the Fermi level coincides with the trap level. Similar quantities are given for holes.

As a good exercise, one can prove that at the steady state, $R_n = R_p$ (or $df_t / dt = 0$), and by solving the trap occupancy f_t, the recombination rate reduces to the more familiar Shockley–Read–Hall recombination rate.

Usually, there is more than one kind of trap in the system, and we add a trap index j in Equations (6.95)–(6.97), such as N_{tj} or f_{tj}. For simplicity we drop the trap index in the discussion while keeping in mind that the theory applies to all other traps.

The simplified situation is that we consider a region with substantial amount of traps (e.g., 1E+17 cm^{-3}). There is only light intentional doping, or it is totally undoped. The free carriers are supplied via injection from nearby contacts. In Appendix C, we will derive the small signal AC current flowing through this region under AC voltage ΔV. The AC current can be expressed as

$$\Delta I = G_{tp} \Delta V + j\omega C_{tp} \Delta V \tag{6.98}$$

where the conductance G_{tp} is given by

$$G_{tp} = \frac{V_{tp} N_t \omega^2 \left(\dfrac{dc_i}{dV}\right)}{\omega^2 + \left(c_n n_0 + c_p p_0\right)^2} \tag{6.99}$$

and the capacitance is obtained as

$$C_{tp} = \frac{V_{tp} N_t \left(\dfrac{dc_i}{dV}\right)\left(c_n n_0 + c_p p_0\right)}{\omega^2 + \left(c_n n_0 + c_p p_0\right)^2} \tag{6.100}$$

where n_0 and p_0 are the steady-state electron and hole concentrations, respectively; V_{tp} is the volume of the trap region; and c_i is a parameter depending on carrier injection.

It is interesting to note that the frequency response of capacitance and conductance are complementary: the capacitance peaks on the lower frequency side, while the conductance peaks on the higher side. Furthermore,

$$C_{tp} + G_{tp} = frequency\_independent\_constant \qquad (6.101)$$

The characteristic frequency for the peaks is given by $c_n n_0 + c_p p_0$, and it is dependent on carrier density and trapping rate. Due to the existence of more than one kind of trap in the system, the AC behavior is not simple to predict.

We demonstrate the AC TCAD simulation using APSYS in the following structure. An AlGaN/GaN diode is built with Al mole fraction of 0.1 for AlGaN ($Al_{0.1}Ga_{0.9}N$). $Al_{0.1}Ga_{0.9}N$ is P doped with doping concentration of 1E+17 cm^{-3} and GaN is N doped with doping concentration of 1E+17 cm^{-3}, as seen in Figure 6.9.

Donor-type traps located at 1.2 eV below conduction band are placed within GaN material with trap concentration of 5E+17 cm^{-3}; the capture cross section is 1E-15 cm^{-2}. The same structure without traps is also simulated as a comparison. A positive DC bias is applied to the cathode (GaN side) for both diode devices.

The capacitance and conductance in Figure 6.10 show the predicted complementary profile with capacitance peaks at the lower frequency side. In a real device, the number of traps and traps levels are rather complex and it is not easy to judge whether or not trap regions are the dominant factor in the device characteristics. Our theory here may help in two ways.

1. If a material is well-known to have deep-level traps (e.g., near the substrate), the TCAD can help determine if such traps would have an impact to the AC capacitance and conductance.
2. If frequency response peaks are measured in the lower end of capacitance or higher end of conductance that cannot be easily explained by simple junction or depletion region theory, then traps may be the cause, and one may use reverse engineering to deduce the existence of traps in one or more regions of the device.

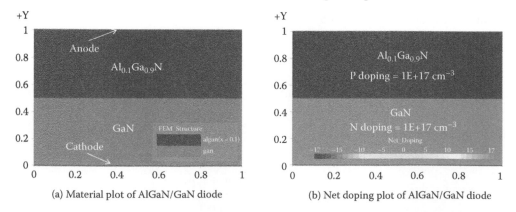

(a) Material plot of AlGaN/GaN diode   (b) Net doping plot of AlGaN/GaN diode

FIGURE 6.9 Simulation structure views of the AlGaN/GaN diode to demonstrate how traps influence device performance.

AlGaN/GaN Diode with Traps

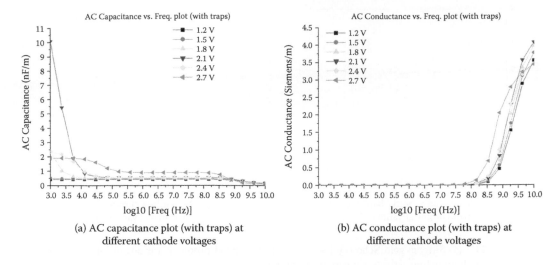

(a) AC capacitance plot (with traps) at
different cathode voltages

(b) AC conductance plot (with traps) at
different cathode voltages

AlGaN/GaN Diode without Traps

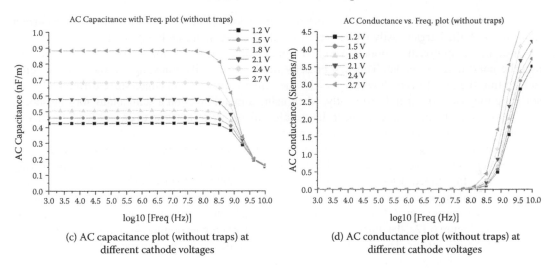

(c) AC capacitance plot (without traps) at
different cathode voltages

(d) AC conductance plot (without traps) at
different cathode voltages

FIGURE 6.10 Simulated AC capacitance and conductance results of AlGaN/GaN diode at different cathode voltages: (a)–(b) With traps; (c)–(d) Without traps.

6.6   QUANTUM TUNNELING

6.6.1   IMPORTANCE OF QUANTUM TUNNELING FOR POWER DEVICES

Intraband quantum tunneling is important when a potential barrier is formed to block the transporting carriers and the barrier thickness is in the order of quantum mechanical wavelengths (typically 10 nm or shorter). For silicon-based power devices, such a situation arises in metal/silicon interfaces where Schottky or Ohmic contacts are formed. For Schottky contacts, the importance of tunneling is obvious since the Schottky barrier forms a significant barrier for transporting carriers.

Ohmic contacts should be regarded as a special case of Schottky contacts with extremely high doping at the semiconductor contact region. As indicated in Figure 6.11a, the potential barrier gets

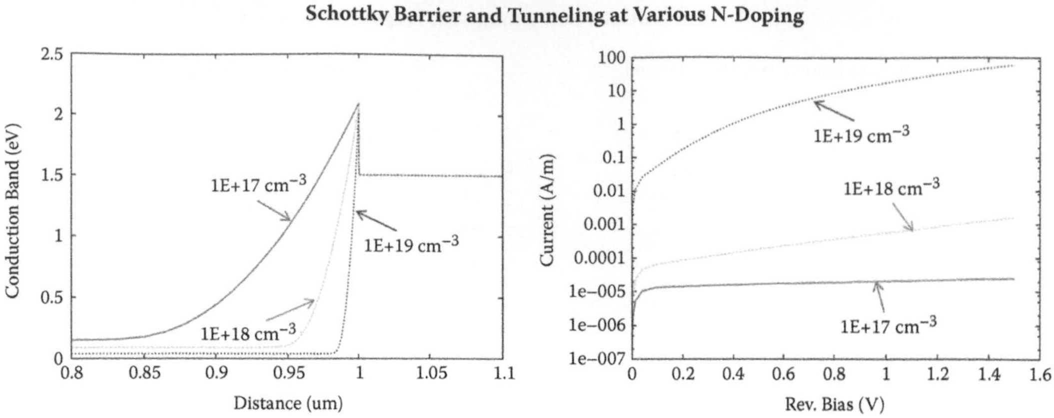

(a) Silicon-gold Schottky barrier conduction
band profile at various N-doping

(b) Reverse bias current with tunneling
activated at various N-doping

FIGURE 6.11 (a) Silicon–gold Schottky barrier conduction band profile at various n-doping; (b) Reverse bias current with tunneling activated at various n-doping.

narrower as doping is increased. In the extreme of high doping, the barrier is so thin that the current flows through the barrier easily with a large tunneling coefficient (see Figure 6.11b).

A common approach to modeling a good Ohmic contact is to assume the barrier does not exist and the band diagram is flat. This can be achieved by artificially adjusting the metal work function so that the metal Fermi level aligns with the semiconductor band edge. Since different TCAD software may treat the metal differently, when using a metal layer on a semiconductor one should always examine the band diagram near the metal/semiconductor junction to ensure the Schottky or Ohmic contacts have the expected band alignment.

For AlGaN/GaN power devices, in addition to the tunneling at Schottky/Ohmic contacts, one must also consider tunneling through a heterojunction or a whole barrier layer. The current flowing through a highly doped heterojunction can get rather complicated since high doping can create a highly depleted layer in the barrier and cause a large voltage drop there unless quantum tunneling is activated. We consider an AlGaN/GaN heterojunction subject to different densities of n-doping. As indicated by Figure 6.12a, the tunneling barrier gets thinner at higher doping, similar to the situation of Schottky/Ohmic contacts. The current computed using tunneling theory is indicated in Figure 6.12b, indicating much higher reverse current for higher doping.

One can show that the barrier height at high doping is dependent on whether there is a composition grading at the heterojunction. The highest barrier is obtained for an abrupt junction while composition grading reduces or even completely removes the potential barrier at high doping. If the details of transport in the highly doped heterojunction are not important and one wishes to avoid dealing with tunneling, a shortcut would be to artificially grade the abrupt heterojunction to remove the high barrier.

6.6.2 BASIC THEORY OF TUNNELING FOR TCAD SIMULATION

As discussed in previous subsections, there are ways to avoid the use of tunneling for highly doped heterojunctions or Ohmic contacts if the details of transports are not important. For barriers where tunneling is of primary importance, we need to formulate the tunneling theory in a simple way that TCAD software can employ.

In quantum mechanics, tunneling depends on carrier distribution as a function of kinetic energy. Since drift-diffusion theory is based on the assumption that carriers assume a Fermi–Dirac

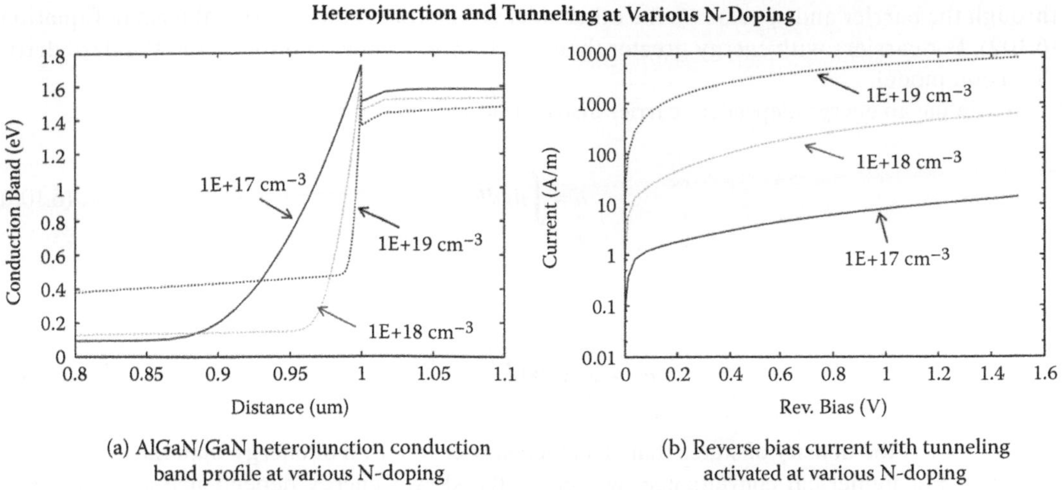

Heterojunction and Tunneling at Various N-Doping

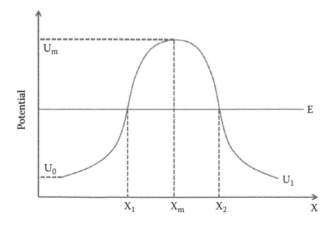

(a) AlGaN/GaN heterojunction conduction
band profile at various N-doping

(b) Reverse bias current with tunneling
activated at various N-doping

FIGURE 6.12 (a) AlGaN/GaN heterojunction conduction band profile at various n-doping; (b) Reverse bias current with tunneling activated at various n-doping.

distribution (quasi-equilibrium), a reasonable approach for quantum tunneling is to assume that carriers incident upon the barrier follow this distribution.

The formulas we use here are based on quasi-equilibrium approximations, and the approach is similar to that used by Grinberg et al. [175]. Since carriers at the top of the barrier are not blocked, it is sufficient we calculate the current flow at the top of the barrier.

We use the schematic in Figure 6.13 and consider the drift current on top of the barrier X_m:

$$J = qvn_m + qvn_m\alpha_T \tag{6.102}$$

where α_T is the tunneling coefficient to be derived. The velocity v is dependent on mobility or the thermionic emission properties at the top of the barrier.

Here the basic understanding is that the quasi-Fermi level is relatively flat around the barrier such that the energy distribution of the carriers can be expressed by a simple Boltzmann function (or exponential function) since the Fermi level is usually far below the top of the barrier. We assume that the carriers with energy between U_0 and U_m are capable of tunneling

FIGURE 6.13 Schematic of the potential barrier for tunneling theory.

through the barrier and appears on the other side, giving rise to the additional term in Equation (6.102). For carriers with energy greater than U_m, they are accounted for by the standard drift-diffusion model.

We define an energy-dependent carrier distribution n_E such that

$$n = \int n_E dE \tag{6.103}$$

Using Boltzmann distribution

$$n_E = g(E)\exp\left(\frac{U_m - E}{kT}\right) \tag{6.104}$$

where $g(E)$ is the density of states that is a much slower function of energy than the exponential function. For numerical convenience, we ignore the slower energy dependence and replace it by an average constant n_{Em} so that Equation (6.103) still holds. As we shall see later, our final results do not depend on the choice of constant n_{Em}.

Since we are able to express the energy distribution as

$$n_E = n_{Em}\exp\left(\frac{U_m - E}{kT}\right) \tag{6.105}$$

the tunneling current can be expressed as

$$J_{tun} = \int_{U_0}^{U_m} qvn_E D_T(E)dE = qvn_{Em}\int_{U_0}^{U_m}\exp\left(\frac{U_m - E}{kT}\right)D_T(E)dE \tag{6.106}$$

where $D_T(E)$ is the energy-dependent tunneling transparency. We note that the carrier density at the top of the potential barrier can be expressed in term of n_{Em}:

$$n_m = \int_{U_m}^{\infty} n_{Em}\exp\left(\frac{U_m - E}{kT}\right)dE = n_{Em}kT \tag{6.107}$$

Therefore, the tunneling current can be written as

$$J_{tun} = qvn_m(kT)^{-1}\int_{E_0}^{U_m}\exp\left(\frac{U_m - E}{kT}\right)D_T(E)dE \tag{6.108}$$

The tunneling coefficient α_{Tm} at the top of the barrier is given by

$$\alpha_{Tm} = (kT)^{-1}\int_{E_0}^{U_m}\exp\left(\frac{U_m - E}{kT}\right)D_T(E)dE \tag{6.109}$$

The main task of computing the tunneling current has been reduced to calculation of the energy-dependent tunneling transparency $D_T(E)$. A commonly used method for computation of tunneling transparency is the transfer matrix method (TMM) or the propagation matrix method, which cuts up an arbitrary potential profile into piecewise constant barriers. For each of the constant barriers, a 2×2 matrix describing forward and backward traveling waves can be constructed. The total transparency can be obtained by multiplying the 2×2 matrices.

The previous formulas are relatively easy to implement in TCAD software and were used by the APSYS software package from Crosslight Software.

6.6.3 INTRODUCTION TO NONEQUILIBRIUM GREEN'S FUNCTION FOR TUNNELING

The previous subsection describes the equilibrium theory of quantum tunneling, which is consistent with the drift-diffusion theory. The latest simulation technology may employ the nonequilibrium Green's function (NEGF) method for tunneling. Due to the complexity of the NEGF, only a conceptual introduction is given here referring to the equilibrium theory we have described previously.

In the quasi-equilibrium theory, the system can be described by the Fermi level and carrier distribution in the whole tunneling region is described by Fermi or Boltzmann functions. The NEGF method does not assume such Fermi or Boltzmann distribution and Fermi levels are not used.

The NEGF solves for the Green's function, a correlation function relating the probability of finding a carrier at one position with another. The Green's function is solved in spatial, momentum, and energy coordinates and can take into account of various scattering mechanisms such as phonon scattering while tunneling. Some advanced TCAD packages (such as APSYS from Crosslight) have implemented NEGF and interested readers should consult the relevant documentations.

6.7 DEVICE SIMULATOR MODELS CALIBRATION

As mentioned in Section 5.9, the accuracy of TCAD simulation depends on how well the models in the process and device simulators are calibrated for a particular set of equipment and fabrication facilities. In this section, a general approach to calibrate the models for a device simulator will be discussed.

Assuming that the process simulator has been calibrated in the first place, the device simulator has to be calibrated such that the electrical characteristics or parameters of the simulated device can be as close as possible to those of the experimental device. For example, in the case of a metal-oxide-semiconductor field-effect transistor (MOSFET), important electrical characteristics for calibration include the drain current versus gate voltage curves at constant drain voltage, the on-state drain current versus drain voltage curves at various gate voltages, and the off-state drain current versus drain voltage curve at zero gate voltage. Electrical parameters of the MOSFET can be extracted from these electrical curves. For example, threshold voltage can be extracted from the drain current versus gate voltage curves. Moreover, on-state resistance, saturation current, and on-state breakdown voltage can be extracted from the on-state drain current versus drain voltage curves. Finally, off-state breakdown voltage and leakage current can be extracted from the off-state drain current versus drain voltage curve. A well-calibrated device simulator is able to predict electrical characteristics at arbitrary bias conditions such that the discrepancy between the simulated and measured device performance is acceptable. However, if the discrepancy persists after considerable effort has been put in calibrating the device simulator, calibration of the corresponding process simulator will have to be reviewed [142].

Calibration of a popular device simulator, Medici [176], for MOSFET simulation is illustrated as an example. It begins with calibrating the simulated drain current versus gate voltage curve with the corresponding measured curve [177]. The drain current in the linear region is mainly determined by the carrier mobility, whereas the threshold voltage is determined by several parameters such as the gate metal work function and the interface charges. All of these parameters can be adjusted in the device simulator. It is followed by calibration of the on-state drain current versus drain voltage curves. The on-state resistance, saturation current, and on-state breakdown voltage are mainly determined by the carrier mobility, saturation velocity, and impact ionization rate, respectively. For the case of a power MOSFET, which operates at high power in the on-state, modeling of thermal effect and lattice heating have to be enabled in the device simulator for a more precise calibration. Finally, the leakage current and off-state breakdown voltage, which can be extracted from the off-state drain current versus drain voltage curve are mainly determined by the band-to-band tunneling and impact ionization rate, respectively. Both of these parameters can be adjusted in the model for the curve calibration.

7 Power IC Process Flow with TCAD Simulation

Before we explore the integrated power devices, it is always better to prepare the reader with some knowledge of how the power devices are fabricated in a smart power IC technology.

7.1 OVERVIEW

While all real process technologies are proprietary, in this chapter, we will present a mock-up process flow with step-by-step technology computer-aided design (TCAD) simulations. This process simulation follows a published [178] 0.18 μm smart power IC flowchart with some modifications. The corresponding process conditions (e.g., implant dose, energy) are arbitrary and not optimized for any real device applications. The purpose of this chapter is to help familiarize the reader with typical process flows and introduce necessary knowledge for the next chapters on device physics and simulation. Please be cautious: blindly following the process steps and recipes listed in this chapter without modifications to account for real manufacturing conditions is not recommended.

7.2 A MOCK-UP POWER IC PROCESS FLOW

7.2.1 PROCESS FLOW STEPS

For smart power IC technology development, we need to keep in mind that a lot of different devices need to be integrated onto a single chip (not a single package). Each device type has its own structure, whose process steps are either shared with other devices or dedicated to it. In smart power IC design, it is desirable to have as many shared process steps as possible to reduce costs. While many process steps like N+ source and drain implant can be easily shared between digital, analog, and power MOSFETs, it is often impossible to share the same thermally grown gate oxide between digital MOSFETs and power lateral double-diffused MOS (LDMOS).

Similar to the process flow found in Figure 4.3, a simplified mock-up flow is created according to [178] and illustrated in Figure 7.1. Some processing steps like silicide formation require a chemical reaction and cannot be simulated by the TCAD simulator; these steps are therefore omitted. Here "Twin-Well" means that N-well and P-well are formed by separate implants and masks. Two gate oxide thicknesses are used. Thick gate oxide (TGOX) is used for LDMOS, whereas thin gate oxide (GOX) is used for analog and digital MOSFETs. For simplicity, only one metal layer is used. Poly-insulator-poly (PIP) or metal-insulator-metal (MIM) capacitors are not included in this simulation. Low-voltage wells are used primarily for highly doped emitter of NPN and PNP BJTs, while high-voltage wells are used for lightly doped N-drift or P-drift regions of LDMOS.

We will discuss all these process steps in detail in the following sections. But before that, we will first take a look at the whole picture of the finished process flow in the next section.

7.2.2 STRUCTURE VIEW OF THE MOCK-UP PROCESS FLOW

In this chapter, a simplified mock-up process flow is built. For illustration purpose, three types of devices are placed side by side and simulated by a TCAD process simulator. For the sake of simplicity, we will include only three important device types: N-type LDMOS; NPN BJT; and a digital PMOS.

FIGURE 7.1   Simplified smart power IC process flow for simulation.

Figure 7.2 is a structure view of the finished smart power IC process flow with three device types arranged side by side: from left to right, an N-type LDMOS, an NPN BJT isolated from the LDMOS by shallow trench isolation (STI) and finally, a digital PMOS placed besides the BJT. Only one layer of metal is applied in this simulation, even though in real fabrication two or more metal layers are always necessary.

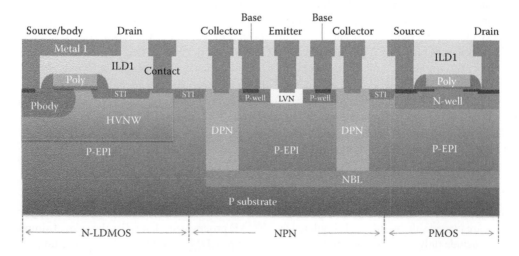

FIGURE 7.2   Structure view of the finished smart power IC process flow.

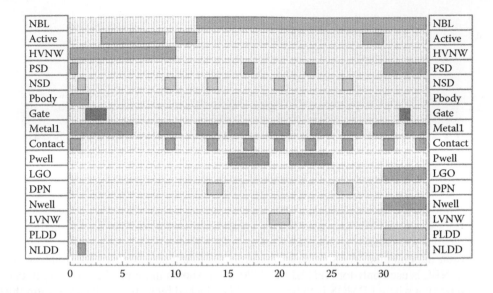

FIGURE 7.3 Mask set view of the smart power IC process flow.

Figure 7.3 is the mask set view of the smart power IC process flow; 16 photomasks are used in this example.

7.3 SMART POWER IC PROCESS FLOW SIMULATION

7.3.1 P+ SUBSTRATE

A P+ substrate is commonly used in smart power IC technology. Highly doped P-type substrates can reduce minority carrier injection and reduce cross-talk between devices.

In this simulation, boron is used as the dopant, with a constant doping concentration of 1E+17 cm^{-3} and wafer orientation of (100) for better silicon surface quality. The substrate resistivity is estimated to be about 0.2 $\Omega \cdot$ cm.

In terms of simulations, a mesh needs to be defined for any simulation tools based on finite element analysis. Mathematically, simulation results using a coarse mesh are not as accurate compared with a fine mesh as this means smooth curves are approximated by piecewise linear ones. However, a very dense mesh requires considerable simulation time and computer memory, so trade-offs are always important: the computing power and engineering time and resources needed to run each version of a design should be considered. A proper mesh should be designed so that the important areas (e.g., contacts, junctions, interfaces, doping gradients) have fine grids while relatively unimportant areas use a coarser grid. Figure 7.4 shows a sample mesh for our P-substrate process; the simulation area is 34 µm wide and 3 µm thick.

7.3.2 N+ BURIED LAYER

The N+ buried layer (NBL) is widely used in smart power IC technology. NBL is used to reduce CMOS latch-up and as isolation between devices, especially in conjunction with deep trench isolation. As a floating layer, it can also provide shielding for high-side applications. In this simulation, the NBL is used as the collector link to create a low resistance current path for the NPN BJT.

The NBL layer is one of the first steps of process flow. It has to undergo a lot of thermal cycles as the process flow moves on. To prevent too much diffusion of the implanted dopants, heavier

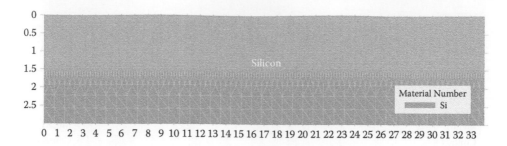

FIGURE 7.4   P-substrate simulation result with mesh plot.

species with lower diffusion coefficients like arsenic (As) and antimony (Sb) are used instead of phosphorus; a screen oxide is also used to keep contaminants away and reduce channeling effects. Channeling allows dopants to travel deeper before colliding with lattice atoms and is an unwanted feature of ion implantation. A photomask is used to define the area for the NBL implant. In this simulation, NBL is used only for NPN BJT and PMOS: as shown in Figure 7.5, the NBL implant in the silicon portion where LDMOS is to be built will be blocked by a photoresist layer 1 μm thick.

In this simulation, screen oxide is thermally grown in H_2O to achieve a thickness of 200 Å. The growth condition is 15 minutes at 1000°C. Arsenic is used as the dopant and implanted with energy of 15 keV and a dose of 1E+14 cm^{-2}; note that in real fabrication, a much higher dose (1E+15 cm^{-2} or above) is commonly used for NBL. After implant, implant damage anneal and drive-in are carried out. In the simulation, the furnace has an initial temperature of 800°C and is ramped up to 1000°C over 10 minutes. The temperature is then kept at 1000°C for 20 minutes before being ramped down to 800°C for another 10 minutes. The process recipe is listed in Table 7.1 and the temperature vs. time curve is plotted in Figure 7.6; the ramp rate in this process step is 20°C/min. In future simulation steps, we may ignore the temperature ramping and just use constant furnace temperature at plateau for simplicity. Please note that the NBL layer will diffuse upward with later thermal cycles such as those for the epi layer and high-voltage wells.

The process simulation result of NBL is illustrated in Figure 7.7. Subsequent process steps will further drive the NBL implant in all directions.

7.3.3   Epitaxial Layer Growth and Deep N Link

An epitaxial layer (or epi layer) of silicon is grown in a special furnace after the NBL implantation: the crystal orientation of that layer will follow that of the substrate material. In fabrication a (100)

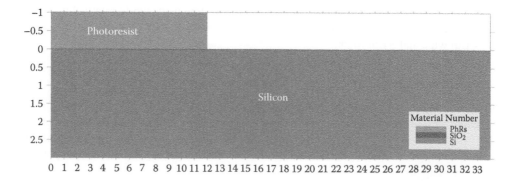

FIGURE 7.5   Photomask for NBL implant.

TABLE 7.1
Diffuse Parameters for NBL Anneal and Drive-in

Process Name	Time (minutes)	Initial Temperature (°C)	Final Temperature (°C)	Condition
Diffuse	10	800	1000	Nitrogen
Diffuse	20	1000	1000	Nitrogen
Diffuse	10	1000	800	Nitrogen

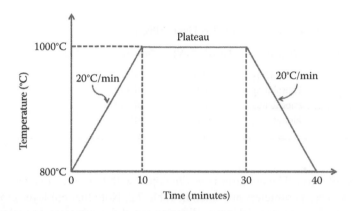

FIGURE 7.6 Furnace anneal with temperature ramp up and down.

Si surface is often used to produce a high-quality epitaxial layer with low defect density. *In situ* doping is necessary to control the epi layer resistivity.

In the simulation, the vertical grid lines will be inherited from the substrate grid, while horizontal grid lines need to be defined for all deposited/grown layers. For simulation purpose, we can simply use the deposition of silicon with constant P-type (boron) doping instead of real silicon epi growth. The thermal budget of the epi layer growth is combined with the previous NBL drive-in step and the deep N-link (DPN) step that follows the epi layer growth.

In this mock-up process flow, we break up the epitaxial growth into two steps. After each growth step, a photomask is used to create deep N-type implanted wells that connect the silicon surface to the NBL layer. These deep N-links are used for NPN BJTs in this simulation. The first epitaxial layer has a thickness of 3 μm, and the second epi layer is thinner with a thickness of 2 μm for a total

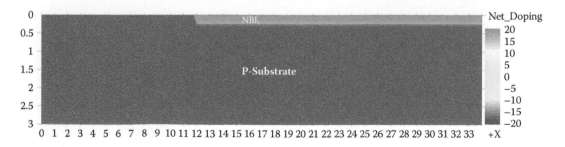

FIGURE 7.7 Net doping plot after NBL anneal.

TABLE 7.2

Epi Layer Process Simulation Parameters

Process Name	Material	Thickness (μm)	Dopants	Doping Concentration (cm⁻³)
Deposit	Silicon	3	Boron	1E+16
Deposit	Silicon	2	Boron	1E+16

TABLE 7.3

Implant Chain Parameters of the First DPN Implant

Process Name	Dopants	Energy (keV)	Dose (cm⁻²)	Angle (°)	Rotation (°)
Implant	Phosphorus	100	3E+12	0	0
Implant	Phosphorus	375	3E+12	0	0
Implant	Phosphorus	675	3E+12	0	0
Implant	Phosphorus	900	3E+12	0	0

epi layer thickness of 5 μm. The epi layer is *in situ* doped with boron to create a constant doping of 1E+16 cm⁻³. The process parameters are listed in Table 7.2. Note that epi layer growth is a thermal process, but in simulation we simply deposit silicon on top of the substrate. In real processes, the epi growth temperature should be designed so that the associated thermal budget is negligible.

The reason that we do not grow the epi layer in one step is because of the DPN implants. A very deep implant greater than a few microns is difficult to realize, and typically very high implant energy levels need to be applied. High energy combined with high dose implants will substantially damage the silicon lattice, which causes undesirable transient enhanced diffusion effects. Even though in practice implant energies can be in the MeV range, we will use maximum implant energy of 900 keV in this simulation.

An implant chain (combining several individual ion implants into a single process step) is necessary to create a smooth and deep implant after drive-in. For the first DPN implant, we use the implant conditions listed in Table 7.3, the photomask step for DPN implant is simulated in Figure 7.8.

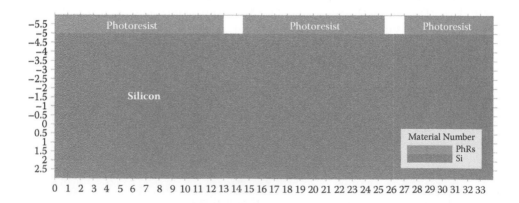

FIGURE 7.8   Photomask for DPN implant.

TABLE 7.4

Implant Chain Parameters of Second DPN Implant

Process Name	Dopants	Energy (keV)	Dose (cm⁻²)	Angle (°)	Rotation (°)
Implant	Phosphorus	150	2E+12	0	0
Implant	Phosphorus	400	1E+12	0	0
Implant	Phosphorus	550	1E+12	0	0
Implant	Phosphorus	700	5E+12	0	0

TABLE 7.5

Diffuse Parameters for Second DPN Anneal and Drive-in

Process Name	Time (minutes)	Initial Temperature (°C)	Final Temperature (°C)	Condition
Diffuse	80	1100	1100	Nitrogen

Since the second epitaxial layer is thinner (2 μm) than the first one (3 μm), we can use lower implant energies; the implant chain parameters are listed in Table 7.4.

A thermal process is necessary to anneal implant damage as well as activate and drive-in the dopants in DPN; the process simulation parameters for this step is listed in Table 7.5. This is a high thermal budget drive-in step to allow both the DPN and NBL layers diffuse to the desired depth and width.

The net doping plot after the DPN anneal/drive-in is illustrated in Figure 7.9.

7.3.4 HIGH-VOLTAGE TWIN-WELL

High-voltage twin-well consists of a high-voltage N-well (HVNW) and a high-voltage P-well (HVPW). They are used to create the drift regions for n-LDMOS and p-LDMOS. For process simulation in this chapter, only n-LDMOS is included so only HVNW is simulated.

HVNW is a relatively lightly doped N-well (about 1E+16 cm⁻³) since it has to support high breakdown voltages. The depth of the well is about 1.5 μm in this simulation. Like the deep N-link, an implant chain must be used to make the well deeper and more uniform. A large thermal cycle is also necessary to allow dopant diffusion, which will also apply to the previously implanted wells (NBL and DPN). Thermal cycles in later process steps will create even further diffusion of the already implanted wells, which is why implant wells that require large thermal budgets are always

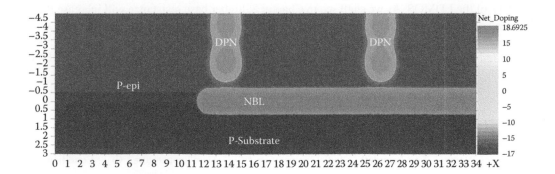

FIGURE 7.9 Net doping plot after DPN implant anneal and drive-in.

TABLE 7.6
HVNW Pad Oxide Simulation Parameters

Process Name	Material	Thickness (μm)
Deposit	Oxide	0.01

TABLE 7.7
HVNW Implant Process Simulation Parameters

Process Name	Dopants	Energy (keV)	Dose (cm^{-2})	Angle (°)	Rotation (°)
Implant	Phosphorus	150	1E+12	0	0
Implant	Phosphorus	350	1E+12	0	0
Implant	Phosphorus	550	1E+12	0	0
Implant	Phosphorus	750	1E+12	0	0

processed first. Well implants that need smaller thermal budgets to prevent dopants from diffusing too much (e.g., source/drain implants) are usually placed at the very end of the process flow.

In real fabrication, a thin layer of sacrificial oxide is usually used to protect the silicon surface from direct implant damage and contaminants and to prevent dopants from escaping the silicon during annealing. For shallow implant, this thin oxide layer can help reduce the channeling effect, which causes implanted dopants to travel too deep in the silicon lattice.

In this simulation, a thin layer of oxide is deposited, but we will not repeat this step for the rest of this chapter for the sake of simplicity. The simulation parameters are listed in Table 7.6, Table 7.7, and Table 7.8. The photomask for HVNW is shown in Figure 7.10.

The process simulation result after implant drive-in is illustrated in Figure 7.11. We see that after the thermal process of HVNW the deep N-link is now connected to the NBL.

7.3.5 P-BODY IMPLANT FOR N-LDMOS

This process step is dedicated to create a P-type implant well for the body region of the n-LDMOS. The photomask for the P-body implant is shown in Figure 7.12.

This well must be deep enough to be useful, and setting the doping concentration can be quite tricky. If the doping level is too low, it may cause punch-through of the depletion region from the drain-drift/P-body junction to the source/P-body junction, and leakage current surge can cause premature breakdown. Also, a lightly doped P-region will increase base resistance of the parasitic NPN BJT of the LDMOS. As we shall see in Chapter 8, higher base resistance will increase the risk of turning on the parasitic BJT, which eventually causes permanent damage to the LDMOS. On the other hand, a heavily doped P-body is not feasible either, since it will greatly increase the threshold voltage and create a longer channel by diffusion. Table 7.9 lists the process condition for P-body implant; boron is used as the dopant.

TABLE 7.8
HVNW Iimplant Drive-in Simulation Parameters

Process Name	Time (minutes)	Initial Temperature (°C)	Final Temperature (°C)	Condition
Diffuse	120	1100	1100	Nitrogen

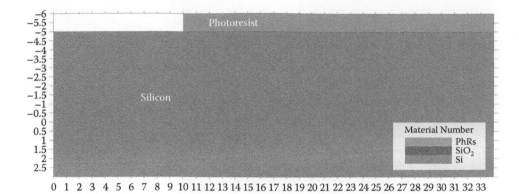

FIGURE 7.10 Photomask for HVNW.

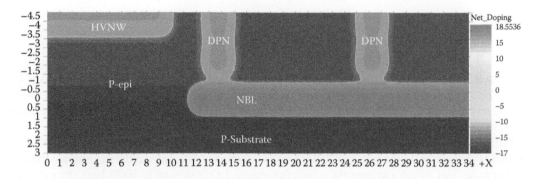

FIGURE 7.11 Net doping after HVNW drive-in.

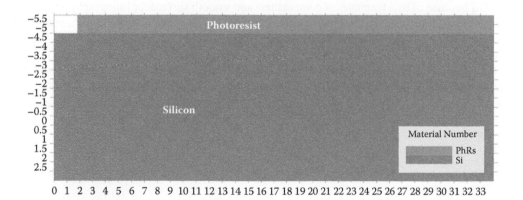

FIGURE 7.12 Photomask of P-body implant.

TABLE 7.9

P-Body Implant Process Simulation Parameters

Process Name	Dopants	Energy (keV)	Dose (cm⁻²)	Angle (°)	Rotation (°)
Implant	Boron	100	5E+13	0	0

TABLE 7.10
P-Body Implant Drive-in Simulation Parameters

Process Name	Time (minutes)	Initial Temperature (°C)	Final Temperature (°C)	Condition
Diffuse	30	1000	1000	Nitrogen

After implant, an anneal step is necessary to activate the dopants and drive-in the implanted well to the desired depth and width. The doping concentration of the well is related to the total thermal budget: engineers should use the total thermal budget after a particular well implant to estimate how much diffusion occurs and how high the doping concentration is at the end of the process. TCAD software provides a useful calculation in this regard. Table 7.10 gives the simulation drive-in time and temperature after P-body implant. Note that photoresist needs to be removed before any diffusion step.

The process simulation result after P-body anneal is shown in Figure 7.13.

7.3.6 ACTIVE AREA/SHALLOW TRENCH ISOLATION (STI)

The phrase *active area* describes regions that perform "active" functions. Basically it means the areas used to build devices and are not used for isolation. However, in power devices like LDMOS, oxide layers can also be used within the device to improve breakdown voltage. In GDSII files, active areas are drawn with oxide layer covered areas defined.

Shallow trench isolation (STI) is an isolation technology used widely in today's CMOS, BiCMOS, and BCD fabrications. In the past, LOCOS (LOCal Oxidation of Silicon) was the mainstream technology, but as devices shrink and enter the submicron regime, LOCOS (with its bird's beak shaped oxide) takes up too much silicon real estate. It also causes an uneven Si surface that is undesirable for photolithography. Most of today's process technology with feature size smaller than 0.25 μm use STI instead of LOCOS because of its advantages of compactness and smooth surface for lithography. In this chapter, only STI will be discussed. Interested readers can read classic books [135] by Drs. Plummer, Deal, and Griffin from Stanford University for a detailed explanation of LOCOS technology.

The STI process simulation starts with pad oxide and nitride deposition. Si_3N_4 (nitride) is used as an etch stop layer for chemical mechanical polishing (CMP) while the pad oxide is used to release the stress that is introduced by the nitride layer. The pad oxide is thermally grown to 400 Å by wet oxidation. The nitride is deposited and has a thickness of 800 Å. The process simulation parameters for the pad oxide and the nitride layer are listed in Tables 7.11 and 7.12.

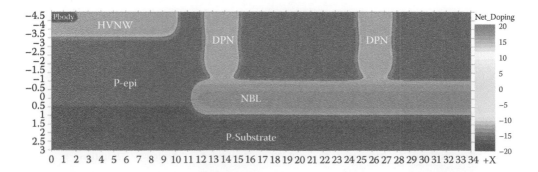

FIGURE 7.13 Net doping plot after P-body implant anneal.

TABLE 7.11

Process Simulation Parameters for STI Pad Oxide Growth

Process Name	Time (minutes)	Initial Temperature (°C)	Final Temperature (°C)	Condition
Diffuse	15	900	900	Wet oxide

TABLE 7.12

Process Simulation Parameters for STI Nitride Layer

Process Name	Material	Thickness (μm)
Deposit	Nitride	0.08

The process simulation result after pad oxide and nitride deposit is shown in Figure 7.14, with a zoom-in view of the top 0.6 μm.

The next step is to etch a shallow trench. In fabrication, photolithography steps using masks for active/STI are used to expose and develop the photoresist and define areas to be covered or to be exposed. Reactive ion etching (RIE) is typically applied to remove the nitride layer and pad oxide and gain access to the silicon in the exposed area. The silicon is then dry-etched to a certain depth to create trenches with very small angles in the sidewalls. In simulation, the depth of the silicon etch is chosen to be 0.3 μm, and sidewall angle is 2 degrees, measured from the plane normal to silicon surface. The process simulation parameters for the STI etching are listed in Table 7.13. Some overetch of nitride and oxide is necessary to make sure they are completely removed from the areas where they are not needed.

Figure 7.15 is the process simulation result after the STI etch.

Before we fill in the etched trench, a dry liner oxide is thermally grown. This liner is important for the STI/silicon interface quality since dry oxide is denser and usually has a more superior quality than deposited oxide. The oxide/silicon interface quality will impact device reliability like hot carrier injection (HCI) for power LDMOS. Table 7.14 illustrates the process simulation parameters for STI liner oxide growth. The process simulation result after the STI liner is shown in Figure 7.16, and an enlarged view of the STI liner is given in Figure 7.17.

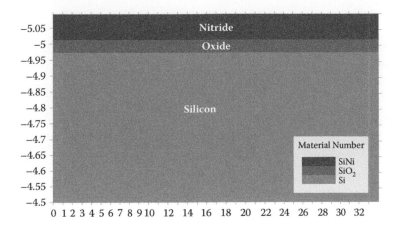

FIGURE 7.14 Process simulation result after pad oxide growth and nitride deposition.

TABLE 7.13

Process Simulation Parameters for STI Etch Steps

Process Name	Material	Thickness (µm)	Angle (°)	Etch Condition
Etch	Nitride	0.1	2	Dry
Etch	Oxide	0.05	2	Dry
Etch	Silicon	0.3	2	Dry

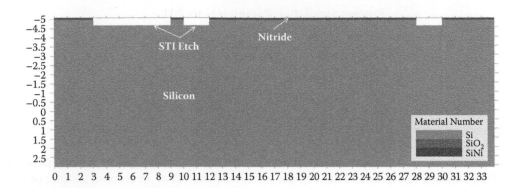

FIGURE 7.15   Process simulation result after STI etch.

TABLE 7.14

Process Simulation Parameters for STI Oxide Liner Growth

Process Name	Time (minutes)	Initial Temperature (°C)	Final Temperature (°C)	Condition
Diffuse	10	1000	1000	Dry oxide

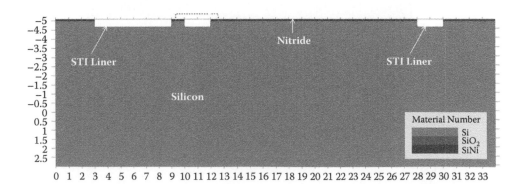

FIGURE 7.16   Process simulation result after STI liner growth.

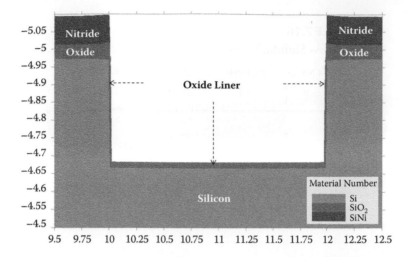

FIGURE 7.17 Enlarged view of oxide liner with Y from −4.5 μm to −5.1 μm and X from 9.5 μm to 12.5 μm.

TABLE 7.15
Process Simulation Parameters for STI Oxide Fill

Process Name	Material	Thickness (μm)
Deposit	Oxide	0.5

With the oxide liner complete, we are ready to fill the trench with deposited oxide. In fabrication, high-density plasma (HDP) oxide is deposited using chemical vapor deposition (CVD). In simulation, oxide is deposited with a layer thickness of 0.5 μm. The process simulation parameters are listed in Table 7.15. Figure 7.18 shows the process simulation result after the oxide fill step.

CMP is performed to get rid of extra oxide and nitride to leave a flat surface. In fabrication, CMP typically stops at a certain designed nitride thickness; the remaining nitride is then stripped off using the hot phosphor method. In simulation, a square etch with designated x1, x2,

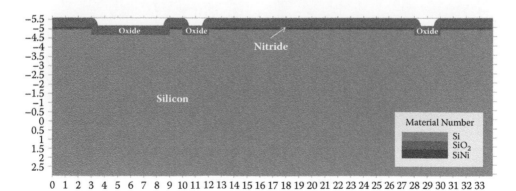

FIGURE 7.18 Process simulation result after oxide fill step.

TABLE 7.16
Process Simulation Parameters for Square Etch

Process Name	Material	(X1,Y1) (μm)	(X2,Y2) (μm)
Square etch	All	(0, –20)	(34, –4.95)

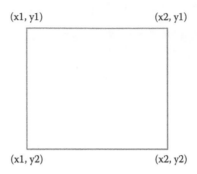

(x1, y1)   (x2, y1)

(x1, y2)   (x2, y2)

FIGURE 7.19 The four coordinates in a square etch simulation.

y1, and y2 coordinates is applied to get rid of all materials within the specified area. Table 7.16 is the process simulation parameters for the square etch. Please note that in the process simulator the positive direction for the Y axis points downward toward the substrate. So the mesh points with y = –20 μm are above those with y = –4.95 μm. Figure 7.19 shows how the four coordinates are arranged in the process simulation. After square etch, the remaining nitride is removed. Figure 7.20 shows the simulation result after CMP and nitride removal.

7.3.7 N-Well and P-Well

N-well and P-well twin wells implant have relatively higher doping concentration. N-well is used for digital PMOS, whereas P-well is used as the base region of the NPN BJT. The process simulation parameters for N-well and P-well are listed in Tables 7.17 and 7.18.

Figure 7.21 shows the photomask of N-well while the photomask of P-well is shown in Figure 7.22.

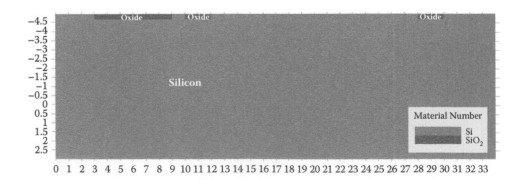

FIGURE 7.20 Process simulation for STI formation after CMP and nitride removal.

TABLE 7.17

Process Simulation Parameters for N-Well Implant

Process Name	Dopants	Energy (keV)	Dose (cm⁻²)	Angle (°)	Rotation (°)
Implant	Phosphorus	50	1E+13	0	0

TABLE 7.18

Process Simulation Parameters for P-Well Implant

Process Name	Dopants	Energy (keV)	Dose (cm⁻²)	Angle (°)	Rotation (°)
Implant	Boron	30	1E+13	0	0

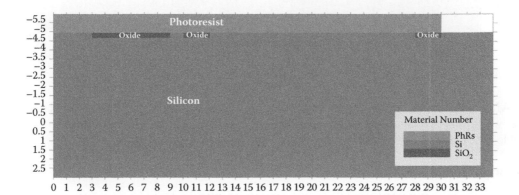

FIGURE 7.21 Photomask of N-well.

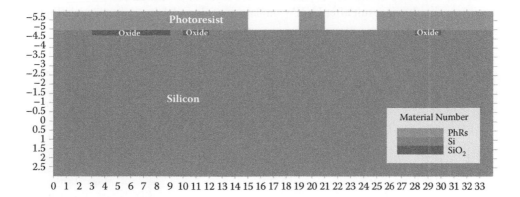

FIGURE 7.22 Photomask of P-well.

TABLE 7.19

Process Simulation Parameters for N-Well and P-Well Drive-in

Process Name	Time (minutes)	Initial Temperature (°C)	Final Temperature (°C)	Condition
Diffuse	10	900	1000	Nitrogen
Diffuse	20	1000	1000	Nitrogen
Diffuse	10	1000	900	Nitrogen

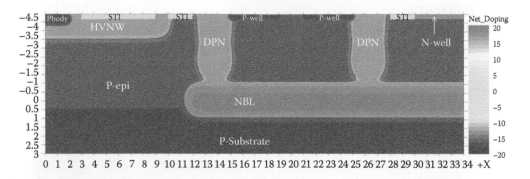

FIGURE 7.23   Net doping plot after N-well and P-well drive-in.

After implant, a thermal process is performed to activate the dopants and drive-in the wells. The process simulation parameters for N-well and P-well drive-in are listed in Table 7.19. Figure 7.23 is the net doping plot after N-well and P-well drive-in.

7.3.8   LOW-VOLTAGE TWIN WELLS

Low-voltage N-well (LVNW) and low-voltage P-well (LVPW) are highly doped wells in the process flow. In our example, LVNW is used as the highly doped emitter region of the NPN BJT. The photomask of LVNW is shown in Figure 7.24.

The ion implantation process simulation parameters for LVNW are listed in Table 7.20. The thermal anneal and drive-in simulation parameters are listed in Table 7.21.

The process simulation result after LVNW implant anneals and drive-in is shown in Figure 7.25.

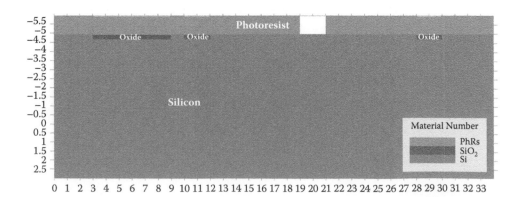

FIGURE 7.24   Photomask for LVNW.

TABLE 7.20

Process Simulation Parameters for LVNW Implant

Process Name	Dopants	Energy (keV)	Dose (cm⁻²)	Angle (°)	Rotation (°)
Implant	Phosphorus	20	5E+14	0	0

TABLE 7.21

Process Simulation Parameters for LVNW Anneal and Drive-in

Process Name	Time (minutes)	Initial Temperature (°C)	Final Temperature (°C)	Condition
Diffuse	10	900	1000	Nitrogen

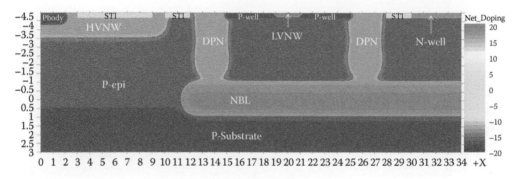

FIGURE 7.25   Net doping after LVNW implant anneal and drive-in.

7.3.9   THICK GATE AND THIN GATE OXIDE

Unlike a CMOS process flow, there are multiple gate oxide thicknesses for smart power ICs. In our example, two gate oxide thicknesses are chosen: a thick gate oxide for power devices like LDMOS; and a thin gate oxide for digital/analog MOSFETs.

The thick gate oxide (240 Å) is grown first. There is no photomask used for this step, and it is grown everywhere on the wafer. The process simulation parameters for the thick gate oxide growth are listed in Table 7.22, and the process simulation results are illustrated in Figure 7.26. For digital and analog MOSFETs, a thinner gate oxide is required; otherwise the threshold voltage will be too high. Our previous step has grown a thick gate oxide everywhere on the wafer active area, so a photomask is now used to expose the locations where a thinner gate oxide is needed. Figure 7.27 is the process simulation view of a thin gate oxide photomask.

TABLE 7.22

Process Simulation Parameters for Thick Gate Oxide Growth

Process Name	Time (minutes)	Initial Temperature (°C)	Final Temperature (°C)	Condition
Diffuse	20	1000	1000	Dry oxide

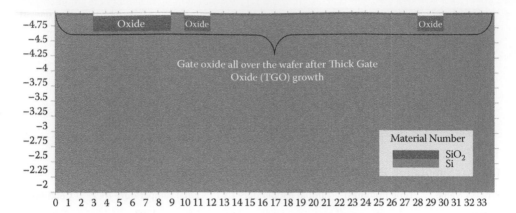

FIGURE 7.26 Process simulation result of thick gate oxide growth (enlarged view).

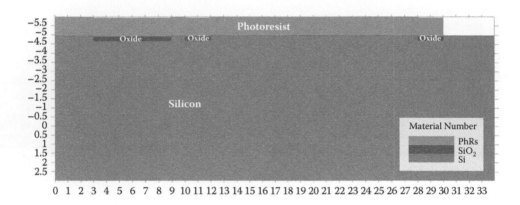

FIGURE 7.27 Photomask of thin gate oxide.

TABLE 7.23
Dry Etch for Thin Gate Oxide Growth

Process Name	Material	Thickness (µm)	Angle (°)	Etch Condition
Etch	Oxide	0.03	0	Dry

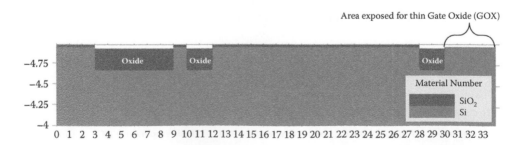

FIGURE 7.28 Enlarged view of the oxide etched region.

TABLE 7.24

Process Simulation Parameters for Thin Gate Oxide Growth

Process Name	Time (minutes)	Initial Temperature (°C)	Final Temperature (°C)	Condition
Diffuse	5	900	900	Dry oxide

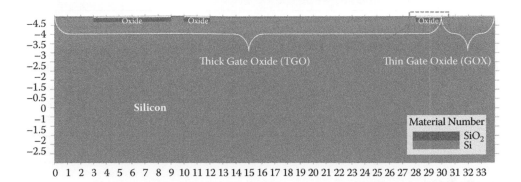

FIGURE 7.29 Process simulation result after thin oxide growth.

In our example, only the PMOS area is exposed by the photomask. The thick gate oxide in the exposed area is subsequently etched with simulation parameters listed in Table 7.23. An enlarged view of the oxide etched portion after etching the thick gate oxide is shown in Figure 7.28.

A second oxidation process is performed to grow thin gate oxide; like our previous step, the thin gate oxide is grown throughout the wafer. The area where the thick gate oxide was etched away has a higher growth rate than those areas still covered by the thicker gate oxide. Table 7.24 lists the process condition for thin gate oxide growth. The final thin gate oxide thickness is about 34 Å, and thick gate oxide is about 250 Å. The process simulation result after thin oxide growth is shown in Figure 7.29, and an enlarged view is provided in Figure 7.30.

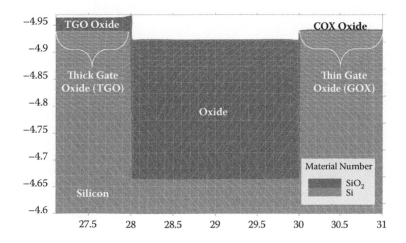

FIGURE 7.30 Process simulation result after thin oxide growth with enlarged view: x from 27 μm to 31 μm and y from −4.6 μm to −5 μm.

TABLE 7.25

Process Simulation Parameters for Poly Deposition

Process Name	Material	Thickness (μm)	Dopants	Doping Concentration (cm^{-3})
Deposit	Poly	0.25	Phosphorus	1E+20

TABLE 7.26

Process Simulation Parameters for Poly Thermal Anneal

Process Name	Time (minutes)	Initial Temperature (°C)	Final Temperature (°C)	Condition
Diffuse	1	850	1000	Nitrogen
Diffuse	1	1000	1000	Nitrogen
Diffuse	1	1000	850	Nitrogen

7.3.10 POLY GATE

Multiple polysilicon (poly) layers are normally used in smart power IC process flow to create, for example, poly-insulator-poly capacitors. In this example, only one poly layer is simulated. The poly layer in the simulation has a thickness of 0.25 μm (2500 Å) and is *in situ* doped with phosphorus. The process parameters for poly deposition are listed in Table 7.25.

After poly deposition, a dry etch is performed and the exposed poly is etched away while the material underneath the photomask is preserved. A thermal step is necessary to activate the dopants in the poly and the process simulation parameters are listed in Table 7.26. In comparison with previous thermal steps, poly anneal has a smaller thermal budget to mimic the widely adopted practice of rapid thermal anneal (RTA).

The finished polysilicon gate simulation is shown in Figure 7.31. Both the LDMOS gate and PMOS gate are created at the same time. An enlarged view (Figure 7.32) of the LDMOS gate corner indicates there is a step-coverage at the STI/silicon boundary. In real fabrication (not in simulation), this corner may have a large impact on device reliability if the corner somehow leads to a thinner gate oxide [179].

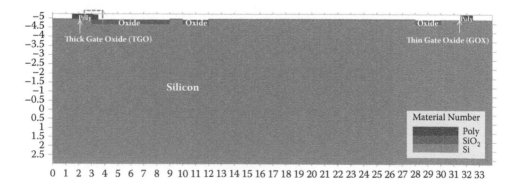

FIGURE 7.31 Process simulation result of poly gates.

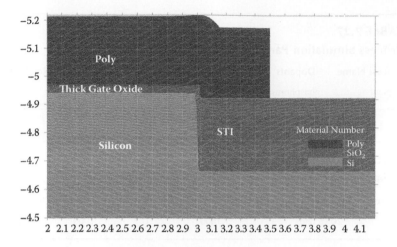

FIGURE 7.32 An enlarged view of LDMOS poly gate at STI corner.

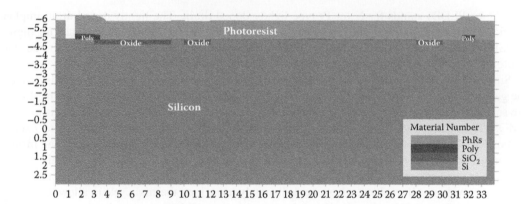

FIGURE 7.33 Photomask for NLDD implant.

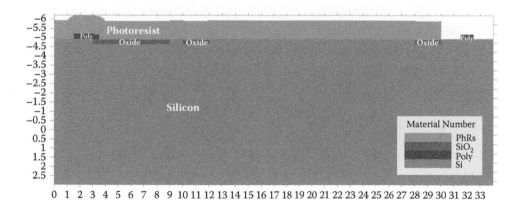

FIGURE 7.34 Photomask for PLDD implant.

TABLE 7.27

Process Simulation Parameters for NLDD Implant

Process Name	Dopants	Energy (keV)	Dose (cm^{-2})	Angle (°)	Rotation (°)
Implant	phosphorus	20	9E+12	0	0

TABLE 7.28

Process Simulation Parameters for NLDD Thermal Anneal

Process Name	Time (seconds)	Initial Temperature (°C)	Final Temperature (°C)	Condition
Diffuse	20 seconds	1000	1000	Nitrogen

7.3.11 NLDD AND PLDD

NLDD (or PLDD) stands for N (or P) type lightly doped drain. The lightly doped drain implant is primarily used to alleviate the high electric field near the heavily doped drain. Since for digital MOSFETs source and drain are interchangeable, this implant applies to both drain and source. The photomasks for NLDD and PLDD are shown in Figures 7.33 and 7.34, respectively.

Similar to a CMOS process flow, NLDD and PLDD are implanted after the gate poly is defined: the gate serves as an implant block for the low energy implant of the LDD, and the implanted wells are self-aligned to the gates. In this simulation, NLDD is used for n-LDMOS, and PLDD is used for PMOS. The NLDD and PLDD process simulation parameters are listed from Tables 7.27 to 7.30.

For NLDD, phosphorus or arsenic can be used, whereas for PLDD BF$_2$ is normally used instead of boron because the heavier ion will create a shallower implant that is desirable for this step. A quick anneal is performed to activate the dopants, with a much smaller thermal budget than the one used for previous well implants where we deliberately use a long thermal cycle to allow dopants to diffuse and form a DPN or P-well. This is because like in a CMOS process, LDD needs little diffusion to create a shallow junction and minimize the short channel effect; in real manufacturing,

TABLE 7.29

Process Simulation Parameters for PLDD Implant

Process Name	Dopants	Energy (keV)	Dose (cm^{-2})	Angle (°)	Rotation (°)
Implant	BF$_2$	40	2E+13	0	0

TABLE 7.30

Process Simulation Parameters for PLDD Thermal Anneal

Process Name	Time (seconds)	Initial Temperature (°C)	Final Temperature (°C)	Condition
Diffuse	30 seconds	1000	1000	Nitrogen

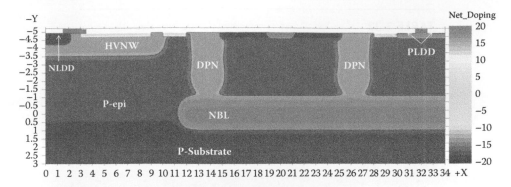

FIGURE 7.35   Process simulation result of NLDD and PLDD implants.

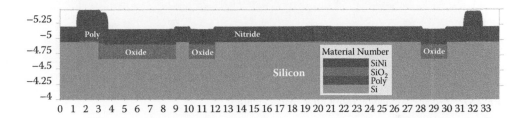

FIGURE 7.36   Process simulation result after spacer deposition (silicon truncated below Y = −4 μm).

an RTA is usually applied but a detailed treatment of RTA is beyond the scope of this book. The process simulation result of NLDD and PLDD is illustrated in Figure 7.35.

The Halo implants with angled implant and rotations (Boron or BF_2 for NMOS and phosphorus or arsenic for PMOS), are popular with standard CMOS technology to reduce the short channel effect for digital MOSFETs. Since our focus is not on advanced CMOS technology, this implant step is omitted.

7.3.12 SIDEWALL SPACER

A silicon nitride layer (Si_3N_4) is then deposited using chemical vapor deposition. Usually a tensile stress comes with the deposited nitride layer, which may (in some cases) be beneficial for advanced CMOS processes that rely on strain-enhanced carrier mobility. However, this is not the purpose for our smart power IC process. The process simulation parameters for nitride deposition are listed in Table 7.31. The simulation result after spacer deposition is given in Figure 7.36.

Now, the Si_3N_4 layer is dry-etched without using a photomask: the anisotropic dry etch will leave a side wall "spacer" along the gate edges so a wet isotropic etch is not the right choice for this step.

TABLE 7.31
Process Simulation Parameters for Spacer Deposition

Process Name	Material	Thickness (μm)
Deposit	Si_3N_4	0.25

TABLE 7.32

Dry Etch of Nitride Spacer

Process Name	Material	Thickness (μm)	Angle (°)	Etch Condition
Etch	Si_3N_4	0.03	0	Dry

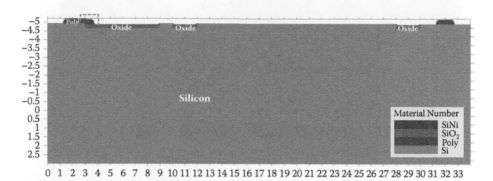

FIGURE 7.37 Process simulation result after spacer etching.

The simulation parameters are listed in Table 7.32. Figure 7.37 is the simulation result after spacer etching, and Figure 7.38 shows an enlarged view with mesh.

7.3.13 NSD AND PSD

N+ (or P+) source and drain implants (NSD and PSD) are typically one of the last implant steps for smart power IC technology. A shallow junction is required by most modern process technologies for source and drain implant, so these steps use the least amount of thermal budget. NSD is used for n-LDMOS, NPN collector, and emitter contacts. PSD is used for PMOS, NPN base contact, and P-body contact for n-LDMOS. The photomasks for NSD and PSD are shown in Figures 7.39 and 7.40, respectively.

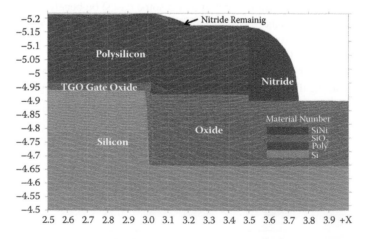

FIGURE 7.38 Process simulation result after spacer etching (enlarged with mesh on).

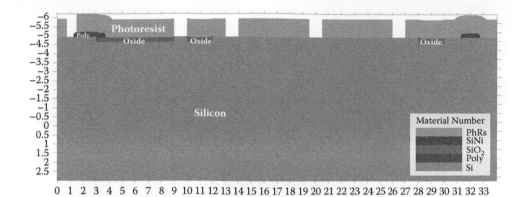

FIGURE 7.39 Photomask for NSD implant.

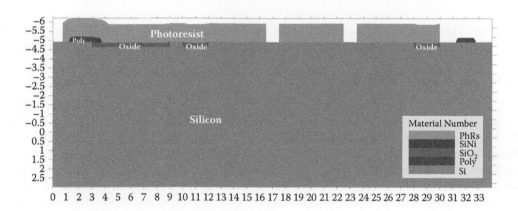

FIGURE 7.40 Photomask for PSD implant.

In practice, a pre-amorphization implant step is generally required to reduce the channeling effect and obtain shallower junctions, and an RTA is performed to activate the dopants. The process simulation parameters are listed from Tables 7.33 to 7.36.

The simulation result after NSD and PSD anneal is shown in Figure 7.41. In practice, after the NSD and PSD steps, there should be a self-aligned silicidation process to reduce the contact

TABLE 7.33

Process Simulation Parameters for NSD Implant

Process Name	Dopants	Energy (keV)	Dose (cm^{-2})	Angle (°)	Rotation (°)
Implant	Phosphorus	25	1E+14	0	0

TABLE 7.34

Process Simulation Parameters for NSD Thermal Anneal

Process Name	Time (seconds)	Initial Temperature (°C)	Final Temperature (°C)	Condition
Diffuse	15 seconds	950	950	Nitrogen

TABLE 7.35

Process Simulation Parameters for PSD Implant

Process Name	Dopants	Energy (keV)	Dose (cm⁻²)	Angle (°)	Rotation (°)
Implant	Boron	15	1E+14	0	0

TABLE 7.36

Process Simulation Parameters for PSD Thermal Anneal

Process Name	Time (seconds)	Initial Temperature (°C)	Final Temperature (°C)	Condition
Diffuse	5 seconds	950	950	Nitrogen

resistance. This means a contact metal like cobalt or titanium is deposited and chemical reactions are used to transform the silicon into a silicide. The process simulation for this step is omitted.

7.3.14 BACK-END OF THE LINE

Back-end of the line typically refers to the process steps that form interlayer dielectric (ILDs), contacts, vias, and metal layers. In the past, aluminum was used extensively for power IC technology, but copper is now more prevalent thanks to its lower resistivity and greater immunity to the electromigration problem. However, unlike aluminum, copper is difficult to selectively etch so that deposition and metal patterning methods are not feasible. Fortunately, engineers have developed a new method called dual damascene, which creates etched patterns on the ILD oxides that are then filled in by copper. A CMP step is subsequently performed to expose the copper interconnect. Since

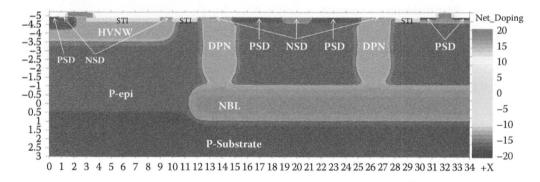

FIGURE 7.41 Net doping plot after NSD and PSD anneal.

TABLE 7.37

Process Simulation Parameters for ILD Deposition

Process Name	Material	Thickness (µm)
Deposit	Oxide	0.8

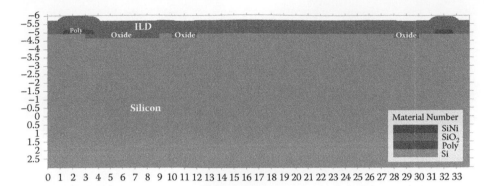

FIGURE 7.42   Process simulation result after ILD deposition.

TABLE 7.38
Process Simulation Parameters for ILD CMP

Process Name	Material	(X1,Y1) (μm)	(X2,Y2) (μm)
Square etch	All	(0, −20)	(34, −5.65)

the thermal cycles during back-end processes are typically done at much lower temperatures, we can neglect the diffusion of dopants in the silicon underneath the ILDs. In this example, the thermal cycles that allow ILDs such as BPSG to flow are omitted.

After silicidation, a layer of ILD is deposited; for simulation purposes we use a 0.8 μm thick layer of oxide as the ILD rather than the more complicated materials used in practice. In manufacturing, an etch stop layer (Si_3N_4) should be inserted for the dual damascene process; for simulation, we can just define the etch depth. The process simulation parameters for the ILD are listed in Table 7.37, and the simulation result after ILD deposition is shown in Figure 7.42.

The top surface after ILD deposition is rough, and CMP is necessary to smooth out the top surface and prepare for the lithography/etch of the contacts and metal layers. In this simulation example, a square etch is performed, the parameters of which are listed in Table 7.38. The definition of (X1, Y1)

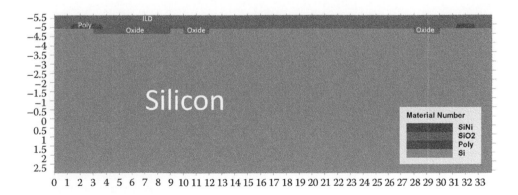

FIGURE 7.43   Process simulation result after ILD CMP.

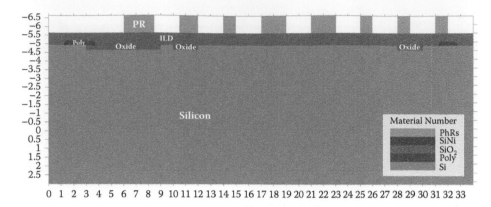

FIGURE 7.44 Photomask for metal layer pattern.

TABLE 7.39
Dry Etch of ILD Oxide

Process Name	Material	Thickness (μm)	Angle (°)	Etch Condition
Etch	Oxide	0.3	0	Dry

and (X2, Y2) are the same as those in Figure 7.19, with (X1, Y1) being the top left corner and (X2, Y2) being the bottom right corner. The process simulation result after CMP is given in Figure 7.43.

In the dual damascene process, metal layer pattern is etched first up to the etch stop layer. For simulation, we just need to define the etch thickness instead of using an etch stop layer. The photomask for the metal layer pattern is shown in Figure 7.44.

With this photomask, a dry etch is performed with simulation parameters shown in Table 7.39. The simulation result after metal layer pattern etch is given in Figure 7.45.

The next step is to etch the contacts. Figure 7.46 is the photomask for contacts etch. The simulation parameters are listed in Table 7.40. Please note that overetch is performed to make sure oxide is completely removed at desired contact locations.

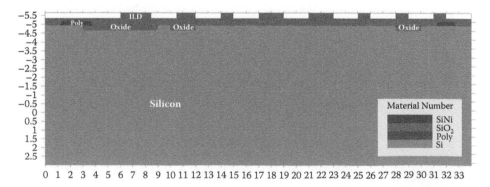

FIGURE 7.45 Process simulation result after oxide etch for metal patterning.

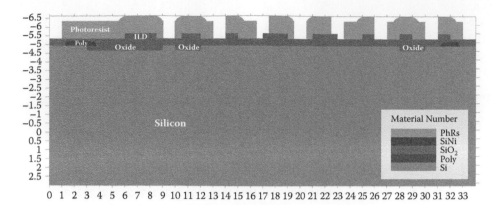

FIGURE 7.46 Photomask for contacts etch.

TABLE 7.40

Dry Etch of ILD Oxide

Process Name	Material	Thickness (μm)	Angle (°)	Etch Condition
Etch	Oxide	1	0	Dry

The process simulation result after contacts etch is shown in Figure 7.47.

In real fabrication, electroplating technology is applied for copper interconnect; older aluminum processes used sputter deposition methods. First, Ta or TaN is used as a barrier layer with thickness of about 75 Å to prevent diffusion of copper into the oxide. Next, a seed copper layer with a thickness between 500 to 1000 Å is deposited with LPCVD before electroplating of the copper layer [180]. TCAD simulations use purely geometrical methods rather than solving the full chemistry of the

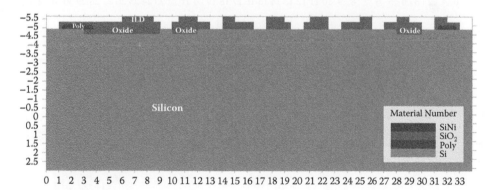

FIGURE 7.47 The simulation result after contacts etch.

TABLE 7.41

Copper Deposition

Process Name	Material	Thickness (μm)
Deposit	Copper	1.2

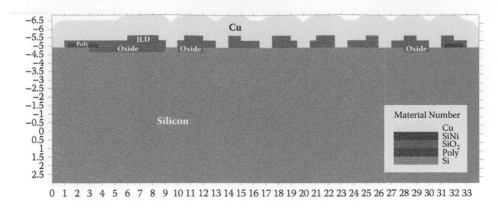

FIGURE 7.48   Process simulation result after copper deposition.

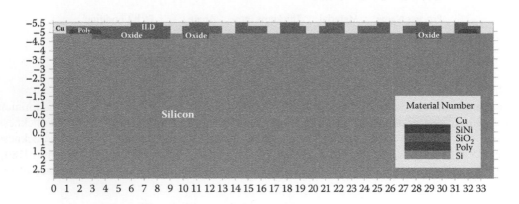

FIGURE 7.49   Process simulation result after copper metal layer CMP.

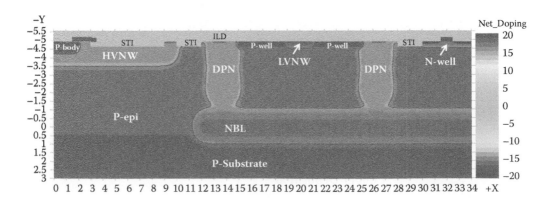

FIGURE 7.50   Final net doping plot after copper metal layer CMP.

Enlarged Views of n-LDMOS, PMOS and NPN Transistors

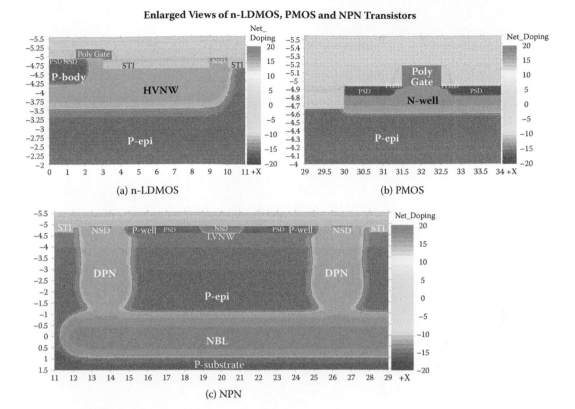

(a) n-LDMOS

(b) PMOS

(c) NPN

FIGURE 7.51 Enlarged net doping views of (a) n-LDMOS, (b) PMOS, and (c) NPN transistors from the final process simulation result.

problem, so a simple deposition command is used instead; the parameters are listed in Table 7.41. The process simulation result after copper deposition is illustrated in Figure 7.48.

Finally, a CMP step is carried out to get rid of extra copper and reveal the metal pattern; CMP will stop at the ILD layer. In our example, only one metal layer is simulated, while in real manufacturing flow multiple metal layers with different thicknesses are often required. Figure 7.49 shows the final simulation result after CMP is done. The net doping plot is illustrated in Figure 7.50.

The enlarged net doping plots are provided in Figure 7.51.

Passivation of the layers must also be done at this point to protect the surface and the layers below. This step and other postprocessing and packaging steps that occur in industry are beyond the scope of this book. The complete CSUPREM code for this chapter is listed in Appendix B.

8 Integrated Power Semiconductor Devices with TCAD Simulation

This chapter explores the devices used in power ICs with the help of technology computer-aided design (TCAD) simulation. For the sake of simplicity, in this chapter, we will use only a device simulator (e.g., APSYS) to define structures and doping concentrations and to perform device electrical and thermal simulations. The next chapter contains examples that are simulated with both process and device simulators (e.g., CSUPREM and APSYS).

8.1 PN JUNCTION DIODES

Diodes are widely used in analog and power ICs, even though synchronous rectification is now replacing diodes for many applications. Since a PN junction laterally integrated diode is a fundamental building block for most power ICs, we will begin with modeling such a diode.

8.1.1 PN JUNCTION BASICS

A PN junction is formed by bringing together a P-type semiconductor with a high concentration of holes and an N-type semiconductor with a high concentration of electrons. By bringing the two regions together, electrons tend to diffuse into the P-type region, while holes will diffuse into the N-type region, which creates diffusion current. However, when charge carriers migrate, they leave behind uncompensated ionized dopants, which form an internal electric field. This electric field drags electrons back to the N-region and holes back to the P-region and creates a drift current. Since the drift current is opposite to the diffusion current, after a very short period of time equilibrium is achieved with a continuous space charge region forming across the PN junction to achieve charge balance. Since the total charge on each side is calculated by multiplying the charge concentration (ionized donor or acceptors) with the charge width, the heavier the doping concentration in the semiconductor, the higher the charge concentration will be in the ionized region (assuming fully ionized dopants). To achieve charge balance, the ionized region with a higher dopant concentration has a narrower width than with lower dopant concentration. In the extreme case when a very heavily doped P- or N-region meets a very lightly doped N- or P-region, the space charge region extends almost entirely into the lightly doped region: this is called a single-sided space charge region.

A space charge region is sometimes referred to as a depletion region, because electrons and holes are "depleted" in this region. Any electrons or holes that enter this region will be swept away by the internal "built-in" electric field. This electric field corresponds to a potential difference, and the relationship between these quantities is governed by Poisson's equation:

$$\frac{d^2V}{dx^2} = -\frac{dE}{dx} = \frac{\rho}{\varepsilon_s} \tag{8.1}$$

where ε_s is the semiconductor permittivity, ρ is the charge density (C/cm^3), E is the electric field, and V is the internal potential. It should be stressed that V does not correspond to the external voltage or bias applied to the device. Even under thermal equilibrium conditions with zero volts of applied bias, the potential in the PN junction has a spatial dependence corresponding to the built-in value. The internal voltage drop of the device should therefore always be plotted as $V(x)|_{bias} - V(x)|_{equilibrium}$, a quantity referred to as the *differential potential*.

From Poisson's equation, it is easy to derive that the depletion width on the P- and N-sides are related by

$$N_a \cdot X_P = N_d \cdot X_N \tag{8.2}$$

where N_a and N_d are acceptor and donor concentrations in the P- and N-regions, respectively, X_P is the depletion width in the P-region, and X_N is the depletion width in the N-region. The built-in potential in the space charge region can be calculated with

$$V_{bi} = V_t \cdot \ln\left(\frac{N_a \cdot N_d}{n_i^2}\right) \tag{8.3}$$

where n_i is the intrinsic carrier concentration (about 1.5E + 10 cm^{-3} at 300 K), and V_t is the thermal voltage (about 26 mV at room temperature) and is related to lattice temperature T and Boltzmann constant k with

$$V_t = \frac{k \cdot T}{q} \tag{8.4}$$

The width of the space charge region W is calculated as the added total of and X_P and X_N. It is found to be related to the built-in potential as

$$W = X_P + X_N = \sqrt{\frac{2\varepsilon_s V_{bi}}{q}\left(\frac{1}{N_a} + \frac{1}{N_d}\right)} \tag{8.5}$$

8.1.2 Lateral PN Junction Diode at Equilibrium

Since this book is about integrated devices, to explore the physics of a PN junction diode a simple lateral (rather than vertical) PN junction diode is simulated using a TCAD device simulator.

Showing only the upper 20 μm thick of the P-type substrate with boron concentration of 5E + 16 cm^{-3}, the diode is created with an N-diffused region on top. The P-region of the device can be a P-type substrate, a P-epitaxial (P-epi) layer, or an implanted P-well. For simplicity, a P-substrate is used. The schematic representation of this simple structure is illustrated in Figure 8.1.

A Gaussian doping profile is used to define the doping concentration, as shown in Figure 8.2. Here, the area between y1 and y2 defines the constant doping at maximum concentration; for a single implant, y1 normally equals to y2, and dy1 and dy2 are the standard deviation of the Gaussian profile. The same definition is applied to the z direction in 3D simulations.

A 2D simulation is performed using the schematic of Figure 8.1. Figure 8.3 explains how the X and Y axes are defined for process and device simulators. Please note that for device simulators (e.g.,

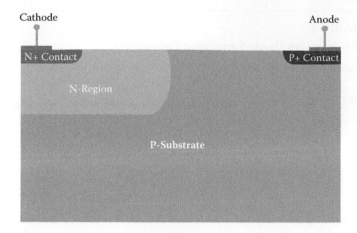

FIGURE 8.1 Schematic representation of a simple lateral PN junction diode.

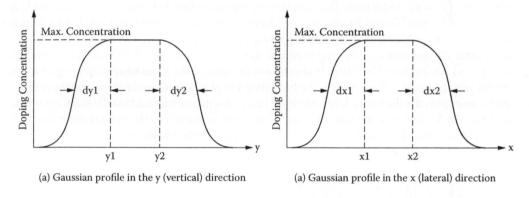

(a) Gaussian profile in the y (vertical) direction

(a) Gaussian profile in the x (lateral) direction

FIGURE 8.2 Gaussian doping profile definition in both x and y directions.

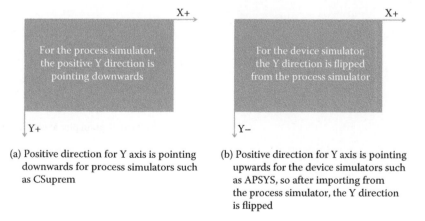

(a) Positive direction for Y axis is pointing downwards for process simulators such as CSuprem

(b) Positive direction for Y axis is pointing upwards for the device simulators such as APSYS, so after importing from the process simulator, the Y direction is flipped

FIGURE 8.3 The polarity of Y direction for the process and device simulators.

TABLE 8.1

PN Junction Diode Simulation Parameters

Parameters for Simulation	Value
X range	0 to 10 μm
Y range	0 to −20 μm
P-substrate (P-base constant doping)	5E + 16 cm^{-3}
N-region max concentration	1E + 17 cm^{-3}
N-region location (x1, x2, dx1, dx2) Gaussian	x = (0, 3, 1, 1); y = (−1, −1, 1, 1)

APSYS), the axis polarity is reversed after the import from the process simulator (e.g., CSUPREM); that is, Y is more negative when going downward.

Going back to the simulation details, the device has a lateral cell pitch of 10 μm (x direction) and a thickness of 20 μm (y direction). A P-type substrate with constant boron doping of 5E + 16 cm^{-3} is assumed. The N-region is defined in the device simulator with Gaussian doping parameters: the maximum doping concentration (1E + 17 cm^{-3}) is found at a single point corresponding to the implant depth (y1 = y2 = −1.0 μm). The simulation parameters for the lateral PN junction diode are listed in Table 8.1 (the N+ and P+ regions for Ohmic contacts are neglected) with doping location parameters (i.e., x1, x2, dx1, dx2) defined according to Figure 8.2.

The lateral PN junction's net doping profile is shown in Figure 8.4a. A 1D cut of the band diagram at y = −1 μm (x from 2 to 6 μm) is also shown in Figure 8.4b. (The band diagram plot is relative to the starting point of the cutline.) The band diagram plot is done under equilibrium conditions as can be seen from the flat Fermi level, and even in the absence of applied bias the built-in potential V_{bi} (approx. 0.8 V) can be observed. Because of the built-in potential, the voltage drop across the diode during the on-state cannot be zero, even if no parasitic series resistance exists.

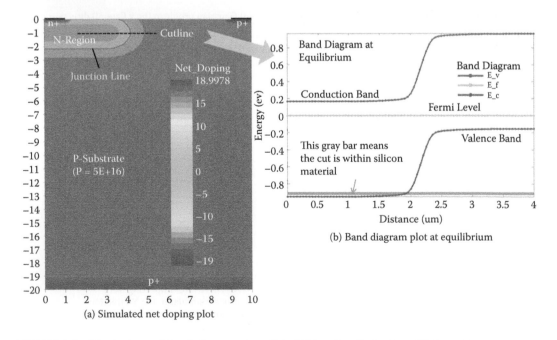

(a) Simulated net doping plot

(b) Band diagram plot at equilibrium

FIGURE 8.4 Net doping and band diagram plots of the PN junction diode at equilibrium.

Band Diagram Plots at Different Anode Voltages

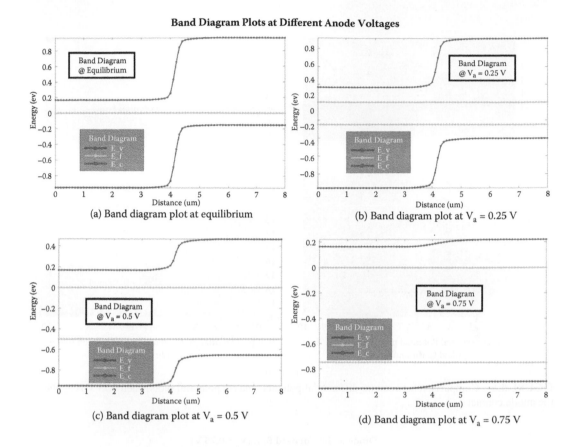

(a) Band diagram plot at equilibrium

(b) Band diagram plot at $V_a = 0.25$ V

(c) Band diagram plot at $V_a = 0.5$ V

(d) Band diagram plot at $V_a = 0.75$ V

FIGURE 8.5   Band diagrams of a forward-biased diode.

8.1.3 FORWARD CONDUCTION (ON-STATE)

Forward biasing the diode means applying a positive voltage on the anode with respect to the cathode. The space charge region width gets narrower as the forward voltage V_a increases:

$$W_{forward} = \sqrt{\frac{2\varepsilon_s (V_{bi} - V_a)}{q}\left(\frac{1}{N_a} + \frac{1}{N_d}\right)} \tag{8.6}$$

The barrier height of the band diagram is decreased, which allows more electrons and holes to diffuse into the P- and N-regions, respectively. Thus, the net drift-diffusion current is no longer zero, and the quasi-Fermi levels for the electrons and holes split. Simulated band diagrams for progressively higher anode voltages are shown in Figure 8.5.

Figure 8.6 is an example of the aforementioned difference between the internal potential V and the differential potential (i.e., the internal voltage drop). For the remainder of this book, all potential plots will show the differential potential since it is more meaningful for device and IC engineers.

Figure 8.7a shows the flow of the total current (electrons and holes) under forward bias at $V_a = 0.75$ V; the I-V curve under forward bias is also plotted in Figure 8.7b. The diode current increases exponentially with applied anode voltage, which can also be expressed in another way: the voltage drop across the anode and cathode is current dependent. That is, the higher the load current, the larger the diode voltage drop and thus the more power that is dissipated in the diode.

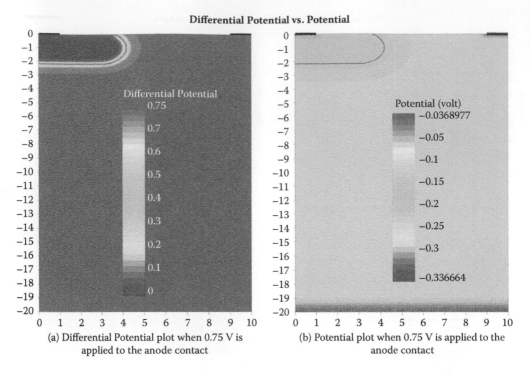

FIGURE 8.6 Comparison between (a) differential potential and (b) potential plots when 0.75 V is applied to the anode contact.

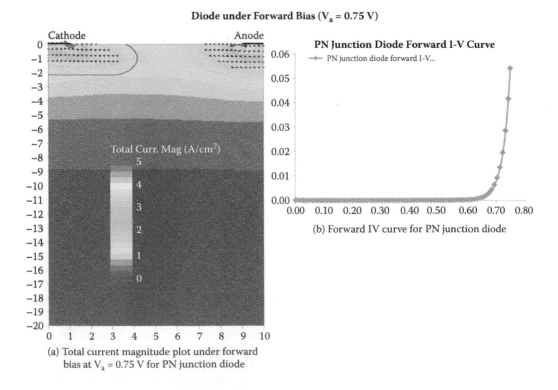

FIGURE 8.7 Forward-bias I-V curve and total current vector plot.

An analytical model of the diode I-V characteristic is given by the diode equation

$$I_a = I_s \cdot \left[e^{\left(\frac{q V_a}{kT} \right)} - 1 \right] = I_s \cdot \left[e^{\left(\frac{V_a}{V_t} \right)} - 1 \right] \tag{8.7}$$

where I_a is the diode anode current, V_a is the voltage applied at the anode, V_t is the thermal voltage, and I_S is the reverse saturation current, which is the leakage current when the diode is reverse biased:

$$I_s = A q n_i^2 \left(\frac{D_p}{L_p N_d} + \frac{D_n}{L_n N_a} \right) \tag{8.8}$$

$$L_p = \sqrt{D_p \tau_p} \tag{8.9}$$

$$L_n = \sqrt{D_n \tau_n} \tag{8.10}$$

In these equations, A is the area of the diode; L_n and L_p are the diffusion lengths for electrons and holes, respectively; τ_n and τ_p are the recombination lifetimes for electrons and holes, respectively; D_n and D_P are the electron and hole diffusivities, respectively. These latter values are related to the electron μ_n and μ_p hole mobilities and through the Einstein relationship

$$D_n = \frac{kT}{q} \mu_n \tag{8.11}$$

$$D_p = \frac{kT}{q} \mu_p \tag{8.12}$$

8.1.4 REVERSE BIAS OF A PN JUNCTION DIODE

Reverse biasing a diode means applying a positive voltage to the cathode terminal or a negative voltage to the anode terminal while keeping the other terminal grounded. The space charge region of the PN junction will be extended under reverse bias. If we define V_r as the absolute value of the reverse-bias voltage being applied, this effect can be seen in the space charge width equation

$$W_{reverse} = \sqrt{ \frac{2 \varepsilon_s (V_{bi} + V_r)}{q} \left(\frac{1}{N_a} + \frac{1}{N_d} \right) } \tag{8.13}$$

The extension of the space charge region provides support for additional voltage drop across the PN junction. Ideally, no current exists during reverse bias, but in fact a small reverse saturation current still exists. At reverse bias, the carriers are swept away by the electric field within the space charge region. Electrons at the edge of the space charge region and P-type neutral region are swept to the N-side, while holes at the edge of the space charge region and N-type neutral region are swept back to the P-side. Thus, no minority carriers exist at the space charge region edge. The minority carriers in both N- and P-type neutral regions that are within diffusion length of the space charge region edge can enter the space charge region and are then swept by the high field to contribute to the reverse saturation current.

The I-V curve for reverse breakdown is shown in Figure 8.8. The default impact ionization model used in this simulation is the Baraff model, which yields a breakdown voltage of around 47 V. A nonparallel junction such as this case usually yields lower breakdown voltage than parallel junction. However, our device also has a graded junction, which means that the depletion region extends into both the N- and P-regions: this is beneficial and helps improve the breakdown voltage compared with an abrupt junction.

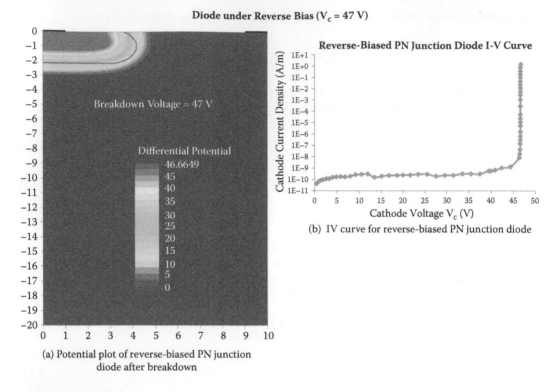

(a) Potential plot of reverse-biased PN junction
diode after breakdown

(b) IV curve for reverse-biased PN junction diode

FIGURE 8.8   Potential plot and I-V curve for reverse-biased PN junction diode.

8.1.5 LATERAL PN JUNCTION DIODE WITH NBL

The previously described simple PN junction diode consists of a diffused N-region within a P-substrate or P-epi layer. In real manufacturing, diode structures can be much more complex, and additional implant steps are often necessary. These diodes maybe derived directly from a bipolar junction transistor (BJT) or lateral double-diffused MOS (LDMOS) transistor processes, a detailed discussion of which is beyond the scope of this book.

For a real lateral device used in analog and power ICs, the substrate of the wafer or die is always grounded. For the aforementioned simple diode structure, a grounded substrate means directly connecting the P-region to ground, which is inappropriate when a positive voltage is applied to the anode (direct short from anode to grounded substrate). So the structure needs to be modified if a substrate contact is included.

In this section, we will add a highly doped N-type buried layer (NBL) to the simple PN junction diode. A buried layer is usually grown before an epitaxial layer at the very beginning of a manufacturing process, as demonstrated in the previous chapter. For simulation purposes only, instead of growing an epitaxial layer, we will simply represent the NBL with an extra Gaussian doping region in our device structure. The simulation parameters are listed in Table 8.2, and the net doping plot is shown in Figure 8.9.

8.1.6 BREAKDOWN VOLTAGE ENHANCEMENT OF THE PN JUNCTION DIODE

Breakdown voltage is a major concern for power semiconductor device engineers. This section demonstrates how to use TCAD simulators to optimize the diode design so that a high breakdown voltage can be achieved. Our original PN junction diode with a 10 μm pitch size is not optimized. One might want to reduce the drift region so that the device will become more compact; in other

TABLE 8.2

Simulation Parameters for a PN Junction Diode with NBL

Parameters for Simulation	Value
X range	0 to 10 μm
Y range	0 to −20 μm
P-substrate (P-base constant doping)	$5E + 16 \text{ cm}^{-3}$
N-region max concentration	$1E + 17 \text{ cm}^{-3}$
N-region location (x1, x2, dx1, dx2) Gaussian	x = (0, 3, 1, 1); y = (−1, −1, 1, 1)
NBL max concentration	$1E + 17 \text{ cm}^{-3}$
NBL location (x1, x2, dx1, dx2) Gaussian	x = (0, 10, 1, 1); y = (−11, −11, 1, 1)

cases, a higher breakdown voltage is desired. For the purposes of this exercise, we would like to keep the cell pitch at 10 μm and increase the breakdown voltage by modifying the P-substrate doping concentration and making smarter use of the NBL region. The doping concentration in the N-region is unchanged.

8.1.6.1   Basic Understanding of How to Improve Breakdown Voltage

Before we do some structure modifications, we need to have some basic understanding of breakdown in semiconductor devices. As we can see from Figure 8.8 for the diode with a P-substrate doping of $5E + 16 \text{ cm}^{-3}$, the potential lines are crowded at the PN junction. This means that the electric field is concentrated at the junction and the depletion edge is a short

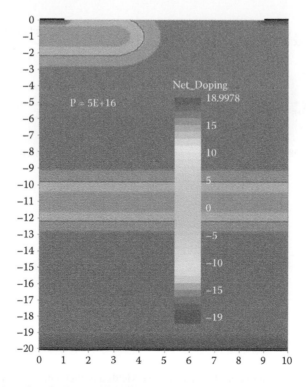

FIGURE 8.9   Net doping plot of a PN junction diode with NBL and P-substrate doping of $5E + 16 \text{ cm}^{-3}$.

distance away from the junction. Other silicon areas within the device are neutral regions without being depleted, so a lot of silicon space is wasted.

As we know, avalanche breakdown is a localized process, where the electric field exceeds the critical field strength and creates electron–hole pairs. These pairs are accelerated within this high field and generate more electron–hole pairs by way of impact ionization. We will observe a sudden increase of current when avalanche breakdown takes place.

To improve the breakdown voltage of a device, we need to improve the utilization of the silicon area to support a larger applied voltage while still preventing impact ionization. Only the depleted region can support the voltage drop. A small depleted area means the device can only support a limited voltage drop and the breakdown voltage is low; increasing the depletion area allows the voltage drop to be spread over a wider range and reduces the average field strength ($E = -dV/dx$). A large depleted area with peak electric field strength lower than the critical field is a more uniform electric field within the depleted area compared with that of a small localized depletion region. This large depleted area can support more voltage drop and yield a higher breakdown voltage, as can be observed in the 1D illustration of Figure 8.10. For a given critical field value, the breakdown voltage is proportional to the shaded area in the Field-Distance plot, so the larger the depletion area, the higher the breakdown voltage. A more uniform electric field will also yield a higher breakdown voltage because it increases the integral of the shaded area. In reality, a complete mesa-like electric field distribution is hard to achieve; a more practical shape of the electric field is a triangle with smaller slope or a combination of triangles, as seen in Figure 8.11. However, a smaller slope means a lower doping concentration, which in many cases compromises the on-state resistance.

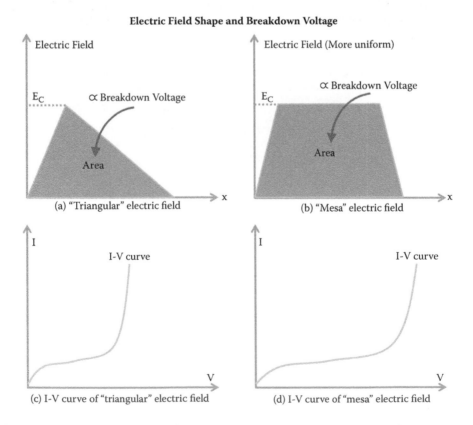

FIGURE 8.10 A more uniform (mesa-like) electric field yields a higher breakdown voltage.

Different Electric Field Slopes

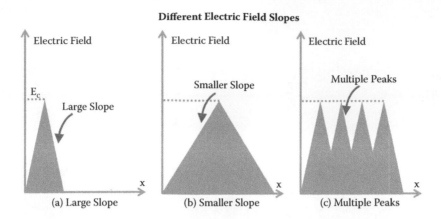

(a) Large Slope   (b) Smaller Slope   (c) Multiple Peaks

FIGURE 8.11 Smaller slope and multiple peaks improve electric field uniformity and boost breakdown voltage.

8.1.6.2 Different P-Substrate Doping for the PN Junction Diode without NBL

To boost the breakdown voltage of the PN junction diode, one common method is to reduce the doping concentration in the N or P or both regions. Since the NBL will affect the breakdown voltage, we will start by first simulating the structure (10 μm pitch size) without the NBL. We will add back this layer later on to discuss its effects.

For this analysis, the N-region doping is held the same as before, and we modify the P-substrate doping concentration. Figure 8.12 compares the results for two P-substrate doping. As can be observed from Figure 8.12b, a lower doping concentration ($5E + 15$ cm^{-3}) in the P-substrate yields a wider spread of equipotential contour lines compared with the higher doping concentration ($5E + 16$ cm^{-3}) in Figure 8.12f. From Figures 8.12c and 8.12g, we also see a wider and more uniform electric field for the low-doping case, which leads to a higher breakdown voltage (92 V vs. 47 V) and better usage of the silicon area for the same pitch size (10 μm).

However, this increase of breakdown voltage does not come without a price. The lower substrate doping also leads to higher parasitic series resistance, which dissipates more power during the on-state of the diode (forward bias). We should also note that lower P-doping does not necessarily bring a higher breakdown voltage either. It depends on how much depletion space is available, and if the P depletion width is restricted for some reasons, it may even bring down the breakdown voltage. An example of this effect is shown in Figure 8.13; as the doping level is reduced the breakdown voltage is increased, but there is an optimal value below which no further improvement can be achieved because the geometry of the device limits the depletion width.

In practice, the available space and doping with which to create the depletion region depends on the process flow being used. For example, if P-well is used, which is likely the case in a BCD technology, the diode may have to share the same P-well process as the NMOS body. Since the P-well cannot be doped too low for the NMOS, the breakdown voltage of the diode is limited by this other part of the process. Even a larger pitch size may not help to increase the breakdown voltage as we have previously seen: much of the silicon space becomes wasted once the pitch size exceeds the depletion width. Of course, there are other possibilities of constructing a lateral PN junction diode. For example, the shallow trench isolation (STI) process may be used to boost the breakdown voltage without reducing the P- or N-doping concentration. This takes advantage of a higher critical field in the oxide, and the purposefully designed peak field in the oxide region will help increase the breakdown voltage. The STI method will be discussed in detail later.

The 3D surface plot of the electric field magnitude is shown in Figure 8.14. The word *surface* refers to the 3D rendering of a physical parameter value such as the electric field magnitude of a 2D plane. From

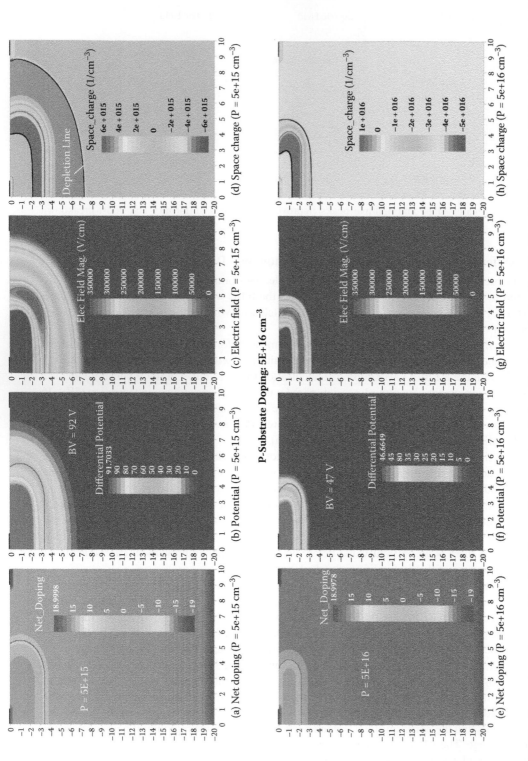

FIGURE 8.12 Comparison of reverse-bias simulation results between two diodes with substrate doping $P = 5E + 15$ cm^{-3} ((a) Net doping; (b) Potential; (c) Electric field; (d) Space charge) and $P = 5E + 16$ cm^{-3} ((e) Net doping; (f) Potential; (g) Electric field; (h) Space charge)).

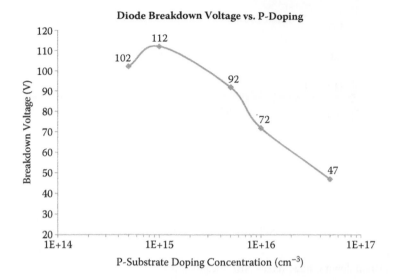

FIGURE 8.13 Breakdown voltage versus P-substrate doping concentration plot for PN junction diodes without NBL.

Figure 8.14, one can observe that in the higher-doping case the field is confined in the upper left corner, whereas the field in the low-doping case is more uniform. Just as we previously showed in our 1D model, the breakdown voltage is proportional to the integrated electric field ($\int EdS$), so having a more uniform field allows the region with a high electric field to contribute over a larger area, which tends to increase the breakdown voltage.

8.1.6.3 Breakdown Enhancement for Diode with NBL

As we have discussed before, without NBL and if the substrate is grounded, forward biasing the diode will instantly cause a high leakage current due to an internal short between the anode P-region

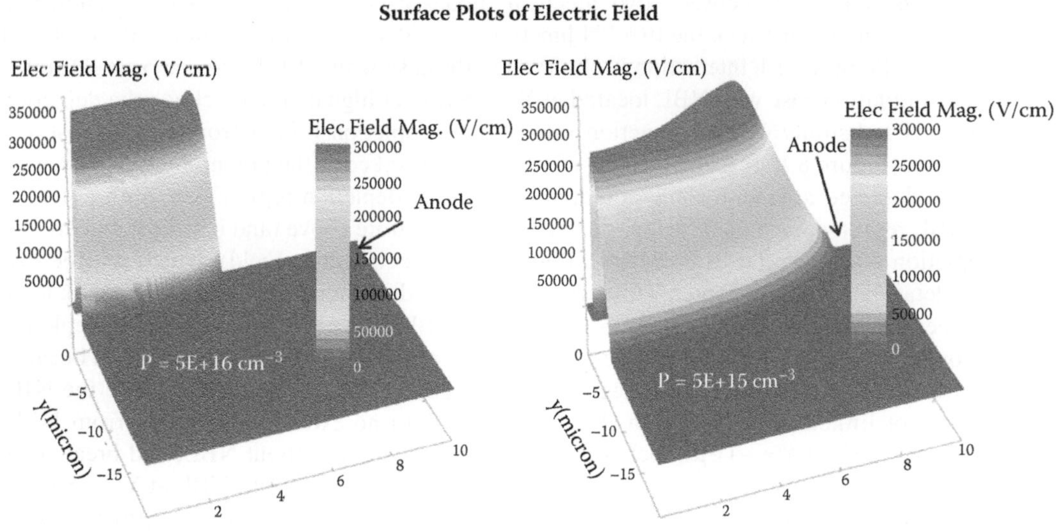

Surface Plots of Electric Field

(a) Surface plot of electric field (P = 5E+16 cm^{-3})   (b) Surface plot of electric field (P = 5E+15 cm^{-3})

FIGURE 8.14 Comparison of electric field 3D surface plots between two P-substrate doping concentrations at breakdown.

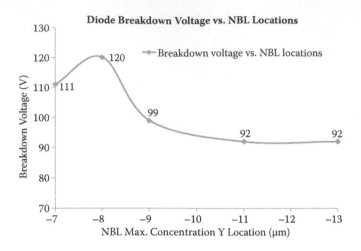

FIGURE 8.15   Breakdown voltage versus NBL locations.

and the P-substrate. NBL isolates the anode from the substrate and allows the substrate to stay at ground potential. The diode structure with NBL may have different reverse bias behavior than without NBL, due to the interaction between the N-region and underlying NBL with the depletion regions. Sometimes such an interaction is desirable since it may help enhance breakdown voltage.

Our previous PN junction diode example with a P-substrate doping of $5E + 15$ cm^{-3} is simulated with the additional NBL layer at different Y locations in the P-substrate. The substrate terminal is connected to ground in these simulations. As can be seen in Figure 8.15, the breakdown voltage is dependent on the position of the NBL. With the NBL, there is an additional PN junction (P-substrate/NBL) besides the main P-substrate/N-region junction. Proper placement of the NBL allows the depletion regions of these two PN junctions to essentially merge under large bias: that is, the P-side depletion of the P-substrate/N-region junction reaches the P-side depletion from the P-substrate/NBL junction.

A comparison is made between two identical diodes except NBL locations. As shown in Figure 8.16, interaction between the two PN junctions is evident in the diode with NBL located at $Y = -8$ μm, while little such interaction is observed in the case with NBL located at $Y = -11$ μm.

For the simulation case with NBL located at $Y = -8$ μm, at high cathode voltage the depletion region from P-substrate/N-region junction touched the depletion region from the P-substrate/NBL junction (Figure 8.16d). The potential of NBL is then linked to the potential of the N-region through the depleted area with some potential drop in the depletion region. The high potential that the NBL region gained then further depletes the P-substrate above (and below) it and creates a 2D depletion effect similar to RESURF (Reduced Surface Electric Field), which will be discussed in detail later in this chapter. This two-dimensional depletion effect boosts the breakdown voltage when properly designed. As for the diode with NBL at $Y = -11$ μm, we see the depletion region from the P-substrate/N-region failed to touch that of the P-substrate/NBL junction because the distance between the NBL and N-region is too large (Figure 8.16h) and therefore the NBL potential is not linked to the N-region (Figure 8.16f). With no extended depletion from NBL, the diode with NBL at $Y = -11$ μm behaves like a standard diode without NBL, and breakdown voltage reached only 92 V compared with 120 V of that from the case with NBL at $Y = -8$ μm.

The surface electric field plots of structures with NBLs at $Y = -8$ μm and $Y = -11$ μm are shown in Figure 8.17. With NBL at $Y = -8$ μm, the electric field has a more uniform field magnitude over a larger portion of the device area than the case with NBL at $Y = -11$ μm.

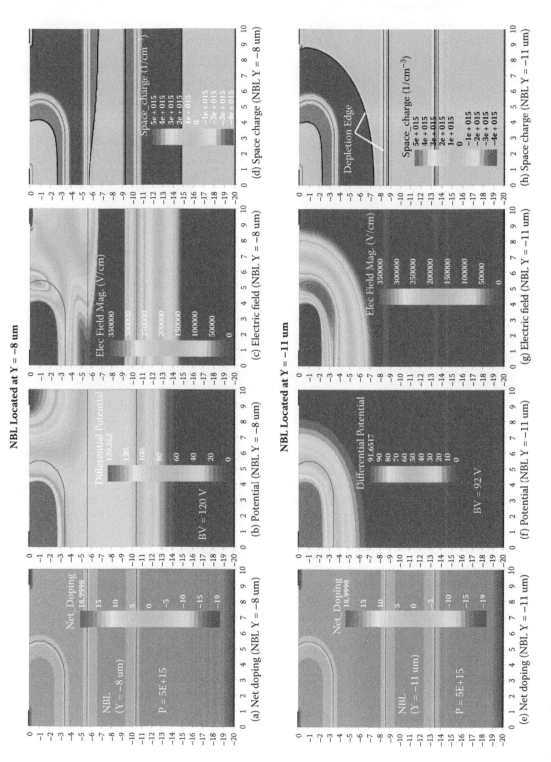

FIGURE 8.16 (See color insert) Comparison plots of different NBL locations with (a) to (d) located at Y = −8 μm and (e) to (h) located at Y = −11 μm.

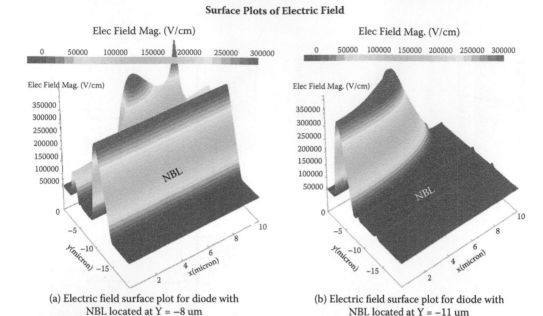

(a) Electric field surface plot for diode with NBL located at Y = -8 um

(b) Electric field surface plot for diode with NBL located at Y = -11 um

FIGURE 8.17　3D electric field surface plots of diode structures with NBL located at (a) Y = -8 μm and (b) Y = -11 μm.

One caution here is that NBL does not always aid in breakdown voltage: it can deteriorate the breakdown voltage if the doping concentrations of the N-region, P-substrate, and NBL are not properly selected. The distance between the N-region and NBL needs to be carefully designed to allow the depletion region from the N-region to touch the depletion region of NBL. If the gap is too large, the two depletion regions fail to touch each other; the potential of NBL cannot be pinned (pulled up) by the N-region. In that case, the NBL is still floating and not capable to help boost breakdown voltage. On the other hand, if the gap is too small, the N-region and NBL region will touch each other, and the depletion region between the two that partially supports the voltage is diminished. For example, when NBL is Y = -7 μm, breakdown voltage drops from 120 V (Y = -8 μm) to 110 V (Y = -7 μm) (Figure 8.15).

8.1.6.4　Reverse Leakage Current Path with Substrate Contact

Leakage current under a reverse-bias condition is structure dependent. With a substrate contact connected to ground and a P-substrate doping concentration of $1E + 16$ cm^{-3}, simulations are done for different NBL positions. As shown in Figure 8.18b, after breakdown the leakage current for the case of NBL located at Y = -13 μm is mostly flowing laterally between the cathode and anode terminals. For the structure with NBL located closer to the N-region (NBL Y = -7 μm), the leakage current flows vertically from cathode to substrate via the NBL (Figure 8.18a).

8.1.7　Reverse Recovery

8.1.7.1　Basic Understanding of Diode Reverse Recovery

For power electronics and power ICs, the diode reverse recovery effect is undesirable since it introduces additional switching power loss and reduces switching speed.

During the turnoff phase of an initially forward-biased diode, the anode current will first drop to zero before becoming negative; during this time, the diode current gradually diminishes to zero.

Leakage Current at Different NBL Locations

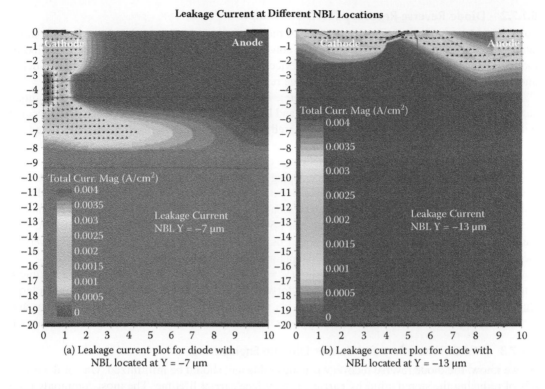

(a) Leakage current plot for diode with
NBL located at Y = −7 μm

(b) Leakage current plot for diode with
NBL located at Y = −13 μm

FIGURE 8.18 Comparison of leakage current path with (a) NBL located at Y = −7 μm and (b) NBL located at Y = −13 μm.

Figure 8.19a illustrates the transient behavior of the diode current during the turnoff period, and Figure 8.19b shows a buck DC/DC converter with a control field-effect transistor (FET) and a diode. Initially, the control-FET is off and the diode is forward biased with current flowing from anode to cathode. When a control signal instructs the control-FET to turn on, the current in the diode starts to drop. Since the diode cathode voltage now equals to input voltage ($V_x = V_{in}$), the diode becomes reverse biased with large power dissipation during the reverse recovery period. The shaded area of the reverse recovery plot in Figure 8.19a indicates charge storage during this period of time.

$$Q_s = \int_{t_1}^{t_2} I_a(t)\,dt \tag{8.14}$$

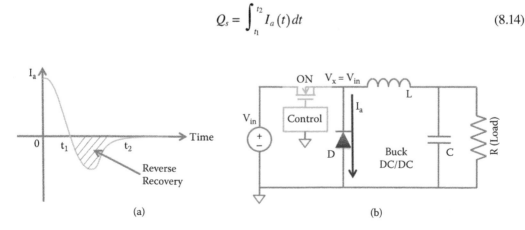

(a)

(b)

FIGURE 8.19 (a) Schematic view of diode reverse recovery; (b) Buck DC/DC converter.

8.1.7.2 Diode Reverse Recovery TCAD Simulation

For power electronics engineers, the effects of diode reverse recovery on the circuit can be simulated with IC simulators (IC CAD tools) that solve compact device models. However, this kind of tool does not reveal the physics within the device. TCAD simulation is necessary for an in-depth understanding of what is going on during diode reverse recovery.

When the diode is forward biased (on-state), both electrons and holes contribute to the diode current. Electrons are injected into the P-side and holes injected into the N-side where they become minority carriers. When the diode is reverse biased (turned off), the minority carriers have finite time to be recombined or removed from either the N- or P-region. This finite time is called carrier lifetime (recombination lifetime). Minority carrier storage effect is more pronounced with lightly doped side.

In this simulation example, the P-substrate doping is $1E + 16$ cm^{-3}. The diode is first forward biased to $V_a = 1.5$ V over a period of 1 ns and then reverse biased to $V_a = -5$ V over a period of 5 ns, allowing sufficient time for the electrons to be removed. As can be seen from the simulation results in Figure 8.20, initially the diode is forward biased with anode voltage of 1.5 V from 0 to 1 ns. Starting from time 1 ns, a reverse bias is applied to the anode, and within another 1 ns, the anode current drops to zero and then reversely increases before finally decreases to reverse saturation leakage current.

Electron concentration plots under forward ($V_a = 1.5$ V, time = 1 ns) and reverse ($V_a = -5$ V, time = 6 ns) bias are shown in in Figure 8.21. Under forward bias right before turning off, there is a large electron concentration in the P-substrate (see PN junction line on the plot); under reverse bias and after the leakage current has stabilized to reverse saturation current, the electrons have been removed.

8.1.7.3 TCAD Simulations for Carrier Lifetime Engineering

As we know, the diode reverse recovery is undesirable and should be minimized. One of the methods of reducing the stored minority carriers is to reduce carrier lifetime. The most commonly used approach is to diffuse gold or platinum or to use electron irradiation to create recombination centers.

In a TCAD simulation, we can simulate this effect by simply modifying the minority carrier lifetimes and observing the results as seen in Figure 8.22. A shorter lifetime reduces the stored charges and allows lower switching power loss for power electronic systems.

8.1.8 SCHOTTKY DIODE

PN junction diode is a bipolar device in which both electrons and holes contribute to current flow. This bipolar action inevitably brings the problem of minority carrier storage and large reverse recovery power loss. Unlike the PN junction diode, a Schottky diode employs a metal-semiconductor junction, and only one type of carrier contributes to the total diode current: a unipolar device. Most

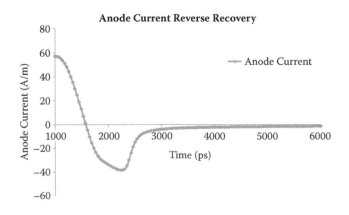

FIGURE 8.20 Transient simulation of diode reverse recovery.

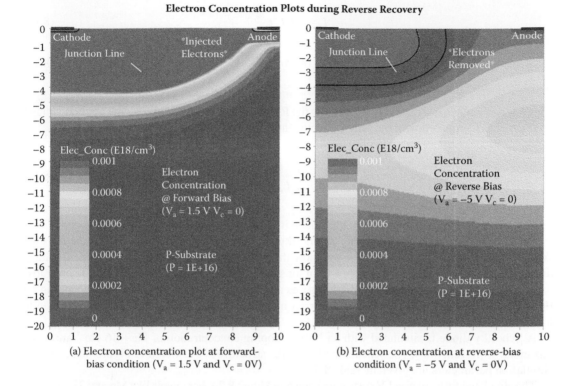

(a) Electron concentration plot at forward-bias condition ($V_a = 1.5$ V and $V_c = 0$V)

(b) Electron concentration at reverse-bias condition ($V_a = -5$ V and $V_c = 0$V)

FIGURE 8.21 Electron concentration plots at (a) forward and (b) reverse bias after the balance is established.

of the time, a metal-to-N-type-semiconductor connection is used to construct the Schottky diode for power electronics and power IC applications.

For sufficiently low doping in the silicon, we neglect tunneling effects and from thermionic emission theory, the I-V relation for the Schottky diode can be derived as

$$I = I_s \left[e^{\left(\frac{V_a}{V_t}\right)} - 1 \right] \tag{8.15}$$

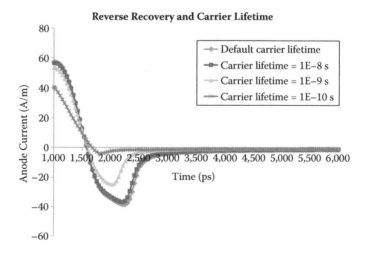

FIGURE 8.22 Reverse recovery by using various carrier lifetimes.

$$I_s = AKT^2 e^{-\frac{V_{bi}}{V_t}} \qquad (8.16)$$

$$K \approx 100 \ A\Big/\left(\frac{cm^2}{K^2}\right) \qquad (8.17)$$

where I_S is the reverse saturation current, K is the Richardson constant, and V_a is the applied voltage on the metal. Compared with the saturation current from the PN junction diode

$$I_{s,PN} = Aqn_i^2\left(\frac{D_p}{L_p N_d} + \frac{D_n}{L_n N_a}\right) \qquad (8.18)$$

The reverse leakage current for a Schottky diode is almost always higher.

One of the most important merits for a Schottky diode is the possibility of reducing the on-state voltage drop. From our previous plots, we know there is always an approximate 0.7–0.8 V voltage drop for a PN junction diode (built-in potential for silicon). This drop can result in a large power loss for a power electronic system if the diode carries a large forward current. However, depending on the metal work function and the resulting barrier height, a Schottky diode may have a significantly lower turn-on voltage than a PN junction diode.

A Schottky diode is simulated with an N-type doping concentration of 5E + 16 cm^{-3} and various metal work functions. The schematic view of the structure is illustrated in Figure 8.23, and the simulation parameters are given in Table 8.3.

The simulated forward bias I-V curves are shown in Figure 8.24. For metals like Mg and Ti, the work functions are too small, whereas Pt has a turn-on voltage almost equal to the PN junction. The metals (Cr and W) in between provide a good balance between turn-on voltage and leakage current for this case.

As an example, two band diagrams, one for Gold (Au) and one for Tungsten (W), are plotted in Figure 8.25 at the Schottky junctions, with the cutline located between points of (X = 0.1 Y = 0) and (X = 0.1 Y = −0.25).

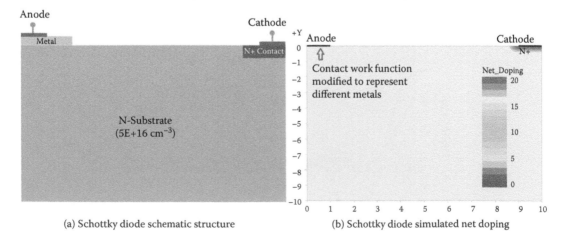

FIGURE 8.23 Schottky diode schematic and simulation structures.

TABLE 8.3
Simulation Parameters for a Schottky Diode

Parameters for Simulation	Value
X range	0 to 10 μm
Y range	0 to −10 μm
P-substrate (P-base constant doping)	5E + 16 cm^{-3}
Metal Schottky contact anode (x1, x2)	(0, 1)
Ohmic contact cathode (x1, x2)	(9, 10)

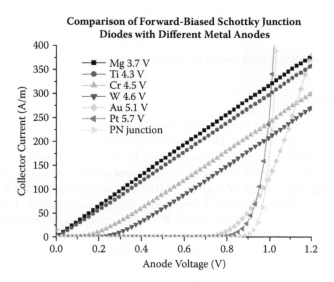

Comparison of Forward-Biased Schottky Junction Diodes with Different Metal Anodes

FIGURE 8.24   Forward-biased Schottky diode with various metal work functions.

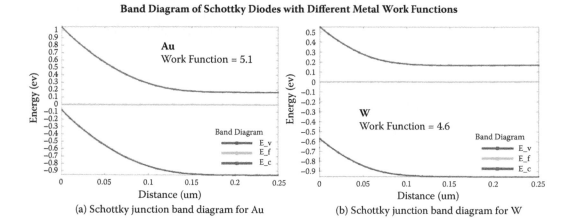

(a) Schottky junction band diagram for Au

(b) Schottky junction band diagram for W

FIGURE 8.25   Schottky junction band diagrams plotted for Au and W.

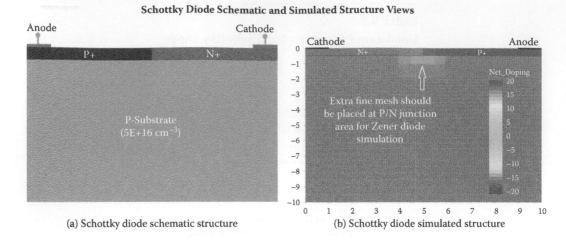

(a) Schottky diode schematic structure   (b) Schottky diode simulated structure

FIGURE 8.26   Zener diode schematic view and net doping plot of the simulated structure.

8.1.9 Zener Diode

When doping concentrations on both sides of the PN junctions are high, another type of breakdown called Zener breakdown occurs. Different from avalanche breakdown, which is due to impact ionization, Zener breakdown happens at low voltages by direct interband (conduction ↔ valence) tunneling of carriers. Zener diodes are useful in many applications for analog and power IC, especially for clamping the gate voltage of a power DMOS. Figure 8.26a shows the schematic view, and Figure 8.26b is the net doping plot of the Zener diode simulated structure. For a successful simulation of a Zener diode, the mesh surrounding the P+/N+ junction should be extra dense to correctly model the tunneling. The simulation parameters are given in Table 8.4.

Shown in Figure 8.27 are linear and log scale of breakdown curves for different doping concentrations. We can see that the Zener breakdown curve is quite "soft," without a sharp corner or steep current increase: this is typical of a tunneling effect. In practice, Zener diodes with breakdown ranges from 3 V to 6 V are available. Zener diodes with a breakdown larger than 6 V are usually not considered a "pure" Zener since they also have significant impact ionization. In real fabrication, junctions formed by very high doping concentrations usually come with large damage and dislocations. This is especially true if both high-dose and high-energy implants are performed.

The band diagrams of Zener diodes with two doping concentration levels ($N_a = N_d = 5E + 19$ cm^{-3} and $N_a = N_d = 5E + 18$ cm^{-3}) at breakdown are shown in Figure 8.28. At larger doping values, the

TABLE 8.4
Simulation Parameters List for Zener Diode

Parameters for Simulation	Value
X range	0 to 10 μm
Y range	0 to −10 μm
P-substrate (P-base constant doping)	5E + 16 cm^{-3}
N+ region max concentration	5E + 19 cm^{-3}
N+ region location (x1, x2, dx1, dx2) Gaussian	X = (0, 5, 0, 0); Y = (−0.5, 0, 0, 0)
P+ region max concentration	5E + 19 cm^{-3}
P+ region location (x1, x2, dx1, dx2) Gaussian	X = (5, 10, 0, 0); Y = (−0.5, 0, 0, 0)

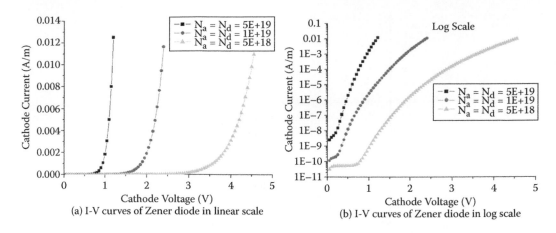

FIGURE 8.27 Zener breakdown comparison with different doping concentrations.

proximity of empty states on the other side of the junction makes it very easy for carriers to tunnel from band to band.

8.1.10 SMALL SIGNAL MODEL FOR PN JUNCTION DIODE

I-V curves of semiconductor devices are generally nonlinear. For example, the forward I-V curve of a diode is nonlinear if we look at a large voltage span. However, if we take an infinitesimal portion of this nonlinear curve, as shown in Figure 8.29, we can linearize the nonlinear device. A small signal model is based on such a linearization around a bias point Q, which is the DC voltage/current level when no signal is applied. Small signal models are very useful tools for analog and power IC design.

Figure 8.30 is the small signal representation of a diode, which can be viewed as a capacitor and resistor connected in parallel: the resistor is the parasitic resistance of the PN junction diode, while the capacitor represents a combination of junction capacitance and diffusion capacitance [167].

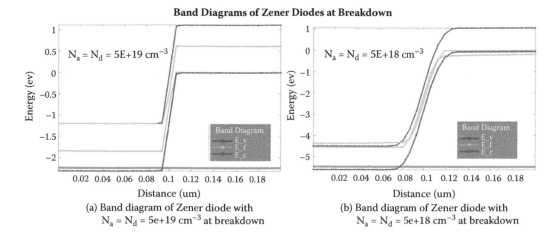

FIGURE 8.28 Band diagram of Zener diode at the PN junction at breakdown with (a) Na = Nd = 5E + 19 cm^{-3} and (b) $N_a = N_d = 5E + 18$ cm^{-3}.

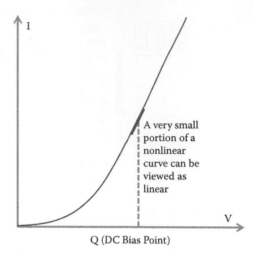

FIGURE 8.29 Small signal analysis.

The small signal conductance and resistance can be calculated as

$$g = \frac{dI}{dV} = \frac{dI_a}{dV_a} = \frac{dI_s \cdot \left[e^{\left(\frac{qV_a}{kT} \right)} - 1 \right]}{dV_a} \approx \frac{dI_s \cdot e^{\left(\frac{qV_a}{kT} \right)}}{dV_a} = \frac{q}{kT} I_s \cdot e^{\left(\frac{qV_a}{kT} \right)} = \frac{q}{kT} I_a = \frac{I_a}{V_t} \tag{8.19}$$

$$R = \frac{1}{g} = \frac{V_t}{I_a} \tag{8.20}$$

There are two capacitance components for a diode. The first one is the junction capacitance, which dominates the reverse bias and small forward-bias region. The other component is called diffusion capacitance and is caused by charge storage effects during the diffusion process; it dominates the large forward-bias region. The total capacitance from both effects can be calculated as

$$C = C_{junction} + C_{diffusion} = \frac{A\varepsilon_s}{W} + \tau_s \frac{dI_a}{dV_a} = \frac{A\varepsilon_s}{\sqrt{\frac{2\varepsilon_s(V_{bi} - V_a)}{q}\left(\frac{1}{N_a} + \frac{1}{N_d} \right)}} + \tau_s \frac{I_a}{V_t} \tag{8.21}$$

This equation should be used with caution because the equation for the junction capacitance is valid only when $V_a < V_{bi}$. If the applied voltage is higher, the total capacitance is dominated by the

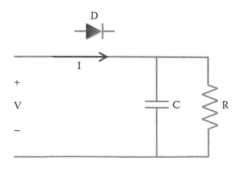

FIGURE 8.30 Small signal representation of a PN junction diode.

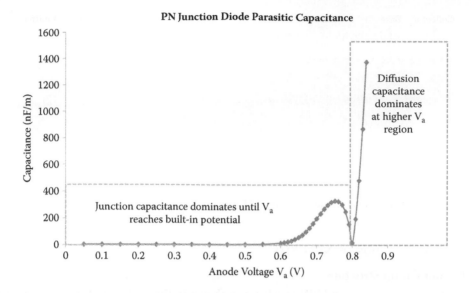

FIGURE 8.31 TCAD simulation of PN junction diode capacitance.

diffusion capacitance, and the junction capacitance should diminish rather than going to infinity as predicted by the equation. The diode capacitance is simulated with TCAD, and the result is shown in Figure 8.31. Junction capacitance dominates when the applied anode voltage is smaller than the built-in potential; after the built-in potential is reached, the junction capacitance drops, and the diffusion capacitance picks up rapidly with applied anode voltage.

8.2 BIPOLAR JUNCTION TRANSISTORS

The BJT was first invented in 1948 at Bell Labs. *Bipolar* refers to the fact that both electrons and holes participate in the transistor operation, contrary to unipolar devices like MOSFETs and Schottky diodes. Before the wide acceptance of CMOS technologies, BJT dominated the semiconductor industry; CMOS gradually replaced BJTs in digital ICs thanks to the latter's advantages in high density, scaling capability, and low power gate drive. BJTs are still preferred in many high-frequency and analog applications for their merits of high speed, low noise, good matching, and high output power.

Two types of BJTs are available: NPN and PNP. NPN BJTs have N+ emitter, P-type base, and N-type collector; PNP BJTs have P+ emitter, N-type base, and P-type collector. Due to the higher mobility of electrons, NPNs are generally preferred over PNPs, so this section discusses only NPN transistors. A simplified NPN BJT structure can be visualized (Figure 8.32) as two back-to-back diodes with a P-region sandwiched between two N-regions.

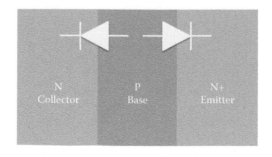

FIGURE 8.32 Basic structure of an NPN BJT.

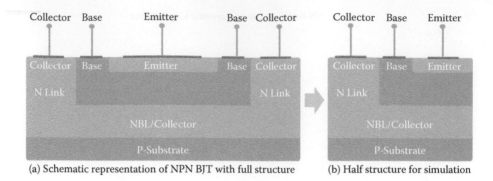

(a) Schematic representation of NPN BJT with full structure   (b) Half structure for simulation

FIGURE 8.33   Schematic representation of NPN BJT with (a) full structure and (b) half structure for simulation.

8.2.1   BASIC OPERATION OF NPN BJTs

8.2.1.1   Simulation Structure

The schematic structure of an NPN transistor is shown in Figure 8.33a. Since this structure is symmetrical, it can be truncated for simulation purposes. By simulating only half of the structure, as seen in Figure 8.33b, results can be obtained faster by using less computer memory and processing power.

In this simulation example, a lateral NPN BJT is created with an NBL connected to the N collector through an N link implant, which can be created by an implant chain of high- and low-energy ion implants. The base P-region can be created either with P-substrate, P-epi layer, or an implanted P-well. For the sake of simplicity, in the simulation a P-substrate is used. The simulation parameters for this NPN BJT are listed in Table 8.5 (N+ and P+ contacts are omitted).

The simulated NPN BJT net doping plot is illustrated in Figure 8.34a with a 3D surface plot of net doping in Figure 8.34b. The surface plot eliminated the substrate below $y = -10$ μm, since the most important device information is located above $y = -10$ μm.

Band diagrams at equilibrium are illustrated in Figure 8.35. Since the BJT has N-P-N structures in both horizontal and vertical directions, a horizontal band diagram is shown with a cutline located from $(x = 0, y = -1)$ to $(x = 10, y = -1)$, and a vertical band diagram is shown with a cutline located from $(x = 9, y = 0)$ to $(x = 9 \ y = -10)$.

TABLE 8.5
NPN BJT Simulation Parameters List

Parameters for Simulation	Value
X range	0 to 10 μm
Y range	0 to −20 μm
P-substrate	$1E + 15$ cm^{-3}
P-base doping	$1E + 17$ cm^{-3}
NBL max concentration	$1E + 16$ cm^{-3}
NBL location (x1, x2, dx1, dx2) Gaussian	$x = (0, 10, 1, 1)$; $y = (-7, -7, 1, 1)$
N-link max concentration	$1E + 16$ cm^{-3}
N-link location (x1, x2, dx1, dx2) Gaussian	$x = (0, 4, 0.25, 0.25)$; $y = (-6, 0, 0.25, 0.25)$
Emitter N-well max	$1E + 19$ cm^{-3}
Emitter N-well location (x1, x2, dx1, dx2) Gaussian	$x = (6.5, 10, 0.25, 0.25)$; $y = (-0.5, -0.5, 0.5, 0.5)$

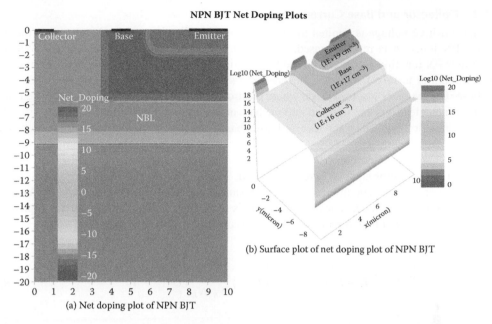

FIGURE 8.34 (a) Simulated net doping plot and (b) 3D net doping surface plot of an NPN BJT.

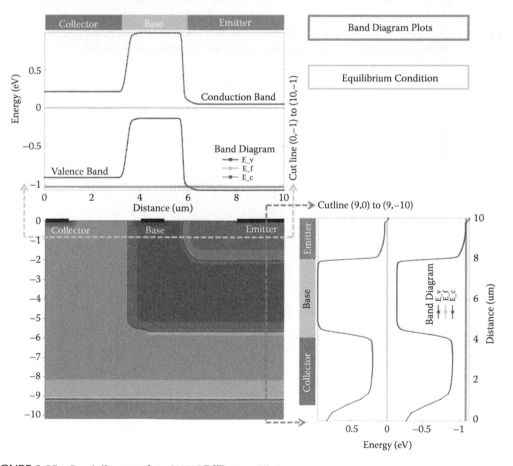

FIGURE 8.35 Band diagram plot of NPN BJT at equilibrium.

8.2.1.2 Collector and Base Current

If a small positive voltage is applied to the collector with all other terminals grounded, the base/collector PN junction is reverse biased, and no current flows except leakage current. When the base/emitter PN junction is forward biased by applying a positive voltage at the base terminal with respect to emitter terminal, holes are injected into the N+ emitter, and electrons are injected into the P-base. Unlike a PN junction diode where electrons enter the P-region and are extracted by the anode terminal, electrons that injected into the base are swept to the collector by the electric field in the depletion region created by the reverse-biased base/collector PN junction. After electrons reach the collector region they will be attracted and collected by the positively biased collector contact. This electron current contributes to collector current I_c, which is independent of the collector bias as long as the base/collector junction is reverse biased.

The collector current is dominated by the diffusion of electrons from emitter to base. Assuming low-level injection (injected minority carrier level is much less than a majority carrier doping concentration) and constant base doping, the collector current I_c can be expressed as [167].

$$I_c = I_s \cdot \left[e^{\left(\frac{V_{BE}}{V_t} \right)} - 1 \right] = A_E q \frac{D_B}{W_B} \frac{n_i^2}{N_B} \left[e^{\left(\frac{V_{BE}}{V_t} \right)} - 1 \right]$$

(8.22)

where I_S is the saturation current, V_{BE} is base to emitter voltage, A_E is the emitter area, n_i is the intrinsic carrier concentration V_t is the thermal voltage, W_B is the base width, and D_B is the minority carrier diffusion coefficient in the base region. At high V_{BE}, the injected minority carrier (electrons in the base region of the NPN transistor) density becomes equal or even larger than majority carrier density (holes in the NPN base region), so the collector current needs some modifications:

$$I_c \propto n_i e^{\left(\frac{V_{BE}}{2V_t} \right)}$$

(8.23)

The base current I_b is created by holes injected into the emitter region. Similar to I_c, I_b, is diffuse dominated and can be expressed as

$$I_b = A_E q \frac{D_E}{W_E} \frac{n_i^2}{N_E} \left[e^{\left(\frac{V_{BE}}{V_t} \right)} - 1 \right]$$

(8.24)

where D_E is the diffusion coefficient of minority carriers (holes in NPN) in the emitter, W_E is emitter width, and N_E is the constant doping concentration in the emitter. Since the emitter is usually heavily doped, it is not easy to reach a high injection level in the emitter.

8.2.1.3 Transistor Gain and Gummel Plot

One of the most important design parameters for BJTs is the DC current gain β, which is defined as

$$\beta = \frac{I_c}{I_b} = \frac{D_B W_E N_E}{D_E W_B N_B}$$

(8.25)

In TCAD simulation, to obtain the current gain a positive bias of 1.0 V is applied to the collector, and then base voltage is scanned from 0 to 1.2 V. The emitter terminal is always grounded. Using the NPN simulation structure described earlier in this chapter, I_c and I_b plots versus V_{BE} are plotted on

BJT Gummel Plot and DC Current Gain

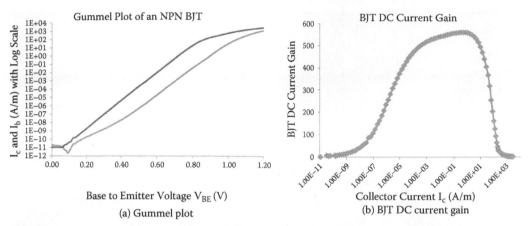

(a) Gummel plot

(b) BJT DC current gain

FIGURE 8.36 Gummel plot and DC current gain plot.

log(Y)-linear(X) scale (Gummel plot) on Figure 8.36a; the resulting DC current gain β is also included (Figure 8.36b).

8.2.2 NPN BJT BREAKDOWN

Since a BJT is a three-terminal device, breakdown can happen between any of the reverse-biased PN junctions or any combination. The four most useful breakdown analyses are BV_{CBO}, BV_{EBO}, BV_{CEO}, and BV_{CES}.

8.2.2.1 BV_{CBO}

Here, BV is short for breakdown voltage, and CBO means breakdown between collector (N-type) and base (P-type) with emitter (N-type) terminal open (floating): BV_{CBO} is the breakdown voltage of the base/collector PN junction with no influence from the emitter, so it is similar to a regular PN junction diode breakdown. Figure 8.37a is the potential plot of BVCBO, we see the breakdown voltage between collector and base with emitter open is 64 V. Figure 8.37b–f plot the electric field, space charge, impact ionization, electron current magnitude, and hole current magnitude. The impact ionization takes place at the collector/base junction close to the surface.

8.2.2.2 BV_{EBO}

EBO means breakdown between emitter (N-type) and base (P-type) with collector terminal open (floating). Similar to BV_{CBO}, BV_{EBO} is just the breakdown voltage of base/emitter PN junction with no influence from the collector. Figure 8.38a is the potential plot of BV_{EBO}, with a simulated breakdown voltage of only 21 V. The low breakdown voltage between emitter and base with collector open is easily understandable since the emitter is a heavily doped region (at about $1E + 19$ cm^{-3}) and the base doping is about $1E + 17$ cm^{-3}. Figure 8.38b–f are plots of electric field, space charge, impact ionization, electron current magnitude, and hole current magnitude. We see the impact ionization happens at the base/emitter junction close to the silicon surface.

8.2.2.3 BV_{CEO}

CEO means breakdown between collector (N-type) and emitter (N-type) with base open (floating). Unlike BV_{CBO} and BV_{EBO}, BV_{CEO} involves two PN junctions. BV_{CEO} may be the most important

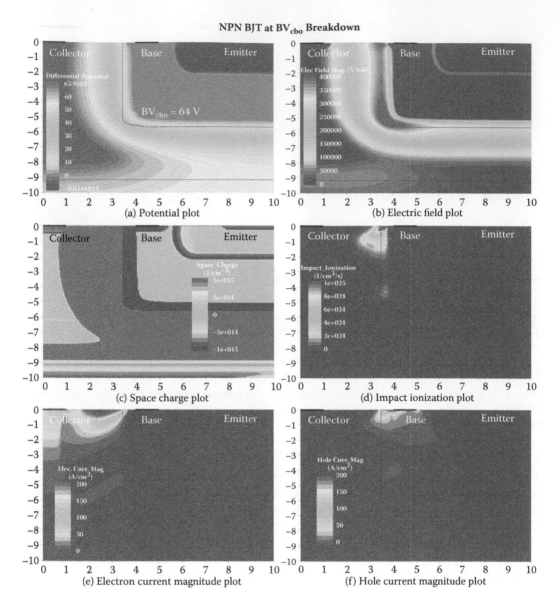

FIGURE 8.37 Simulation plots at BV_{CBO} breakdown.

breakdown voltage in BJTs: it usually has a low breakdown voltage value due to the BJT current amplification effect. When a positive bias is applied to the collector and the emitter is grounded, the collector/base junction is always reverse biased. The potential of the emitter is pinned to zero by grounding the emitter contact. Since the base contact is floating, the base potential is linked to the collector potential minus the voltage drop across the depletion region. If somehow the base potential is higher than the built-in potential of the base/emitter PN junction, the base to emitter junction is then forward biased. Due to the current amplification effect of a BJT transistor, the device operates with current gain that increases with increasing collector current. This positive feedback mechanism produces a collapse in the voltage that can be supported by the transistor and leads to a lower breakdown voltage [158]. Figure 8.39a shows the potential plot of BV_{CEO} with a simulated breakdown voltage of 29 V. Figures 8.39b–f plot the electric field, space charge, impact ionization, electron current magnitude, and hole current magnitude. We note that the impact ionization takes place at the

NPN BJT at BV$_{ebo}$ Breakdown

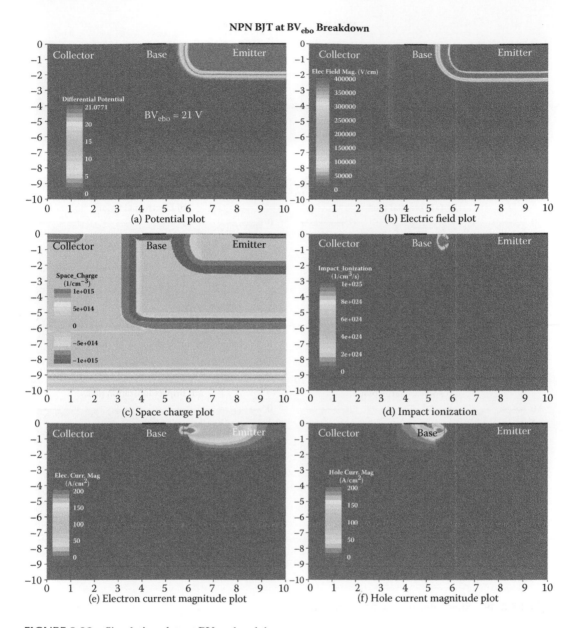

FIGURE 8.38 Simulation plots at BV$_{EBO}$ breakdown.

collector/base junction. An interesting observation is that the potential in the base region is about 0.86 V and that the base/emitter PN junction is forward biased. The hole current is seen flowing from base to emitter, and the current amplification effect is being triggered.

8.2.2.4 BV$_{CES}$

CES means breakdown between collector (N-type) and emitter (N-type) with base terminal shorted to emitter and ground. The collector to base junction is reverse biased, and the base to emitter junction is shorted through the contacts. Ideally, the base to emitter PN junction should not be forward biased, and breakdown voltage should be the same as BV$_{CBO}$. However, because of the parasitic series resistance in the lightly doped base region, leakage current flowing through this resistance will create a voltage drop that can be large enough to cause forward bias in the base/emitter junction. Once this junction

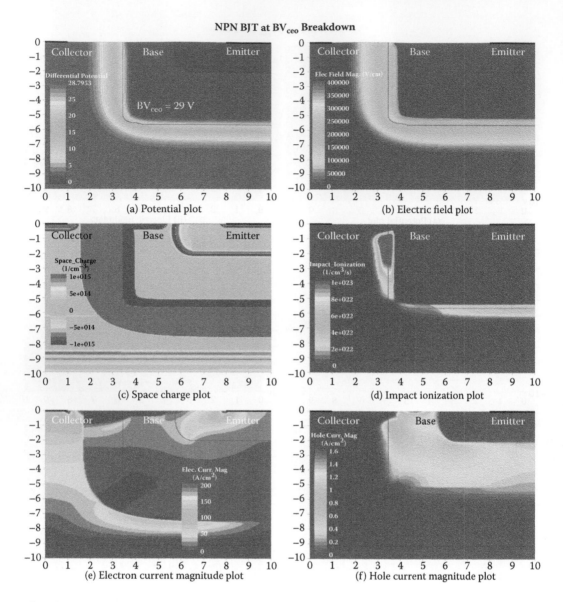

FIGURE 8.39 Simulation plots at BV_{CEO} breakdown.

is forward biased, the leakage current gets amplified, just like the case of BV_{CEO}. However, a properly designed P-type base with sufficiently high doping will reduce the resistance and minimize this effect. Figure 8.40a is the potential plot with a breakdown voltage of 64 V, the same as BV_{CBO} because the breakdown occurs once again between the collector and grounded (shorted) base. Figures 8.40b–f plot the electric field, space charge, impact ionization, electron current magnitude, and hole current magnitude. These plots are very similar to the plots found in Figure 8.37 for BV_{CBO}. Compared with BV_{CES}, the base potential is at 0 V, which means that thanks to the high doping in the P-base region the base/emitter parasitic diode is not turned on. The hole current flow is induced by impact ionization only and is not flowing from base to emitter.

8.2.2.5 Comparison of the Four Types of NPN Breakdown

The simulation result for BV_{CES} is similar to BV_{CBO} because with high P-base doping (about $1E + 17$ cm^{-3}), the base to emitter PN junction is not turned on by the leakage current flowing through the collector to

NPN BJT at BV_{ces} Breakdown

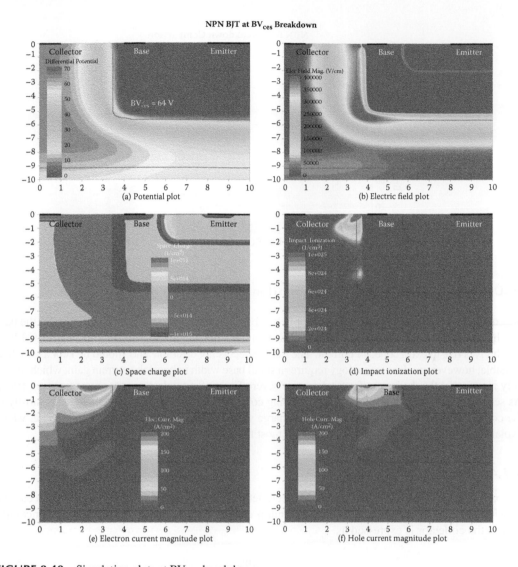

FIGURE 8.40   Simulation plots at BV_{CES} breakdown.

the base. BV_{CEO} simulation yields a 29 V breakdown voltage, which is due to the aforementioned current amplification effect. The BV_{EBO} in our simulation case has the lowest breakdown voltage of only 21 V. This is caused by the highly doped and abrupt emitter/base PN junction: if the emitter doping concentration were more graded, the BV_{EBO} breakdown voltage would be higher. A comparison of breakdown voltages for our NPN BJT is given in Figure 8.41.

8.2.3   BJT I-V FAMILY OF CURVES

Unlike MOSFETs that are turned on by applying a constant gate voltage, BJTs are driven by constant base current. The I_C-V_{CE} curves with different base bias current (I_B) are shown in Figure 8.42. For simulation purposes, the base current is first scanned to a certain value, and then the collector voltage is ramped up while the base current held constant. Incremental base current (I_B) generates a series of I-V curves. We can see these curves are not exactly flat and they are dependent on collector voltage (V_{CE}); this dependence is called the Early effect, named after Dr. James Early. Put simply, a higher voltage on

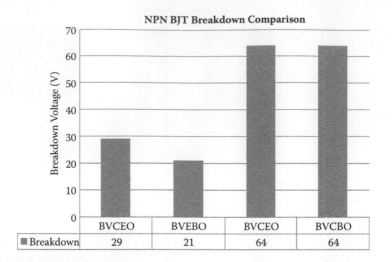

FIGURE 8.41   NPN BJT breakdown comparison chart.

the collector causes the depletion region in collector/base PN junction to widen, which squeezes the base width (calculated using the base metallurgical width minus base depletion width). This decrease of base width gives rise to extra collector current. The Early effect is not desirable and should be kept as small as possible; however, modern technology requires a small base width to boost the current gain, which invariably means that the devices suffer from a more severe Early effect unless proper precautions are taken. As seen in Figure 8.43, with a high P-base doping concentration the depletion region spreads mostly in the collector region: only a very small space charge width increase can be observed in the base, which explains why the simulated I_C-V_{CE} curves are almost flat.

8.2.4   KIRK EFFECT

To explain the Kirk effect (base push-out effect), let us first look at the relationship between current density and doping concentration. The maximum current density under certain doping concentration in the collector is expressed as

$$J_c = qnv_{sat} \qquad (8.26)$$

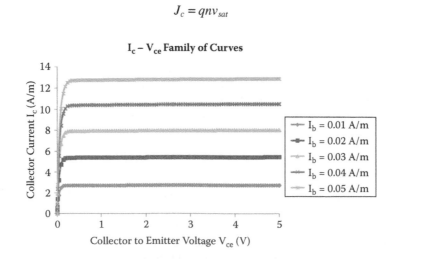

FIGURE 8.42   Simulated BJT I-V family of curves with I_b from 0.01 A/m to 0.05 A/m.

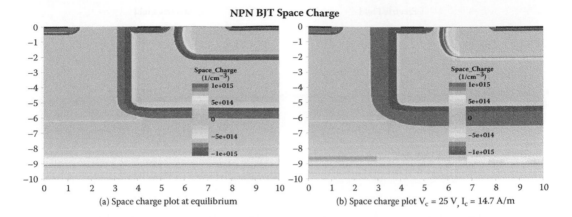

(a) Space charge plot at equilibrium

(b) Space charge plot $V_c = 25$ V, $I_c = 14.7$ A/m

FIGURE 8.43 Space charge region width comparison with (a) at equilibrium and (b) at $V_c = 25$ V and $I_c =$ 14.7 A/m.

where J_c is the collector current density, v_{sat} is the saturation velocity, and n is the carrier (electrons in this case) density. As previously discussed, the charge density in the vicinity of a PN junction is calculated as

$$\rho = qN_c - qn = qN_c - \frac{J_c}{v_{sat}} \tag{8.27}$$

and from Poisson's equation

$$\frac{dE}{dx} = \frac{\rho}{\varepsilon_s} \tag{8.28}$$

With high current density J_c, dE/dx may get a negative slope, as shown in Figure 8.44 [167]. When the collector current is small, $dE/dx > 0$ and the depletion region in the collector has a triangular shape that does not touch the collector N+ region (Figure 8.44a). With increasing collector current, the electric field begins to flatten out, and when $dE/dx = 0$ the ionized dopants and electrons cancel each other: the electric field is now flat in the depletion region until it reaches the N+ collector as shown in Figure 8.44b. Further increases of collector current flip the sign of dE/dx, and now minority carriers outnumber the ionized dopants in the collector as shown in Figure 8.44c. With a large enough collector current, the depletion region in the collector moves away from the base-collector metallurgical junction and the effective base width is "widened" as shown in Figure 8.44d: this base push-out effect is also known as the Kirk effect. The electric field and depletion region in the base and collector N+ regions are omitted, for the sake of simplicity.

The Kirk effect is an important effect for power BJTs since these devices typically operate at high collector current densities. At such high current densities, the lightly doped collector may fail to support the collector current even if the dopant-induced electrons move at saturation velocity [167]. With too few dopant atoms and too much current, the sign of the charge density in the depletion region is reversed [167], which will degrade the device performance under high current conditions. Figure 8.45 shows the diminishing collector depletion region with increasing collector current. From Figure 8.45d, we see a clear reversal of the space charge polarity in the depletion region of the collector (from positive to negative).

8.2.5 BJT Thermal Runaway and Second Breakdown Simulation

BJT has a unique device failure mechanism called second breakdown. Unlike avalanche breakdown, the second breakdown is a purely thermal effect.

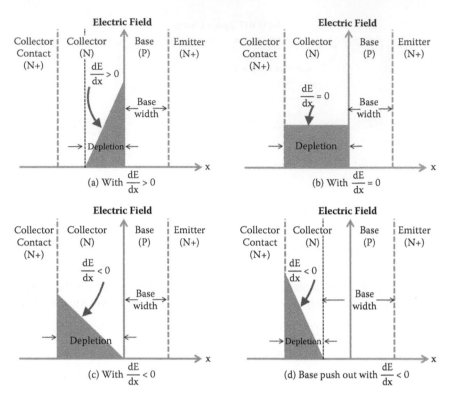

FIGURE 8.44 Modification of depletion region width and Kirk effect.

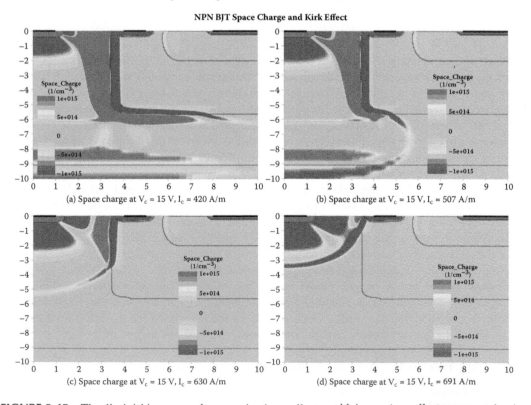

FIGURE 8.45 The diminishing space charge region in a collector with increasing collector current density.

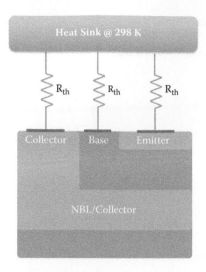

FIGURE 8.46 Schematic of thermal simulation setup.

Suppose a BJT is biased at a forward active region with high collector-emitter current flow. The device is locally heated up by this current. Due to bipolar action, higher temperature causes a higher generation and recombination rate and thus induces an even higher current level, which leads to a positive feedback at some local "hot spots" of the device. This positive temperature and current feedback phenomenon is called thermal runaway and can lead to a catastrophic device failure called second breakdown.

Second breakdown is of course undesirable, and proper design is necessary to minimize this problem. TCAD simulation is done to mimic the thermal runaway and second breakdown and to provide some physical insights during the second breakdown. To be more realistic, heat sink is connected to the contacts through thermal resistors (R_{th}); a schematic of the thermal simulation setup is shown in Figure 8.46.

For this simulation, a 2D simulation is performed so currents are in A/m and power flows are in W/m. Thus, the thermal conductance (R_{th}^{-1}) has units of $(W/m)/K$ rather than W/K. The missing z-depth is omitted. Here, we assume that a thermal conductance of 60 $(W/m)/K$ connects every contacts of the BJT (Figure 8.46) to a heat sink with a constant temperature of 298 K. The BJT base current is first ramped up to 1E-4 A/m, and then the collector voltage is increased to 60 V while the emitter is grounded. Finally, the collector current is ramped up to a high value (about 4200 A/m). From the I-V curve in Figure 8.47, we noticed that breakdown happens at about 65 V but that voltage still increases beyond this point due to internal resistance. After about 75 V is reached, a current snapback takes place: the collector current increases while the voltage decreases. At such high current levels, thermal instability happens, and the device is virtually destroyed. Illustrated in Figure 8.47 are the simulated I-V curves (in linear and log scales) with the self-heating model turned on.

The maximum lattice temperature in the simulated BJT is plotted versus collector voltage and current in Figure 8.48. When the self-heating model is turned off (Figure 8.49), we see a much smaller snapback effect, which highlights the importance of self-heating when simulating high-power devices. With the self-heating model turned on, the temperature rises rapidly after 600 K (327°C): even though the temperature reaches 1400 K in the simulation, a real device would have been destroyed well below this temperature. Readers should also keep in mind that devices are often destroyed by a few hot spots that have been subjected to positive feedback.

This behavior also explains why it is difficult to connect power BJTs in parallel. Since it is almost impossible to make the resistivity of the current paths exactly the same among devices connected in parallel, the one with lowest resistivity will always carry the highest current load and will be more

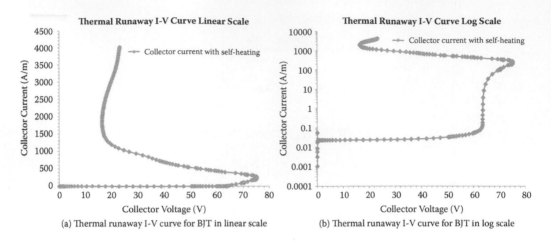

FIGURE 8.47   I-V curves (a) linear scale and (b) log scale of thermal runaway simulation.

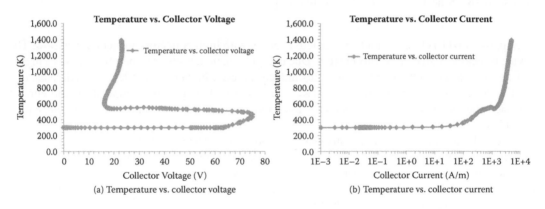

FIGURE 8.48   BJT maximum lattice temperature is plotted against (a) collector voltage and (b) collector current.

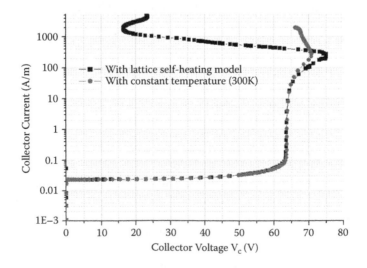

FIGURE 8.49   With lattice self-heating model turned on, snapback effect is more pronounced than with constant temperature.

likely to suffer thermal runaway as it is subject to higher temperatures. For this reason, unipolar devices such as an LDMOS are preferred when connecting power devices in parallel since carrier mobility degradation at elevated temperatures causes a negative temperature-current feedback loop instead of a positive one.

Figure 8.50 is a collection of plots showing the lattice temperature and current magnitude at different points of the I-V curve in Figure 8.47a. Figures 8.50a and 8.50b plot the lattice temperature

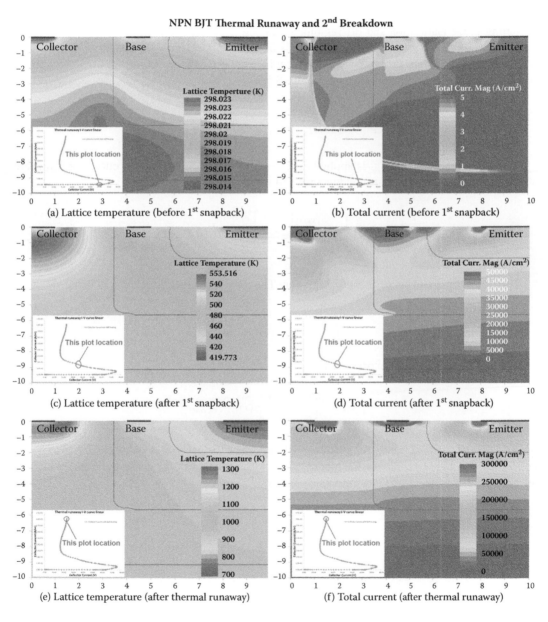

FIGURE 8.50 Lattice temperature and current magnitude plots under different collector voltage, collector current, and temperatures.

and total current before the first snapback (60 V), at room temperature (298 K), and with low current magnitude. Figures 8.50c and 8.50d are at the location after first snapback (V_c = 36 V), with elevated maximum temperature (550 K) and current magnitude. Figures 8.50e and 8.50f are located after the device's second breakdown and thermal runaway. At a very high simulated temperature (1400 K), this device should already have been destroyed; the current magnitude is also very high with relatively low collector voltage (22 V).

8.2.6 BJT SMALL SIGNAL MODEL AND CUTOFF FREQUENCY SIMULATION

A small signal model is useful for analog and power IC design. Unlike a power BJT that works in switching mode (saturation and off region), an analog BJT usually works as an amplifier (active region) and is used to amplify small AC signals. If we regard the NPN BJT as a two-port network with a grounded emitter, the base as the input, and the collector as the output, we can draw a simple small signal model for a BJT, as seen in Figure 8.51.

$$I_c = I_s \cdot \left[e^{\left(\frac{V_{BE}}{V_t} \right)} - 1 \right] \approx I_s \cdot e^{\left(\frac{V_{BE}}{V_t} \right)} \quad \text{when } V_{BE} \text{ is not close to 0} \tag{8.29}$$

When a small AC signal V_{BE} is applied to base emitter junction, a collector current $g_m V_{BE}$ is created where g_m is called the transconductance and can be expressed as [167]:

$$g_m = \frac{dI_c}{dV_{BE}} = \frac{d}{dV_{BE}} \left[I_s \cdot e^{\left(\frac{V_{BE}}{V_t} \right)} \right] = \frac{I_s \cdot e^{\left(\frac{V_{BE}}{V_t} \right)}}{V_t} = \frac{I_c}{V_t} = \frac{I_c}{26 mV \, (@300K)} \tag{8.30}$$

This is a simple and yet very useful equation. The input resistance is calculated using

$$\frac{1}{r_\pi} = \frac{dI_B}{dV_{BE}} = \frac{1}{\beta} \frac{dI_c}{dV_{BE}} = \frac{g_m}{\beta} \tag{8.31}$$

and the input capacitance is expressed as

$$C_\pi = \frac{dQ_F}{dV_{BE}} = \frac{d}{dV_{BE}} \tau_F I_c = \tau_F g_m \tag{8.32}$$

where τ_F is the forward transit time that can be viewed as the delay created as carriers travel from collector to emitter. To complete the model, a junction capacitance C_{JBE} needs to be added in parallel with the diffusion capacitance previously defined.

$$C_\pi = \tau_F g_m + C_{JBE} \tag{8.33}$$

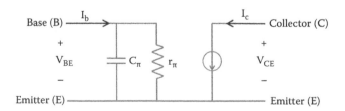

FIGURE 8.51 Simple small signal model for BJT.

A forward-biased base-emitter junction usually has a very small junction capacitance. So it is safe to omit the junction capacitance at this time.

8.2.6.1 Cutoff Frequency

The cutoff frequency is defined as the frequency at which the AC current gain β_{ac} ($= i_c/i_b$) falls to unity. It is related to τ_F as [167]:

$$f_t = \frac{1}{2\pi(\tau_F + \frac{C_{JBE}V_t}{I_c})} \approx \frac{1}{2\pi\tau_F} \tag{8.34}$$

Since for a forward-biased base-emitter PN junction, the junction capacitance is quite small, the cutoff frequency expression can be simplified and is approximately inversely proportional to forward transit time τ_F. Since τ_F can be measured only in experiment, f_t provides a useful way to calculate τ_F.

The cutoff frequency f_t and AC current gain are simulated under different collector current I_c levels, as shown in Figure 8.52. With a fixed collector to emitter voltage $V_{CE} = 1$ V, V_{BE} is increased from 0.6 V to 1.0 V, and I_c varies accordingly. Intuitively, we may think that with increasing I_c, forward transit time will be smaller and f_t will increase. In fact, this relationship is valid only for smaller I_c values. With high collector current, the base width is increased due to the Kirk effect, which increases the base transit time and thus the total forward transit time τ_F. f_t is observed to drop after reaching a maximum value. Designers tend to bias BJT at the peak of f_t to achieve better high-frequency results.

8.3 LDMOS

An LDMOS transistor is one of the most important power devices in power ICs. Most of the design efforts in power IC technology are concentrated on how to improve the performance of LDMOS devices. Like a vertical double-diffused MOS (VDMOS) transistor, the channel region is created by the difference between diffusions of body and source N+ (or P+) implants. This is a self-aligned process so that large lithographic feature sizes can be tolerated. Unlike a VDMOS where the silicon epi-layer is used as drain-drift region and a heavily doped substrate is used as drain contact, an

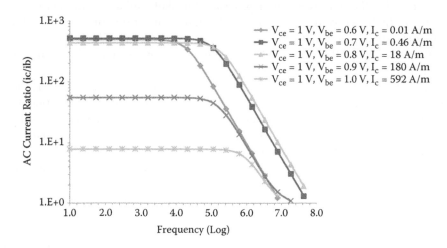

FIGURE 8.52 AC current gain and cutoff frequency comparison under a different collector current.

LDMOS drain contact is at the silicon surface (same side as source and body). For DMOS, lateral or vertical indicates whether current flows laterally or vertically from drain contact to source/body contact.

A well-designed LDMOS should have a high breakdown voltage, low on-state resistance, and a small gate to drain capacitance. This section focuses mostly on how to use a TCAD tool to improve the device breakdown voltage without jeopardizing the on-resistance.

For 0.35 μm technology or above, LOCOS (LOCal Oxidation of Silicon) is usually the isolation technology of choice. As power ICs move toward smaller feature sizes, STI is also becoming popular. In this section, we limit ourselves to LDMOS using STI technology to reflect this trend.

Eight examples of LDMOS (I to VIII), with slightly different structures, are first discussed. The section examines in detail electric field distribution and how to optimize various doping concentrations to improve the breakdown voltage and reduce on-resistance. For power IC applications, the breakdown voltage requirements for LDMOS range from 20 to 1000 V; power ICs that can sustain over 300 V are generally referred to as high-voltage IC (HVIC). They are usually used in applications that require high voltage but low current. In this section, LDMOS with breakdown voltage less than 150 V is discussed.

8.3.1 Breakdown Voltage Improvement

This section presents the LDMOS in a step-by-step fashion. Eight different LDMOS structures are introduced, with a focus on how to use TCAD tools to improve the breakdown voltage and reduce the on-state resistance.

8.3.1.1 Basic LDMOS Structure with N-Type Epitaxial Layer (LDMOS I)

LDMOS with a breakdown voltage less than 150 V are popular for power ICs in consumer electronics and transportation applications. A standard LDMOS structure without STI in the drain-drift region is introduced first. A schematic representation of an LDMOS with an N-type epitaxial layer (or an N-well) is illustrated in Figure 8.53a. The simulation cell pitch for this LDMOS is set to be 10 μm with a depth of 20 μm. (To get a larger view, the depth of plotted structures is truncated to 10 μm, but the simulation size is 20 μm deep.) The source and body terminals are connected together for enhanced breakdown capability. The poly gate is 0.4 μm thick by 2 μm long, is phosphorus-doped at $1E + 19$ cm^{-3}, extends over the P-body region, and covers part of the drain-drift region. Since electrons accumulate underneath the gate when a positive voltage is applied to that electrode, this extension may help reduce the on-resistance. The gate oxide thickness is chosen to be 200 Å to reach an appropriate threshold voltage while maintaining oxide robustness under high gate voltages. The channel length is about 0.8 μm. The total drain length (measured from P-body edge to device right edge) is about 7.5 μm. The P-body has a depth of about 2.75 μm. The N+ and P+ regions (NSD and PSD) each have a width of 0.75 μm; no substrate contact is included in this structure.

Figure 8.53 also illustrates some results from TCAD simulation. Figure 8.53b is the material plot. Three materials—silicon, polysilicon, and oxide—are included in this simulation. Please note that in the simulation contacts are simply boundary conditions and not a type of material; thus, they are plotted for illustration purposes only. The net doping is shown in Figure 8.53c, and Figure 8.53d is the surface plot (with absolute value) of the net doping.

For this simple LDMOS structure, we use N-epi with constant doping as the drain. Suppose we want to design an LDMOS with breakdown voltage greater than 100 V. We can start by checking the single-sided parallel plane breakdown voltage listed in Table 6.6. As a starting point, an N-epi layer doping concentration of $5E + 15$ cm^{-3} may be appropriate, since it has a P+/N breakdown voltage close to 100 V. The N-epi layer will act as the drain-drift region, which will support most of the breakdown voltage. (Part of the breakdown voltage is supported within P-body due to graded junction.) The simulation parameters are listed in Table 8.6.

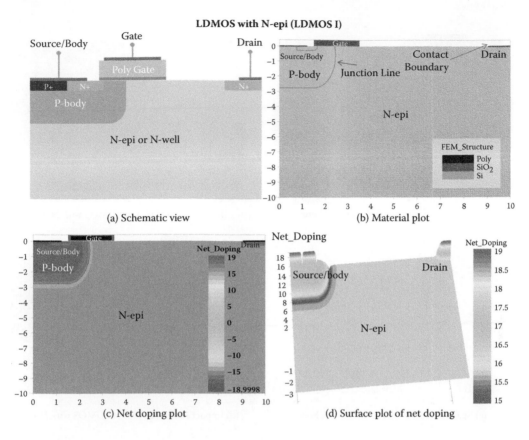

FIGURE 8.53   Schematic reprentation of an LDMOS with N-epi.

The breakdown simulation is performed with source/body contact and gate contact grounded while the potential at the drain contact is increased until leakage current increases rapidly (avalanche breakdown). The substrate contact is not included in this simulation. The simulation results after device breakdown are shown in Figure 8.54. From Figure 8.54a we see the equipotential lines after breakdown extend primarily within the N-epi region, which means that the N-epi supports most of the voltage applied to the drain. Unfortunately, the simulated breakdown voltage is found

TABLE 8.6
Simulation Parameters for LDMOS I

Parameters for Simulation	Value
X range	0 to 10 μm
Y range	0 to −20 μm
N-epi	$5E + 15$ cm^{-3}
P-body max concentration	$2E + 17$ cm^{-3}
P-body location (x1, x2, dx1, dx2) Gaussian	X = (0, 1.5, 0.35, 0.35); y = (−0.6, −0.6, 0.8, 0.8)
Oxide thickness	200 Å
Poly gate thickness	0.4 μm
Poly gate length	2 μm
Poly gate doping	Phosphorus, $1E + 19$ cm^{-3}
Channel length	0.8 μm (approx.)

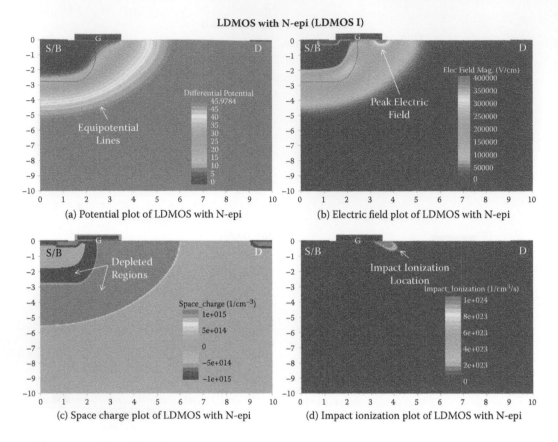

FIGURE 8.54 (See color insert) Simulation results after device breakdown for LDMOS with N-epi (LDMOS I) with (a) potential, (b) electric field, (c) space charge, and (d) impact ionization.

to be just about 46 V, much less than we expected of 100 V. From our earlier simplified PN junction model we know that the curvature of the junction degrades the breakdown voltage to about 80% of that of parallel planes, but this effect alone does not justify our simulation results.

If we take a further look at the simulation, we find that in the electric field plot of Figure 8.54b the peak field is right underneath the gate edge, which also corresponds to a peak of the impact ionization rate in Figure 8.54d. This strongly suggests that the critical field close to the gate edge is triggering impact ionization and reducing the breakdown voltage of our device.

This is easily understood from an enlarged view of the potential plot in Figure 8.55a. Since the poly gate is grounded in the breakdown simulation, the potential changes abruptly from 0 V at the gate edge to a high potential (> 30 V) in the silicon, which leads to a peak in the electric field close to the gate corner. Figure 8.55b is the 3D surface plot of electric field and only the silicon is plotted for clarity.

The breakdown voltage I-V curve is shown in Figure 8.56.

8.3.1.2 LDMOS with Shielding Plate (LDMOS II)

A simple method to alleviate the high field around the gate corner is to use the shielding effect from a field plate. A schematic view is given in Figure 8.57a: the metal layer connecting to the source/body is extended beyond the edge of the gate to relieve the electric field in the silicon underneath the gate corner. The same LDMOS device (LDMOS I) with added ILD and metal layer are simulated with the source field plate extended beyond the gate edge (LDMOS II). The thickness of ILD oxide is 0.7 μm, and that of the aluminum is 0.25 μm.

LDMOS with N-epi (LDMOS I)

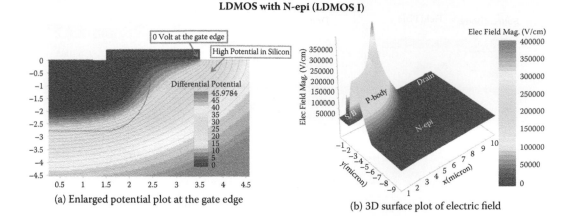

(a) Enlarged potential plot at the gate edge

(b) 3D surface plot of electric field

FIGURE 8.55 (a) Enlarged potential plot and (b) 3D electric field surface plot of LDMOS with N-epi (LDMOS I).

Figure 8.57b shows the simulated result of a material plot. Four materials are used—silicon, poly-silicon (poly), SiO_2 (oxide), and aluminum (al)—with aluminum metal plate extending over the gate. The length of the metal gate will affect the breakdown voltage, which is discussed later. A field plate length of 6 μm (measured from the simulation edge) was chosen as a starting point. Figures 8.57c and 8.57d are an the net doping plots. The simulation parameters are listed in Table 8.7.

A breakdown simulation is performed by applying a high voltage on the drain while keeping the source/body and gate contacts grounded. Figure 8.58 shows the simulation results after device breakdown. (Silicon area below Y = –10 μm is truncated for enlarged view, since that area has little information of interest.) The equipotential lines illustrated in Figure 8.58a show a dramatic improvement compared with those of LDMOS I in Figure 8.54a: the field plate has successfully extended the equipotential lines (Figure 8.58b), and we also see evidence of an increased depletion width in Figure 8.58c. With the field plate, the peak electric field has been successfully shifted away from the gate edge and into the oxide ILD; oxide has a much higher (> 10x) critical electric field than silicon, so it is unlikely to provoke the avalanche effect. Additionally, the impact ionization (Figure 8.58d) now mostly takes place at the PN junction in the bulk rather than at the surface. With

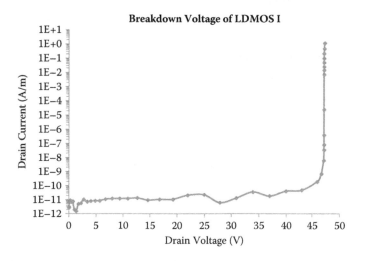

FIGURE 8.56 Breakdown voltage I-V plot of the simple LDMOS (LDMOS I).

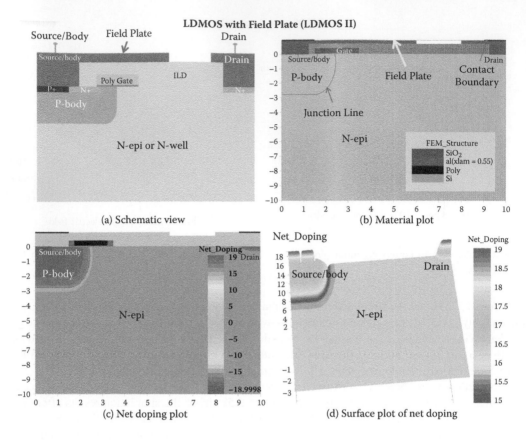

FIGURE 8.57 LDMOS structure with field plate (LDMOS II).

the addition of the field plate, the breakdown voltage has been increased from 47 V to 89 V, but we are still shy of our design goal of 100 V.

The surface plot of an electric field for LDMOS II is shown in Figure 8.59 compared with LDMOS I. With a field plate, a second peak is observed at the PN junction in the silicon bulk. As in previous surface plots for LDMOS I, the electric field is shown only in the silicon material, and Y range is from −10 μm to 0 μm.

TABLE 8.7
Simulation Parameters for LDMOS with Field Plate

Parameters for Simulation	Value
X range	0 to 10 μm
Y range	0 to −20 μm
N-epi	5E + 15 cm^{-3}
P-body max concentration	2E + 17 cm^{-3}
P-body location (x1, x2, dx1, dx2) Gaussian	X = (0, 1.5, 0.35, 0.35); y = (−0.6, −0.6, 0.8, 0.8)
Oxide thickness	200 Å
Poly gate thickness	0.4 μm
Poly gate length	2 μm
Channel length	0.8 μm (approx.)
Interlayer dielectric thickness	0.7 μm
Source metal shield plate length	6 μm (from source) or 2.5 μm (from gate edge)

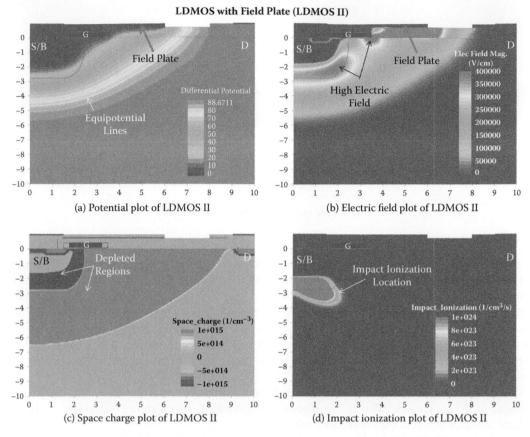

FIGURE 8.58 (See color insert) Simulation results after device breakdown for LDMOS with N-epi and field plate (LDMOS II) with (a) potential, (b) electric field, (c) space charge, and (d) impact ionization.

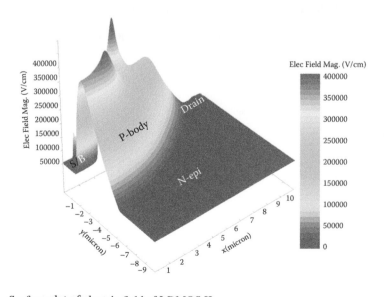

FIGURE 8.59 Surface plot of electric field of LDMOS II.

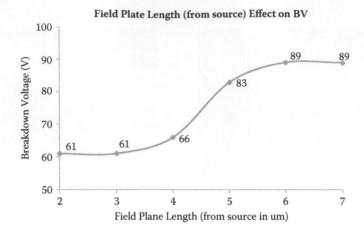

FIGURE 8.60 Breakdown voltage variation with different field plate lengths from the source side.

Different field plate lengths will change the breakdown voltage, so a proper design is necessary to ensure optimal results. The breakdown voltage variation with different field plate lengths is simulated and shown in Figure 8.60. When the field plate is short (e.g., 3 μm measured from the source edge), breakdown voltage is only 61 V; a sharp increase of breakdown voltage is seen when the field shielding plate length is longer than 4 μm because the field plate extends beyond the gate edge, which diverts the peak field into the ILD oxide. With a field plate length of 6 μm, the breakdown voltage increases to 89 V and then saturates; no further increase of breakdown voltage is observed with longer field plates.

Another possible approach for the field plate is to connect it from the drain side (going left) instead of the source side. This is investigated in Figure 8.61, which shows that longer drain field plates are detrimental to the breakdown voltage.

This unusual behavior can be explained by noting that when the field plate is connected to the grounded source, its potential is zero, which flattens out the equipotential lines underneath the plate. However, when the field plate is connected to the drain side, it acquires a high voltage from the drain, which tends to push back the equipotential lines underneath it and makes them more crowded. A comparison of the equipotential lines spacing for source/drain field plates of different lengths is shown in Figure 8.62. The longer field plate from the source side yields the highest

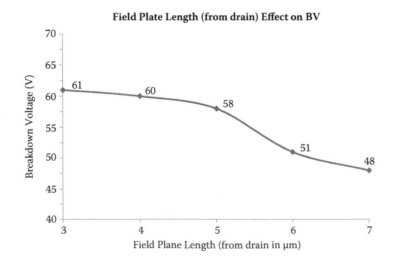

FIGURE 8.61 Breakdown voltage variation with different field plate lengths from the drain side.

LDMOS with Field Plate (LDMOS II)

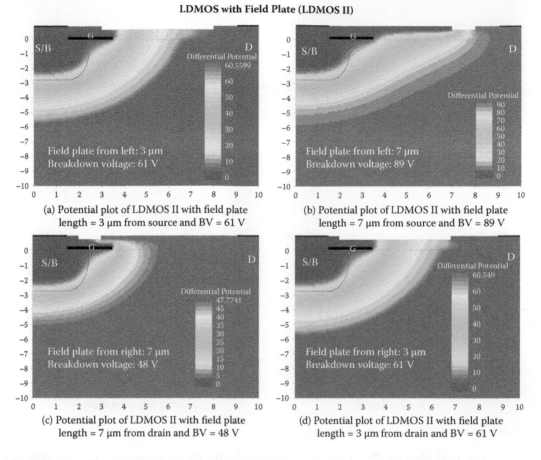

(a) Potential plot of LDMOS II with field plate
length = 3 μm from source and BV = 61 V

(b) Potential plot of LDMOS II with field plate
length = 7 μm from source and BV = 89 V

(c) Potential plot of LDMOS II with field plate
length = 7 μm from drain and BV = 48 V

(d) Potential plot of LDMOS II with field plate
length = 3 μm from drain and BV = 61 V

FIGURE 8.62 Comparison of equipotential lines for field plates connected from the source side and drain side: (a) 3 μm from the source (left) side; (b) 7 μm from the source (left) side; (c) 7 μm from the drain (right) side; (d) 3 μm from the drain (right) side.

breakdown value (89 V), while the same plate from the drain side dramatically lowers the breakdown voltage to 48 V. Short field plates on either side yield a similar breakdown voltage of 61 V, as shown in Figures 8.62a and 8.62d.

8.3.1.3   LDMOS with STI in the Drain-Drift Region (LDMOS III)

A common way to improve breakdown voltage is to divert the peak electric field into the oxide; depending on process technology, either LOCOS or STI is used for this purpose. Since the growing trend is to use STI to replace LOCOS in most process technology below 0.25 μm, we discuss only STI in this section.

The STI trench is located in the drain-drift region but should also have a certain overlap with poly gate. A schematic of our sample structure is shown in Figure 8.63a (LDMOS III) with a material plot in Figure 8.63b. The oxide trench (STI) is within the drain-drift region with a small overlap with the gate (0.5 μm); an edge slope (3 degree angle) is also used to mimic the real manufactured shape. Net doping plots are given in Figures 8.63c and 8.63d. For a 3D surface plot, Y ranges from −10 μm to 0 μm (the simulated device has a total depth of 20 μm and poly gate is located above 0 μm). The simulation parameters for this LDMOS are listed in Table 8.8.

The simulated equipotential lines at breakdown are shown in Figure 8.64a. The breakdown voltage of the LDMOS with STI is about 87 V, which is close to the value achieved for LDMOS II

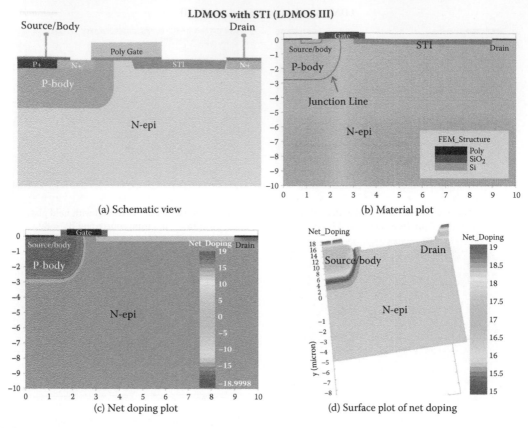

FIGURE 8.63   LDMOS structure with STI (LDMOS III).

using a long field plate on the source side. Compared with the equipotential lines in LDMOS I (Figure 8.54a), using STI in the drain enables a wider spread of the equipotential lines and a more uniform electric field along the PN junction (Figure 8.64b). The peak electric field underneath the gate's right edge is now within the STI oxide material; as before, the oxide has a much higher critical electric field, so it does not trigger impact ionization. Figure 8.64c shows how the space charge

TABLE 8.8
Simulation Parameters for LDMOS with STI

Parameters for Simulation	Value
X range	0 to 10 μm
Y range	0 to −20 μm
N-epi	5E + 15 cm^{-3}
P-body max concentration	2E + 17 cm^{-3}
P-body location (x1, x2, dx1, dx2) Gaussian	X = (0, 1.5, 0.35, 0.35); y = (−0.6, −0.6, 0.8, 0.8)
Oxide thickness	200 Å
Poly gate thickness	0.4 μm
Poly gate length	2 μm
Channel length	0.75 μm (approx.)
STI depth	0.35 μm with etch angle of 3 degrees
Overlap of STI with gate	0.5 μm
STI length	6 μm (x.from = 3, x.to = 9)

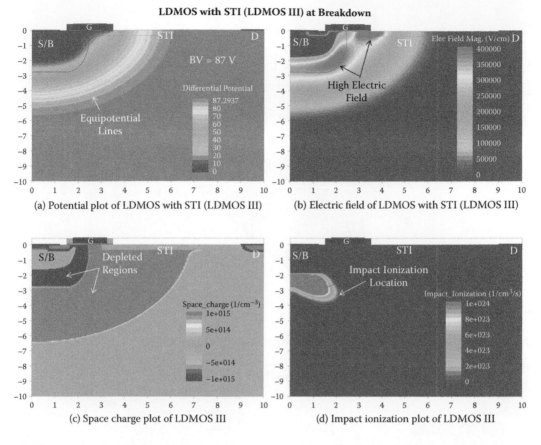

LDMOS with STI (LDMOS III) at Breakdown

(a) Potential plot of LDMOS with STI (LDMOS III)

(b) Electric field of LDMOS with STI (LDMOS III)

(c) Space charge plot of LDMOS III

(d) Impact ionization plot of LDMOS III

FIGURE 8.64 (See color insert) Simulation results at device breakdown for LDMOS with N-epi and STI (LDMOS III) with (a) potential, (b) electric field, (c) space charge, and (d) impact ionization.

extends into the drift region. As seen in Figure 8.64d, the region of maximum impact ionization for this structure has been shifted away from the silicon surface and into the bulk of the device.

For a better view of the electric field distribution within the device, a 3D surface plot is shown in Figure 8.65. Compared with the LDMOS I case (Figure 8.55b), the added STI region has increased the electric field in the bulk of the device and the PN junction, making the electric field more uniform and enabling a higher breakdown voltage.

8.3.1.4 LDMOS with Both STI Oxide and Field Plate (LDMOS IV)

The next step of our design is to combine the concepts of LDMOS II (source field plate) and LDMOS III (STI); both seem to increase the breakdown voltage, but neither method on its own has met our design target value of 100 V.

A schematic view of this structure (LDMOS IV) is shown in Figure 8.66a, along with the material plot in Figure 8.66b. The field plate length is 6 µm just as in LDMOS II, while the oxide trench (STI) has the same physical parameters as in LDMOS III. Figures 8.66c and 8.66d illustrate the net doping plot of LDMOS IV. As with our previous LDMOS structures, the substrate contact is not included. The simulation parameters for our LDMOS IV structure are listed in Table 8.9.

The equipotential lines for this structure are plotted in Figure 8.67a; the breakdown voltage is only 88 V, and no significant improvement is seen over either LDMOS II or III. In fact, both field plate and STI oxide methods employ the same idea of diverting the peak electric field away from silicon at the gate edge and into an oxide layer (either STI oxide or ILD oxide). There seems to be

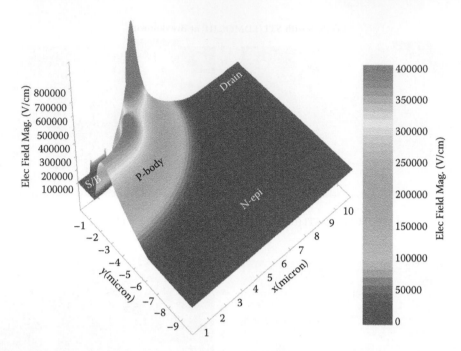

FIGURE 8.65 Surface plot of electric field of LDMOS with STI (LDMOS III).

LDMOS with STI and Field Plate (LDMOS IV)

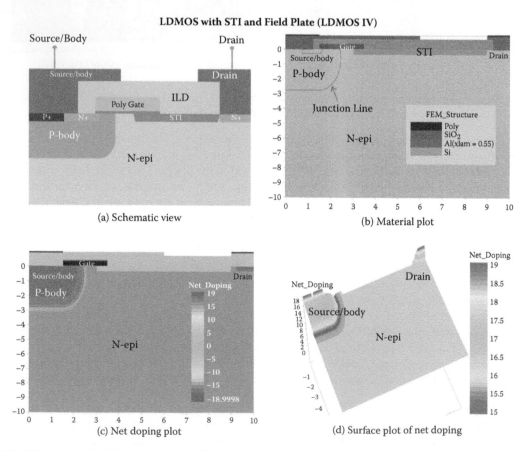

FIGURE 8.66 LDMOS structure with STI and field plate (LDMOS IV).

TABLE 8.9

Simulation Parameters for LDMOS with STI and Field Plate (LDMOS IV)

Parameters for Simulation	Value
X range	0 to 10 μm
Y range	0 to –20 μm
N-epi	5E + 15 cm^{-3}
P-body max concentration	2E + 17 cm^{-3}
P-body location (x1, x2, dx1, dx2) Gaussian	X = (0, 1.5, 0.35, 0.35); y = (–0.6, –0.6, 0.8, 0.8)
Oxide thickness	200 Å
Poly gate thickness	0.4 μm
Poly gate length	2 μm
Channel length	0.8 μm (approx.)
STI depth	0.35 μm with etch angle of 3 degrees
Overlap of STI with gate	0.5 μm
STI length	6 μm (x.from = 3, x.to = 9)
Interlayer dielectric thickness	0.7 μm
Source metal shield plate length	6 μm (from source) or 2.5 μm (from gate edge)

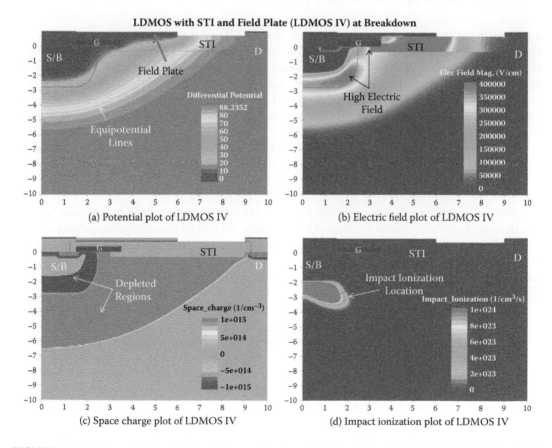

FIGURE 8.67 **(See color insert)** Simulation results after device breakdown for LDMOS with STI and field plate (LDMOS IV) with (a) potential, (b) electric field, (c) space charge, and (d) impact ionization.

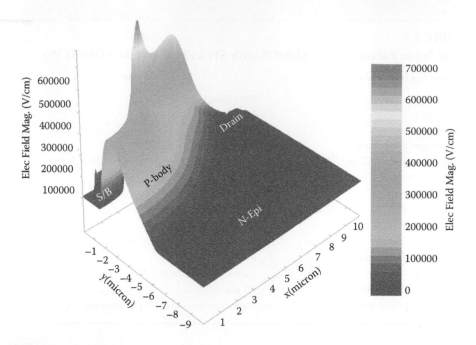

FIGURE 8.68   Surface plot of electric field of LDMOS IV.

a limit to this idea, so thinking outside the box is required to get further improvements. As usual, Figure 8.67b shows the electric field distribution, which is similar to that of LDMOS II and LDMOS III: electric field peaks in the oxide; and increased electric field levels in the silicon bulk at the PN junction. Figure 8.67c is the space charge plot, and Figure 8.67d is the impact ionization plot. The latter happens at the same location as LDMOS II and LDMOS III: as before, moving the impact ionization away from the surface has the added benefit of lowering the gate oxide stress and improving the long-term reliability of the oxide.

The surface plot of the electric field for LDMOS IV is shown in Figure 8.68. We note the similarities between LDMOS II, III, and IV in terms of the electric field distribution within the device.

8.3.1.5   LDMOS with RESURF from P-epi Layer (LDMOS V)

We previously observed that diverting the peak electric field from the silicon to the oxide is effective in improving the breakdown voltage. Unfortunately, the breakdown voltage we achieved is still below our target value of 100 V. In fact, a basic rule of thumb in the power semiconductor industry is to design with a 20 to 30% safety margin in breakdown voltage to account for device fabrication tolerance and long-term degradation. So for a 100 V design we actually need at least 120 to 130 V of breakdown voltage.

If we take a look at the 3D electric field surface plots of all the LDMOS that we have discussed so far, they all feature a triangular-shaped field close to the source/body region, so a large portion of silicon that can support the drain voltage is wasted. From Poisson's equation, we can associate the breakdown voltage with the volume under the 3D surface plots of the electric field: a larger volume leads to higher breakdown voltage. Unfortunately, a *critical electric field* limits how high the electric field level can go. Therefore, the only way to increase the breakdown voltage is to make the surface electric field as uniform as possible so that this peak value can contribute to the integral over a wider range.

To achieve this, we can employ the concept we explored for the PN junction diode with NBL. In that device, the breakdown voltage was greatly improved when merging the depleted regions of the N-region with that of the NBL; the potential gained by of the NBL further depletes the silicon

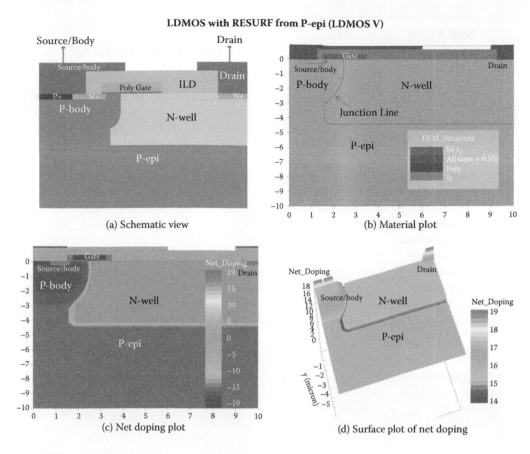

FIGURE 8.69 LDMOS structure with P-epi (LDMOS V).

area above and below, creating a more uniform distribution of electric field. If we take a look at our LDMOS, we find that we can create a similar effect: since we are using an N-epi (or N-well) rather than P-epi (or P-well), a PBL (P-type buried layer) could be used under the P-body so that the interaction of the P-body and PBL depletion regions creates a similar effect as a PN junction diode with NBL.

However, in real manufacturing P-epi is more prevalent than N-epi, and the drain-drift region is created by implantation and diffusion of an N-well. The P-body is typically connected with P-epi in many LDMOS designs: this means the P-epi region under the N-well can gain a low potential from either a grounded substrate or the P-body. We now choose to adjust our design; instead of using PBL, P-epi is used to further deplete an N-well located above it. Our new structure will be called an LDMOS with P-epi (LDMOS V), a schematic of which is given in Figure 8.69a.

Figure 8.69b is the material plot of LDMOS V with added junction lines. For this device structure, we include the same field plate in the simulation as that of LDMOS II (6 μm from the source side), but we do not include the STI from LDMOS III. From the net doping plot of Figures 8.69c and 8.69d we can see the connection between P-body and P-epi.

Table 8.10 lists the simulation parameters of LDMOS V. In this design, doping concentration in the P-epi is chosen to be $1E + 15$ cm^{-3}, and the N-well has a doping concentration of $5E + 15$ cm^{-3}, the same doping concentration as the N-epi we previously used; all other parameters are kept the same.

The simulation is performed with source/body and gate contacts grounded, while the drain voltage increased until breakdown. With the same N-well doping of $5E + 15$ cm^{-3} as before, the breakdown voltage now increases by more than 50% to a value of 147 V. Figure 8.70a is the potential plot, which shows good spacing of the equipotential lines, while the electric field plot is given in

TABLE 8.10

Simulation Parameters for LDMOS with Field Plate and P-epi RESURF (LDMOS V)

Parameters for Simulation	Value
X range	0 to 10 μm
Y range	0 to −20 μm
P-epi	$1E + 15$ cm^{-3}
P-body max concentration	$2E + 17$ cm^{-3}
P-body location (x1, x2, dx1, dx2) Gaussian	x = (0, 1.5, 0.35, 0.35); y = (−0.6, −0.6, 0.8, 0.8)
N-well max concentration	$5E + 15$ cm^{-3}
N-well location (x1, x2, dx1, dx2) Gaussian	x = (2, 10, 0.25, 0.25); y = (−6, 0, 0.25, 0.25)
Oxide thickness	200 Å
Poly gate thickness	0.4 μm
Poly gate length	2 μm
Channel length	0.8 μm (approx.)
Interlayer dielectric thickness	0.7 μm
Source metal shield plate length	6 μm (from source) or 2.5 μm (from gate edge)

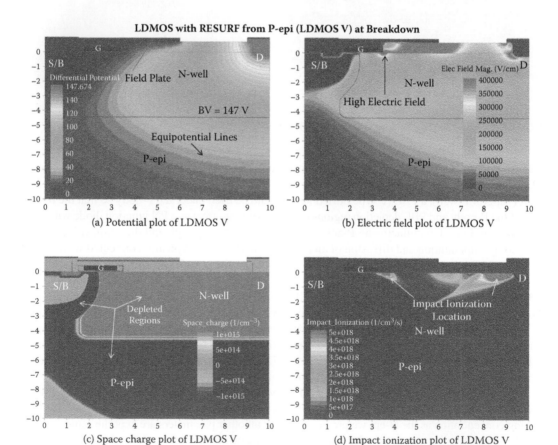

FIGURE 8.70 (See color insert) Simulation results at device breakdown for LDMOS with P-epi RESURF (LDMOS V) with (a) potential, (b) electric field, (c) space charge, and (d) impact ionization.

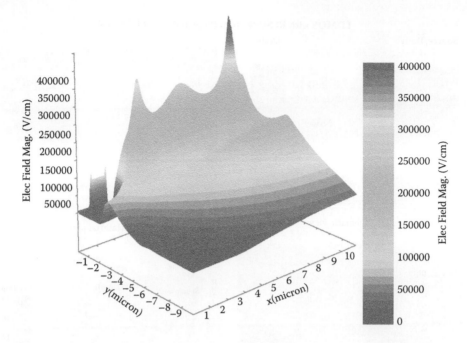

FIGURE 8.71 Surface plot of electric field of LDMOS V.

Figure 8.70b. A more uniform electric field within the N-well is established, and an additional depletion region is formed in the P-epi layer, as is illustrated by the space charge chart of Figure 8.70c. The impact ionization location has been shifted to the drain side, as seen in Figure 8.70d.

The depletion action from the P-epi is regarded as a RESURF effect; research in this field dates back to at least 1979 [181]. Here, the horizontal depletion from the P-body/N-well junction is coupled with a vertical depletion from the P-epi/N-well junction to provide a two-dimensional depletion effect, which allows the electric field to be more uniformly distributed across the drain-drift region. If we take a close look at the 3D surface plot of the electric field (Figure 8.71) and compare it with previous ones with N-epi only, LDMOS V shows a much wider and more uniform electric field distribution.

From the space charge plot in Figure 8.70c, we know that the N-well region is completely depleted by the combined effect of the vertical and horizontal depletions. From the electric field plot in Figure 8.70b and impact ionization plot in Figure 8.70d, we also note that peak field is close to the silicon surface.

8.3.1.6 LDMOS with P-epi RESURF and STI (LDMOS VI)

Our previous design shows a marked improvement but suffers from a flaw similar to that of LDMOS I in that the peak electric field is close to the silicon surface. However, we also know from our earlier LDMOS III design that we can use STI to shunt the peak field away from the silicon and into the oxide to further increase the breakdown voltage. This concept will be the key in the next iteration of our design, LDMOS VI, a schematic view of which is shown in Figure 8.72a.

Figure 8.72b is the material plot for this structure. It is the same as LDMOS V except for an extra STI region (identical to that of LDMOS III) in the drain. Figures 8.72c and 8.72d show the net doping plots.

The list of simulation parameters is given in Table 8.11 and is consistent with our previous designs.

As illustrated in Figure 8.73a, the breakdown voltage has again improved dramatically from 147 V to almost 180 V thanks to the diversion of the peak electric field into the oxide layer (Figures 8.73b). The N-well is fully depleted, and the impact ionization takes place near the STI edges at the drain side (Figures 8.73c and 8.73d). A 3D surface plot of the electric field is given in Figure 8.74.

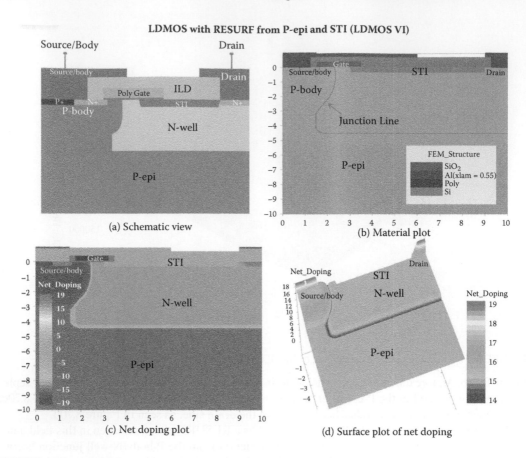

FIGURE 8.72 LDMOS structure with P-epi and STI (LDMOS VI).

TABLE 8.11
Simulation Parameters for LDMOS with P-epi RESURF and STI (LDMOS VI)

Parameters for Simulation	Value
X range	0 to 10 μm
Y range	0 to −20 μm
P-epi	1E + 15 cm^{-3}
P-body max concentration	2E + 17 cm^{-3}
P-body location (x1, x2, dx1, dx2) Gaussian	x = (0, 1.5, 0.35, 0.35); y = (−0.6, −0.6, 0.8, 0.8)
N-well max concentration	5E + 15 cm^{-3}
N-well location (x1, x2, dx1, dx2) Gaussian	x = (2, 10, 0.25, 0.25); y = (−6, 0, 0.25, 0.25)
Oxide thickness	200 Å
Poly gate thickness	0.4 μm
Poly gate length	2 μm
Channel length	0.8 μm (approx.)
STI depth	0.35 μm with etch angle of 3 degrees
Overlap of STI with gate	0.5 μm
STI length	6 μm (x.from = 3, x.to = 9)
Interlayer dielectric thickness	0.7 μm
Source metal shield plate length	6 μm (from source) or 2.5 μm (from gate edge)

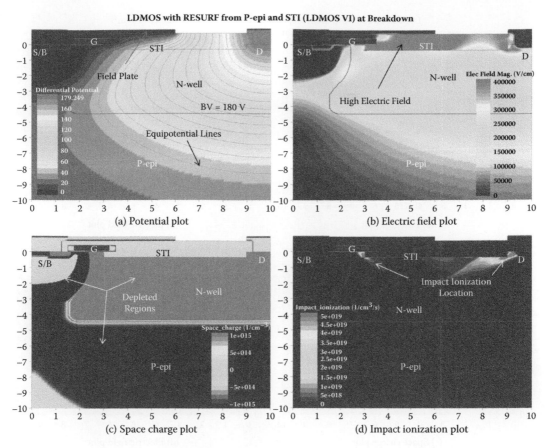

FIGURE 8.73 (See color insert) Simulation results at device breakdown for LDMOS with P-epi RESURF and STI (LDMOS VI) with (a) potential, (b) electric field, (c) space charge, and (d) impact ionization.

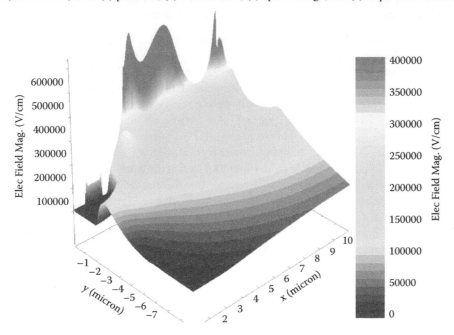

FIGURE 8.74 Surface plot of electric field of LDMOS VI.

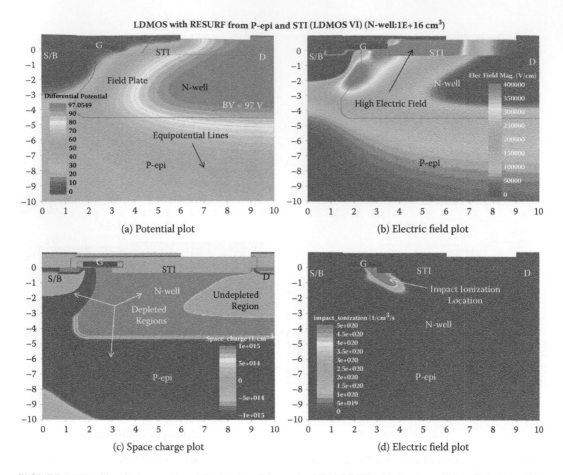

FIGURE 8.75 Simulation results at device breakdown for LDMOS VI with increased N-well doping (1E + 16 cm-3) with (a) potential, (b) electric field, (c) space charge, and (d) impact ionization.

The high breakdown voltage (180 V) seems like overkill for a device designed to handle only 100 V (130 V with safety margin). However, decreasing the on-resistance to achieve better power efficiency often comes at the expense of a lower breakdown voltage, so the large value we achieved gives us a lot of breathing room to make adjustments. To reduce the on-resistance, we need to increase the doping concentration in the N-well; unfortunately, increasing the N-well doping from $5E + 15$ cm^{-3} to $1E + 16$ cm^{-3} while keeping all other parameters the same brings down the breakdown voltage to 97 V.

Shown in Figure 8.75 are the simulation results of LDMOS VI with increased N-well doping concentration ($1E + 16$ cm^{-3}). We note from Figure 8.75a that the equipotential lines are not spreading fully inside the N-well. From Figure 8.75b we see that the electric field is crowded at the source side rather than at the drain side. From the space charge plot of Figure 8.75c, a large portion of the N-well region is not depleted, and the impact ionization also seems to take place close to the source side. The reason behind the degradation in breakdown voltage now seems straightforward: the doping concentration in the N-well is too high to be fully depleted by the combined action of the P-body and P-epi regions.

8.3.1.7 LDMOS with Double RESURF (LDMOS VII)

By increasing the N-well doping, we are able to reduce the on-resistance, but the breakdown voltage is degraded and drops below our requirements. We need to think of ways to improve the breakdown voltage with the increase in N-well doping.

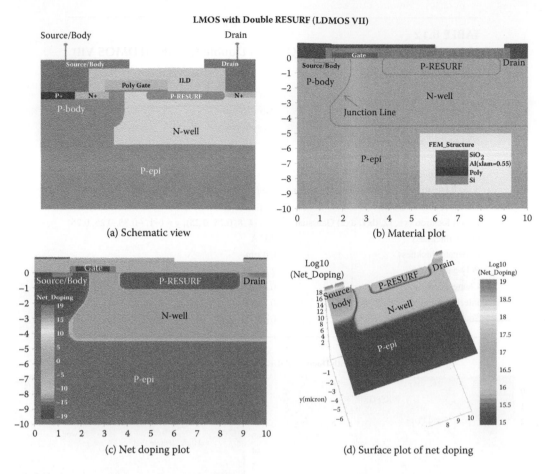

FIGURE 8.76 LDMOS with double RESURF structure and net doping plot. (a) Schematic view (b) Material plot (c) Net doping plot (d) Surface plot of net doping.

The improvements we achieved in LDMOS V were due to the RESURF effect to deplete the N-well from below. Since this is no longer sufficient to fully deplete the region once the N-well doping is increased, a natural development of this idea is to use an additional RESURF region to deplete the N-well from above. This concept is called *double RESURF*, and its origins can be traced back to 1982 [182]. In this book, we call this additional P-region P-RESURF, and it is left floating (not connected to any contacts or the source/body region). A schematic view of this structure is shown in Figure 8.76a, material plot is shown in Figure 8.76b, while Figures 8.76c and d illustrate the net doping.

Note that in this structure (LDMOS VII) we get rid of the STI region introduced in LDMOS VI for two reasons. First, in real practice the STI process may not be compatible with the double-RESURF P implant step; second, the STI in the drain will increase the total on-resistance. The main differences between this structure and LDMOS V are the doping concentration of the N-well (increased to $1E + 16$ cm^{-3}) and the additional P-RESURF region; the channel length is also reduced to 0.7 μm due to the increased doping in the N-well. The simulation parameters are listed in Table 8.12.

The equipotential lines are shown in Figure 8.77a. With double RESURF, the breakdown voltage increases from 97 V to 157 V with an N-well doping concentration of $1E + 16$ cm^{-3}. From the electric field plot of Figure 8.77b, we see multiple high field locations within the device, which is desirable so that the electric field does not concentrate too much in one location. The N-well and P-RESURF regions are fully depleted, as seen from the space charge plot in Figure 8.77c. Impact ionization takes place at the gate edge and at the drain side, as shown in Figure 8.77d.

TABLE 8.12

Simulation Parameters for LDMOS with Double RESURF (LDMOS VII)

Parameters for Simulation	Value
X range	0 to 10 μm
Y range	0 to −20 μm
P-epi	$1E + 15$ cm^{-3}
P-body max concentration	$2E + 17$ cm^{-3}
P-body location (x1, x2, dx1, dx2) Gaussian	x = (0, 1.5, 0.35, 0.35); y = (−0.6, −0.6, 0.8, 0.8)
N-well max concentration	$1E + 16$ cm^{-3}
N-well location (x1, x2, dx1, dx2) Gaussian	x = (2, 10, 0.25, 0.25); y = (−6, 0, 0.25, 0.25)
P-well double RESURF	$1.5E + 16$ cm^{-3}
P-well location (x1, x2, dx1, dx2) Gaussian	X = (4, 8, 0.25, 0.25); y = (−1, −0.35, 0.25, 0.25)
Oxide thickness	200 Å
Poly gate thickness	0.4 μm
Poly gate length	2 μm
Channel length	0.7 μm (approx.)
Interlayer dielectric thickness	0.7 μm
Source metal shield plate length	6 μm (from source) or 2.5 μm (from gate edge)

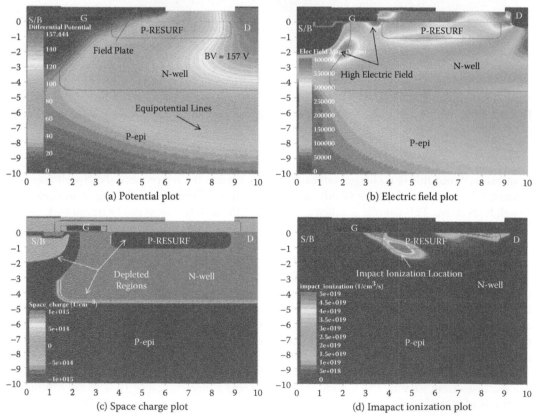

LDMOS with Double RESURF (LDMOS VII) at Breakdown

(a) Potential plot

(b) Electric field plot

(c) Space charge plot

(d) Imapact ionization plot

FIGURE 8.77 **(See color insert)** Simulation results after device breakdown for LDMOS VII with increased N-well doping ($1E + 16$ cm^{-3}) and double RESURF: (a) Potential; (b) Electric field, (c) Space charge; (d) Impact ionization.

The key to design the right P-RESURF region is to achieve dose balance in both the P and N-regions: both regions should be completely depleted to achieve optimal results. It is often difficult to make calculations by hand to achieve the right dose balance since the N-well is depleted by a total of three regions: P-body; P-epi; and P-RESURF. TCAD simulation with design of experiments (DoE) is an easy method to achieve the right dose and doping concentrations. In this LDMOS structure, it has been found that for a P-RESURF maximum doping of 1.5E + 16 cm^{-3}, a dose balance can be achieved.

The downside of this double RESURF technology is the addition of the P-RESURF, which will increase the on-resistance and introduce process complexity. In real manufacturing, even a slight process variation may lead to large degradation of breakdown voltage. Also, in practice, the P-RESURF layer is most likely a special purpose layer not shared with other process steps, so an additional mask is required.

A 3D surface plot of electric field is shown in Figure 8.78. We see multiple electric field peaks in the silicon region, which is beneficial for high breakdown voltage.

If we try to get a further reduction of the on-resistance to account for the presence of this additional P-RESURF region, we again encounter the problem of an undepleted N-region. For example, if we merely increase the N-well doping from 1E + 16 cm^{-3} to 1.5E + 16 cm^{-3} while keeping all other parameters the same, the breakdown voltage will drop to 79 V, as illustrated in Figure 8.79.

In Figure 8.79a the equipotential lines are crowded close to the P-body; Figure 8.79b shows the peak electric field locations; from the space charge plot of Figure 8.79c we can easily see that a large portion of the N-well is not depleted. The undepleted area is not able to support the drain voltage, which causes the breakdown voltage to drop. Figure 8.79d shows the location of impact ionization.

As we previously observed, the key to double RESURF is that the N-well is simultaneously being depleted from three sides: the P-body; P-epi; and P-RESURF. With higher N-well doping, the undepleted area is close to the drain, which means the depletion regions from P-epi and P-RESURF are not able to meet each other. If the higher N-well doping is a necessity, then the obvious solution that allows these shallower depletion regions to meet is to reduce the N-well depth.

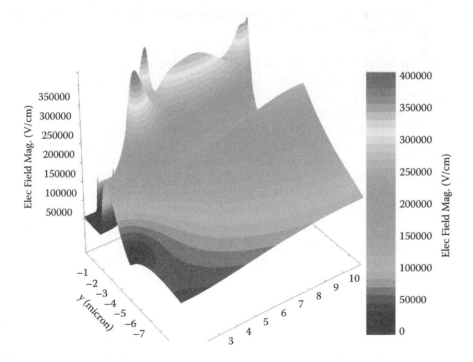

FIGURE 8.78 Electric field surface plot of LDMOS with double RESURF (LDMOS VII).

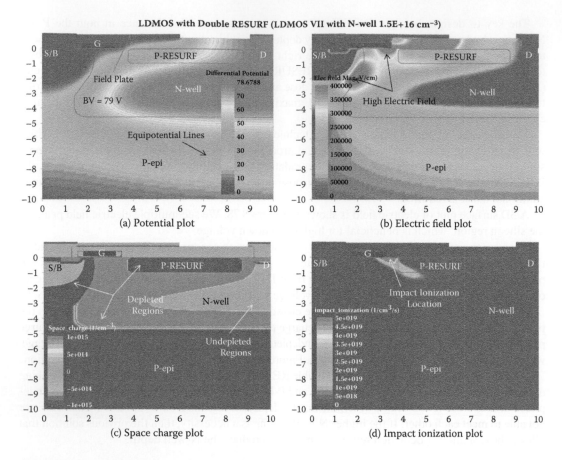

FIGURE 8.79 LDMOS VII with N-well doping concentration increased to 1.5E + 16 cm–3: (a) Potential; (b) Electric field; (c) Space charge; (d) Impact ionization.

If we reduce the N-well depth from about 4.5 μm to 3.35 μm, we can get the breakdown voltage back to 144 V, as illustrated in Figure 8.80. The N-well is almost fully depleted except the top right corner of the drain region: this small pocket area explains why this design is still a few volts lower than the breakdown voltage of our original LDMOS VII design.

Previously when we increased the doping concentration of the N-well from 1E + 16 cm⁻³ to 1.5E + 16 cm⁻³, we broke the dose balance between the N-well and P-regions (P-body, P-epi, and P-RESURF), which caused undepleted regions and lowered the breakdown voltage. By reducing the N-well depth, we are also lowering the total dose of the N-well, which creates a more balanced dose in the device. Unfortunately, the on-resistance is dependent on the total dose of the N-well (rather than doping concentration). A simple explanation of this relationship is given as

$$R = \rho \frac{L}{t} = \frac{1}{q\mu_n N_d} \frac{L}{t} = \frac{1}{N_d t} \frac{L}{q\mu_n} = \frac{1}{D} \frac{L}{q\mu_n} \tag{8.35}$$

$$D = N_d t \tag{8.36}$$

where R is the drift region resistance in 2D, ρ is the N-well resistivity, μ_n is the electron mobility, N_d is the doping concentration in the N-well, L is the N-well length, t is the N-well thickness, and D is the total dose of the N-well. From this equation, we find the resistance in the drain-drift region is inversely proportional to the total dose in the N-well.

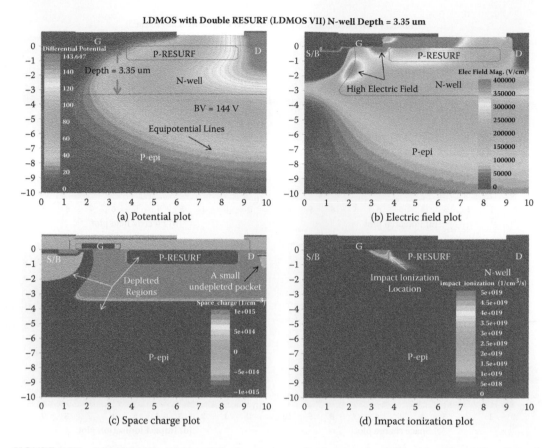

LDMOS with Double RESURF (LDMOS VII) N-well Depth = 3.35 um

(a) Potential plot

(b) Electric field plot

(c) Space charge plot

(d) Impact ionization plot

FIGURE 8.80   LDMOS VII with N-well doping concentration increased to 1.5E + 16 cm⁻³ and N-well depth reduced to 3.35 μm: (a) Potential; (b) Electric field; (c) Space charge; (d) Impact ionization.

By reducing the depth of the N-well, we lowered the total dose in the N-well, which leads to a higher on-resistance. This increase can easily offset the benefits we get from increasing the doping concentration in the N-well; thus, once again, improving the breakdown voltage comes at the expense of the on-resistance.

Since the reduction of the N-well depth can backfire, we need to think of other ways to improve the design. Our goal is to keep the increased N-well dose (doping = 1.5E + 16 cm⁻³) and default thickness (4.5 μm) to maintain a low on-resistance while achieving a breakdown voltage that can meet our specifications (100 V + 30% margin = 130 V). One possibility is to increase the dose of P-RESURF so that it can balance the dose increase in the N-well. Modifying the doping concentrations in the P-body and P-epi might also be possible, but this is more difficult in practice since it will inevitably affect other devices on the same chip that share the same process.

To get a rough idea of the P-RESURF dose required to achieve dose balance, we take N-well doping concentration multiplied by N-well depth, so increasing the doping concentration from 1E + 16 cm⁻³ to 1.5E + 16 cm⁻³ with the same N-well depth of 4.5 μm results in an dose increase of 2.25E + 12 cm⁻². So for a P-RESURF depth of 1 μm, the doping concentration of the P-RESURF should increase by 2.25E + 16 cm⁻³ from 1.5E + 16 cm⁻³ to 3.75E + 16 cm⁻³.

We will use this new P-RESURF dose for the next iteration of our design. As we can see, the simulated breakdown voltage for this structure is only 64 V (Figure 8.81a), which is even worse than for our previous case with a P-RESURF doping of 1.5E + 16 cm⁻³. The reason behind this degradation is that the depletion region in the N-well from the P-RESURF side fails to reach the depletion region from the P-epi side, as seen in the space charge plot in Figure 8.81c. The high doping of the

LDMOS with Double-RESURF (LDMOS VII P-RESURF 3.75+16 cm^{-3})

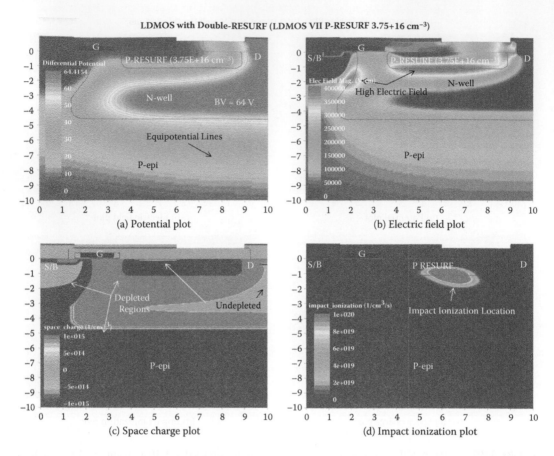

(a) Potential plot (b) Electric field plot

(c) Space charge plot (d) Impact ionization plot

FIGURE 8.81 LDMOS VII with N-well doping concentration increased to 1.5E + 16 cm–3 and P-RESURF changed to 3.75E + 16 cm^{-3}: (a) Potential; (b) Electric field; (c) Space charge; (d) Impact ionization.

P-RESURF also prevents this floating island from being fully depleted. Due to high doping concentrations in both the P-RESURF and N-well, a high electric field is present at this junction (Figure 8.81b) which causes impact ionization as shown in Figure 8.81d.

8.3.1.8 LDMOS Multiple RESURF (LDMOS VIII)

From our previous design (LDMOS VII), we have learned that the N-well in a double RESURF structure cannot be fully depleted if we want a high N-well doping (e.g., 1.5E + 16 cm^{-3}), unless we reduce the N-well depth. But this is not useful because the total dose in the N-well is also reduced and the on-resistance depends on the total N-well dose rather than the N-well doping concentration.

An ingenious idea to work around this problem was proposed by engineers (circa 2000): the P-RESURF does not necessarily have to reside on top of the N-well [183], [184], [185]. This floating region can instead stay floating in the middle of the N-well as a RESURF island. The idea is that the island of P-RESURF now separates the N-well into upper and lower portions. This, in effect, reduces the required depletion thickness so that it is easier to deplete the entire N-well.

We call this implementation of RESURF multiple RESURF, as it has multiple RESURF junctions in the N-well and is more effective in mutual depletions than double RESURF or single RESURF. A schematic of this structure is given in Figure 8.82a.

Figure 8.82b shows the material plot, while Figures 8.82c and 8.82d plot the net doping in the device. The P-RESURF is now floating somewhere in the middle of the N-well but not at the exact center; we will discuss this a little bit later. The doping concentration of the N-well is 1.5E + 16 cm^{-3}, while the maximum doping in the P-RESURF is chosen as 4.5E + 16 cm^{-3}. The location and doping

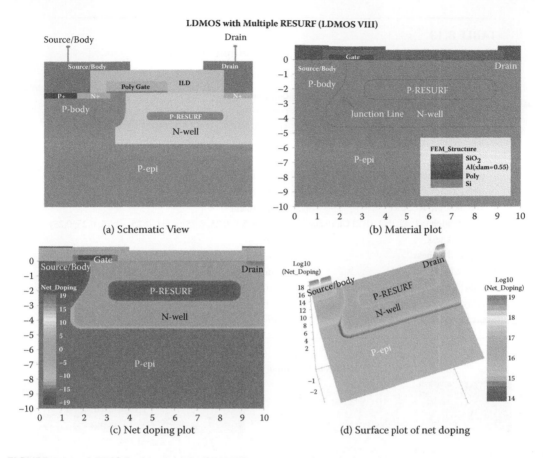

FIGURE 8.82   LDMOS with multiple-RESURF structure and net doping plot.

of the P-RESURF region were chosen according to estimation and finalized with TCAD simulation results and DoE analysis.

The simulation parameters are listed in Table 8.13 for the multiple RESURF LDMOS (LDMOS VIII). The main changes in the table from double RESURF are the maximum doping concentration in the P-RESURF and its location. The maximum doping concentration location (centered at Y = −2.0 μm) of the P-RESURF is made slightly above the middle of the N-well (Y = −2.25 μm) because the lower portion of the N-well is already partially depleted by the P-epi. The channel length also becomes 0.6 μm due to the N-well push-out with increased doping concentration. The field plate length has been shortened to 4 μm since it seems to give better results in this structure.

A satisfactory breakdown voltage of 149 V is obtained with Figure 8.83a showing the equipotential lines in the device after breakdown. The P-RESURF floating island acquires the potential from the reverse-biased PN junction and helps deplete the surrounding N-well. From Figure 8.83b, we find the peak electric field resides at the PN junction of the P-RESURF and N-well close to the drain: this is not surprising since both the P-RESURF and N-well have high doping concentrations. The peak impact ionization is also located in this high field region. Taking a closer look at the space charge plot, we can also observe a small pocket of undepleted material within the P-RESURF. It is left to interested readers to find ways of eliminating this region and squeezing a few extra volts of breakdown voltage out of this design.

A 3D surface plot of electric field of LDMOS VIII is given in Figure 8.84. Multiple electric field peaks help make the electric field distribution more uniform.

TABLE 8.13

Simulation Parameters for LDMOS with Multiple RESURF

Parameters for Simulation	Value
X range	0 to 10 μm
Y range	0 to –20 μm
P-epi	1E + 15 cm⁻³
P-body max concentration	2E + 17 cm⁻³
P-body location (x1, x2, dx1, dx2) Gaussian	x = (0, 1.5, 0.35, 0.35); y = (−0.6, −0.6, 0.8, 0.8)
N-well max concentration	1.5E + 16 cm⁻³
N-well location (x1, x2, dx1, dx2) Gaussian	x = (2, 10, 0.25, 0.25); y = (−5.5, 0, 0.25, 0.25)
P-well multiple-RESURF	4.5E + 16 cm⁻³
P-well location (x1, x2, dx1, dx2) Gaussian	X = (3.5, 8.5, 0.25, 0.25); y = (−2.25, −1.75, 0.25, 0.25)
Oxide thickness	200 Å
Poly gate thickness	0.4 μm
Poly gate length	2 μm
Channel length	0.6 μm (approx.)
Interlayer dielectric thickness	0.7 μm
Source metal shield plate length	4 μm (from source) or 0.5 μm (from gate edge)

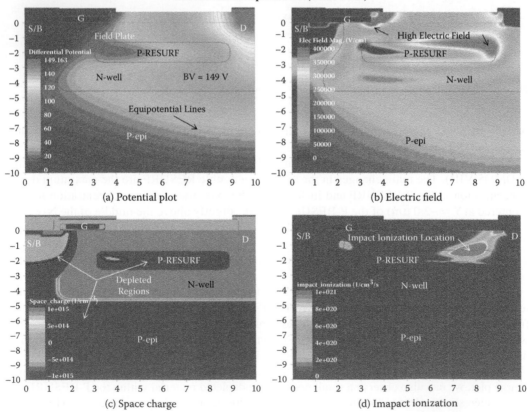

LDMOS with Multiple RESURF (LDMOS VIII)

(a) Potential plot

(b) Electric field

(c) Space charge

(d) Imapact ionization

FIGURE 8.83 **(See color insert)** Simulation results after device breakdown for LDMOS VIII with multiple RESURF: (a) Potential; (b) Electric field; (c) Space charge; (d) Impact ionization.

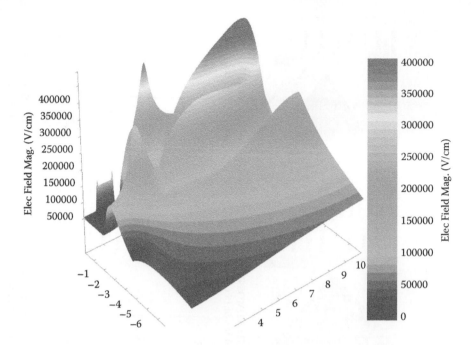

FIGURE 8.84 Electric field surface plot of LDMOS with multiple RESURF (LDMOS VIII).

8.3.1.9 Comparison of 3D Surface Plot of Electric Field

A grand comparison is made among the eight LDMOS structures, as shown in Figure 8.85. LDMOS structures with N-epi (LDMOS I to IV) have considerably inferior breakdown voltages than LDMOS structures with P-epi RESURF (LDMOS V to VIII). LDMOS structures with N-epi have a triangle-shaped electric field compared with the more uniform and multiple peaked electric field distribution exhibited by LDMOS with P-epi RESURF. A well-designed LDMOS with double RESURF or multiple RESURF can handle high breakdown voltage with low on-resistance, thanks to the increased N-well doping that is made possible by the RESURF effect.

LDMOS designs have evolved over the past decades: many structures including multiple RESURF have been developed, and several are covered by patents. New ideas and novel structures continue to be published, and designers are wise to keep up to date with the latest concepts as this book cannot cover the entirety of this topic. Useful academic conferences and journals for this topic include the IEEE International Symposiums on Power Semiconductor Devices and ICs and the International Electron Devices Meeting.

8.3.2 Parasitic NPN BJTs in LDMOS

Unlike a digital or analog MOSFET, a power LDMOS has the source/body shorted to form a simple PN junction diode between drain and source/body. One of the reasons for doing this is because there is a parasitic N (drain-drift) – P (P-body) – N (N+ source) BJT, as shown in Figure 8.86. A floating body region (body P+ contact removed) might cause a BV_{CEO} type of breakdown if a large voltage drop within the P-body region were to trigger forward biasing of the P-body/N+ source junction (base-emitter junction of the parasitic NPN).

As an example, the LDMOS with STI (LDMOS III) is used with body contact removed (contact covers only the N+ source). Shown in Figure 8.87b, the floating body leads to a BV_{CEO} type of breakdown and greatly reduces the breakdown voltage from 87 to 35 V, and a large leakage current flows from the drain (collector in the parasitic NPN BJT) through the P-body (base of the parasitic NPN BJT) to the source (emitter of the parasitic NPN BJT). Figure 8.87a is a comparison of the same

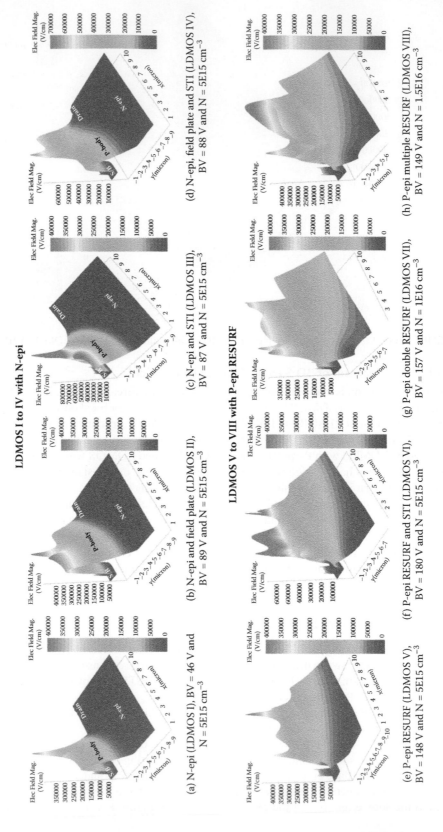

FIGURE 8.85 **(See color insert)** Comparison of electric field surface plots of all LDMOS structures discussed.

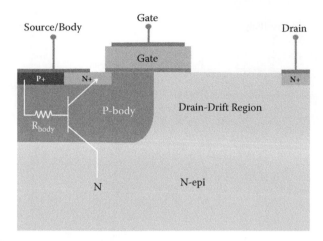

FIGURE 8.86 Parasitic NPN BJT in LDMOS.

LDMOS (LDMOS III) with the body contact connected to the source (standard). The drain voltage is biased to 35 V, and no noticeable leakage current exists.

In fact, even with the body connected to the source, the parasitic BJT may still be turned on when carrying a large leakage current flow (BV_{CES} type of breakdown), since the voltage drop in the body region may be higher than the built-in potential of the body (base) to source (emitter) PN junction. Forward biasing of the base-emitter of the parasitic NPN BJT may cause permanent damage to the device with a second breakdown (thermal runaway): this phenomenon will be discussed later.

8.3.3 LDMOS On-State Resistance

Most of the time power LDMOS are working in either on- or off-states. During the on-state, the gate is biased so that the device is fully turned on: current flows from drain terminal, through the drain-drift region, and into the channel region before finally reaching the source/body terminal.

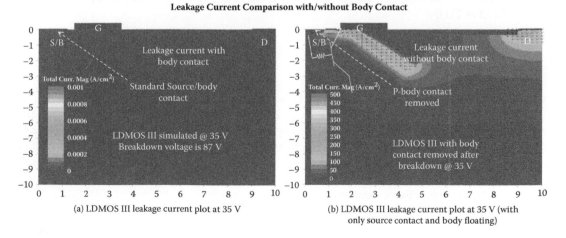

FIGURE 8.87 If body is left floating, BV_{CEO} type of breakdown occurs, resulting in large current flowing through the parasitic BJT.

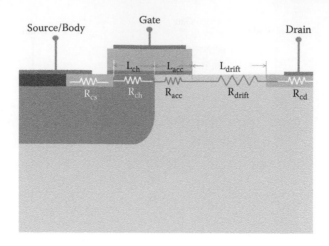

FIGURE 8.88 LDMOS on-resistance.

The on-resistance of the LDMOS can be analyzed with several contributing resistances connected in series. We will use the simple LDMOS with N-epi (LDMOS I) as an example. Figure 8.88 shows the resistance components of an LDMOS. The total on-resistance is thus the sum of these components:

$$R_{on} = R_{cs} + R_{ch} + R_{acc} + R_{drift} + R_{cd} \tag{8.37}$$

where R_{cs} and R_{cd} are the contact resistances of source and drain, which are generally quite small and will be omitted in this book. R_{ch} is the channel resistance, which can be calculated using the standard equation that is also used for digital and analog MOSFETs:

$$R_{ch} = \frac{L}{2\mu_i C_{ox}(V_G - V_{th})} \tag{8.38}$$

where μ_i is the inversion layer mobility for electrons (n-LDMOS) and holes (p-LDMOS), V_G is the gate voltage, V_{th} is the threshold voltage, and L is the channel length. C_{ox} is the gate capacitance and is calculated in 2D by

$$C_{ox} = \frac{\varepsilon_{ox}}{t} \tag{8.39}$$

where ε_{ox} is the dielectric constant of oxide, t is the gate oxide thickness.

The accumulation resistance R_{acc} is unique for DMOS (LDMOS and VDMOS). It represents the resistance in the silicon area that is overlapped by gate and drain. The word *accumulation* means that the overlapped area is in accumulation mode when the gate is turned on. For an n-LDMOS device during on-state, this happens when a sufficiently high positive voltage is applied to the gate: the positive voltage attracts electrons to the overlapped area and lowers the resistance for drain current flow. The accumulation resistance can be calculated by [186].

$$R_{acc} = \frac{L_{acc}}{4\mu_a C_{ox}(V_G - V_{th})} \tag{8.40}$$

where L_{acc} is the accumulation region length, which is from the body/drain PN junction to the gate edge in the drain, and μ_a is the electron or hole mobility in the accumulation layer. Because of the local field, electrons generally have a higher mobility in the accumulation layer than in the inversion layer.

The drift region resistance R_{drift} is calculated using

$$R_{drift} = \rho \frac{L_{drift}}{d_{eff}} = \frac{L_{drift}}{q\mu_n N_d d_{eff}} = \frac{L_{drift}}{q\mu_n D_{eff}} \tag{8.41}$$

$$D_{eff} = N_d d_{eff} \tag{8.42}$$

where L_{drift} is the drift length, d_{eff} is the effective N-well thickness due to unequal distribution of current flow, N_d is the doping concentration in the drain, μ_n is the electron mobility in the drift region, and D_{eff} is the effective total dose in the drain, with n-LDMOS assumed.

8.3.3.1 Specific On-Resistance

To compare resistance of LDMOS with different technology or vendors, it is convenient to use specific on-resistance, $R_{on, sp}$ or R_{sp}, instead of pure resistance; the difference is that specific on-resistance takes device pitch size into consideration. The pitch size is calculated from body contact edge to drain edge. Note that in real devices body and drain are shared with neighboring devices: in that case, the middle point of the body and drain contacts should be used. As an example, most simulated LDMOS in this book use a pitch size of 10 μm. The unit used in this book for specific resistance is $m\Omega \cdot cm^2$.

8.3.3.2 On-Resistance Contribution

To find out how much contribution the channel region and drift region have for the total resistance, we can simulate a truncated area of the LDMOS. For example, since accumulation resistance is relatively small compared with drift resistance, we can combine them into a simple channel resistance. Figure 8.89 is the test structure for on-resistance simulation from channel and accumulation regions only: a test contact is added under the edge of the gate in the drain-drift region to inject carriers in the simulation. The drain region is very short, so it is ignored and small amounts of N+ doping are

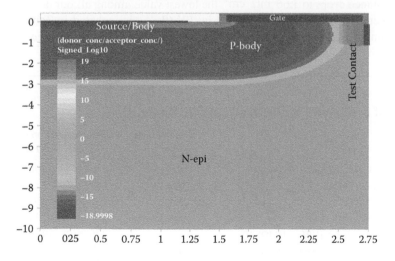

FIGURE 8.89 Test structure for LDMOS on-resistance.

added to create an Ohmic boundary for our test contact. The total length of this test structure is 2.75 μm, including the drain region, which is only 0.25 μm.

For the simple LDMOS with an N-epi doping concentration of 5E + 15 cm^{-3} (LDMOS I), a pitch size of 10 μm, and a gate voltage bias of 5 V, the simulation result shows the channel resistance is only 0.2 mΩ · cm^2 (pitch size of 10 μm is used in calculation of specific on-resistance for comparison), while the total resistance for the LDMOS I with standard drain-drift length is 2.78 mΩ · cm^2. Therefore, we can conclude that the majority of the on-resistance comes from the drain-drift region. However, for low breakdown voltage LDMOS (< 25 V), the drain length is much shorter and the doping concentration is much heavier; therefore, the on-resistance from the channel can be greater than that from the drain.

8.3.3.3   Comparison of Breakdown Voltage and On-Resistance

A comparison is made among the LDMOS discussed before, as illustrated in Table 8.14. The simple N-epi LDMOS (LDMOS I) has the lowest breakdown voltage of 46 V and a relatively low specific on-resistance of 2.78 mΩ · cm^2. With the addition of field plate (LDMOS II), the breakdown voltage increases significantly to 89 V while on-resistance stays the same. With added STI (LDMOS III), the breakdown is similar to LDMOS with a field plate (LDMOS II), but the on-resistance of LDMOS III increases by 40% over LDMOS I: this is caused by the presence of oxide in the drain-drift region, which reduces the accumulation length and makes part of the silicon surface unavailable for current flow. When the field plate and STI are combined (LDMOS IV), the breakdown voltage stays the same, while the on-resistance is similar to that of LDMOS III. In fact, further optimizing the field plate or STI will only show small changes of breakdown voltage.

With P-epi RESURF (LDMOS V), breakdown voltage increases significantly for the same N-well doping concentration (5E + 15 cm^{-3}) and specific on-resistance is similar to LDMOS I at 2.79 mΩ · cm^2. The added STI combined with increased N-well doping (1E + 16 m^{-3}) in LDMOS VI causes the breakdown voltage to drop back down to the level of LDMOS IV while the on-resistance is reduced to 2.09 mΩ · cm^2. To further increase breakdown voltage to satisfy the design spec, double RESURF (LDMOS VII) is used: a much higher breakdown voltage (157 V) is obtained with N-well doping concentration of 1E + 16 cm^{-3}, but the on-resistance rises to 3.07 mΩ · cm^2 due to the reduced total dose in the N-well and the presence of the P-RESURF layer. (P-RESURF has a depth of about 1 μm, while the depth of STI is only 0.35 μm.) Finally, the multiple RESURF method is applied (LDMOS VIII) to create a floating P-RESURF island to help deplete the entire N-well region. Even with a higher doping concentration of 1.5E + 16 cm^{-3}, the breakdown voltage is increased to 149 V and the on-resistance drops to 1.66 mΩ · cm^2, the lowest value among all our designs.

Figure of merits (FOM), which is defined by $BV^2/R_{on, sp}$, is included in Table 8.14. This FOM allows us to compare LDMOS structures in a more straightforward way. We see LDMOS I has the

TABLE 8.14
Comparison of Breakdown Voltage and Specific On-Resistance

LDMOS Structures (Types)	N-well doping (cm^{-3})	Breakdown (V)	Ron,sp (mΩ cm^2)	FOM (BV^2/Ron,sp)
N-epi (LDMOS I)	5E + 15	46	2.78	762
N-epi + field plate (LDMOS II)	5E + 15	89	2.64	3000
N-epi + STI (LDMOS III)	5E + 15	87	3.94	1920
N-epi + field plate + STI (LDMOS IV)	5E + 15	88	3.91	1980
P-epi + field plate + single RESURF (LDMOS V)	5E + 15	146	2.79	7630
P-epi + field plate + STI + single RESURF (LDMOS VI)	1E + 16	97	2.09	4500
P-epi + field plate + double RESURF (LDMOS VII)	1E + 16	158	3.07	8140
P-epi + field plate + multiple RESURF (LDMOS VIII)	1.5E + 16	149	1.66	13400

lowest FOM (762) while LDMOS VIII has the highest (13400). For LDMOS with N-epi, LDMOS II is a better choice with higher FOM than that of LDMOS III or IV.

8.3.4 LDMOS THRESHOLD VOLTAGE

For LDMOS, the threshold voltage design is similar to analog or digital MOSFET. However, unlike analog or digital MOSFET, an LDMOS P-body region is created by diffusion of an implant underneath the gate, which means that a doping concentration gradient exists in the channel region. Since higher P-body doping concentration yields higher threshold voltage, the threshold voltage is determined by the peak P-body doping concentration in the channel, near the source side.

For the gate oxide design, a thicker oxide yields a higher threshold voltage. For power devices with high gate voltage supplies, a thicker oxide is necessary to reduce gate leakage current and improve long-term oxide reliability (e.g., TDDB). A properly designed threshold voltage is always important for IC designers (neither too low nor too high). Oftentimes, multiple levels of threshold voltages are necessary for digital, analog, and power MOSFETs, and multiple oxide thicknesses in the same process flow is desired.

The LDMOS I with N-epi is simulated, and the threshold voltage is found to be close to 0.83 V. The threshold voltage is extracted using the peak transconductance (G_m) method [39]:

1. Plot I_D-V_G at low V_D value (e.g. $V_D = 0.1$ V).
2. Plot the slope of I_D-V_G to get the transconductance curve.
3. Select V_G at maximum transconductance and use the V_G value to find the ($V_{G,Gm.max}$, $I_{D,Gm.max}$) point on the I_D-V_G curve.
4. Draw a tangent line through the ($V_{G,Gm.max}$, $I_{D,Gm.max}$) point and let it intercept the voltage axis of the I_D-V_G plot. This intercept is defined as V_{th}.

The threshold voltage is plotted in Figure 8.90, with the transconductance G_m-V_G curve and tangent line to extract threshold voltage. For analytical calculations, threshold voltage is calculated using

$$V_T = \phi_{ms} - \frac{Q_i}{C_i} - \frac{Q_d}{C_i} + 2\phi_f \tag{8.43}$$

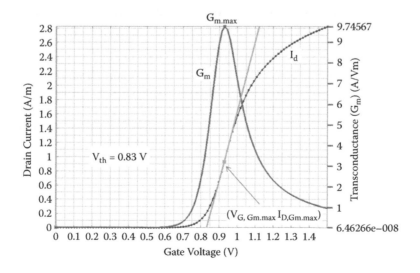

FIGURE 8.90 Threshold voltage plot of simple LDMOS (LDMOS I).

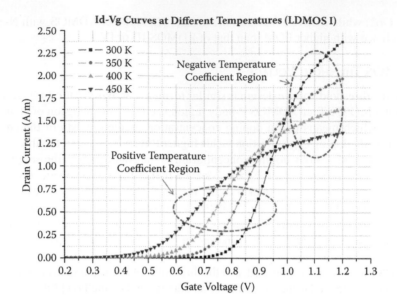

FIGURE 8.91 I_d-V_d curves with different temperatures (LDMOS I).

where Q_i is the interface charge, Q_d is the depletion charge, and ϕ_{ms} is the work function potential difference between the gate and semiconductor:

$$\phi_{ms} = \phi_m - \phi_s \qquad (8.44)$$

ϕ_f is the Fermi level:

$$\phi_f = \frac{kT}{q} \ln\left[\frac{N_A}{n_i}\right] \qquad (8.45)$$

In the power device and power IC industry, sometimes it is more convenient to define threshold voltage using a certain drain current value.

Plots of I_d-V_g curves at different temperatures are illustrated in Figure 8.91. It has been observed that beyond a certain gate voltage range the drain current decreases with increasing temperature due to reduced mobility at elevated temperatures. This negative temperature coefficient is desirable for power devices to reduce the risk of thermal runaway. Unfortunately, below certain gate voltage range, the current increases with higher temperature: this small region of positive temperature coefficient is caused by lower threshold voltage at elevated temperatures. This means there is a small region where positive feedback loops similar to those of BJT are possible, and it is recommended to avoid operating devices in this region.

8.3.5 LDMOS with Radiation Hardening Design

For applications involving outer space, high-altitude flights, or operations around nuclear reactors, power devices should be designed to withstand ionizing radiation such as high-energy protons, alpha particles, x-rays, and gamma rays. Most semiconductor devices are susceptible to radiation damage,

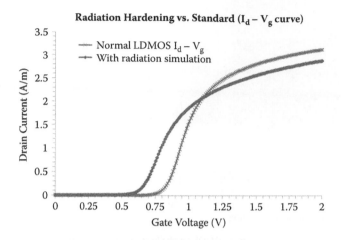

FIGURE 8.92   Radiation hardening simulation compared with normal LDMOS I_d-V_g curve.

and radiation hardening design refers to designs that reduce the susceptibility of semiconductor devices to radiation damage.

Radiation damage comes in many forms and the exact details of the defects created by a radiation source depend on the particle's mass and energy; experimental measurements through deliberate irradiation or Monte Carlo simulation with tools such as GEANT [187] are needed to get this kind of information. However, one commonly observed consequence of ionizing radiation is the creation of positive fixed sheet charges and acceptor-like deep level traps at the oxide/silicon interface. The effect of deep-level acceptors near the silicon channel is similar to an increase of channel P-doping, which would shift the threshold voltage of an n-LDMOS upward. On the other hand, adding a positive fixed charge to the channel interface has an effect similar to n-doping, which shifts the threshold voltage downward [39].

Radiation hardening simulation is carried out using the basic LDMOS (LDMOS I), as seen in Figure 8.92. For the purposes of simulation, we assume additional positive fixed charges with a surface density of 8E + 11 cm^{-2} and deep acceptor traps of 2E + 11 cm^{-2} located at 0.4, 0.6, and 0.8 eV, respectively, below the conduction band. The net effect of the radiation is a downward shift of the threshold voltage from 0.83 to 0.55 V.

8.3.6   LDMOS I-V Family of Curves

The I-V family of curves is a set of drain voltage-current curves at various applied gate voltages: these curves are especially important for analog amplification applications. We will use LDMOS with N-epi (LDMOS I; see net doping plot in Figure 8.93a) as a test structure. The I-V curves under various gate voltages (1.5 V, 2.0 V, 2.5 V, 3.0 V, 3.5 V, 4.0 V, and 4.5 V) are given in Figure 8.93b.

From Figure 8.93b, we notice the I-V curves for gate voltages above 3.5 V exhibit a significant drain current saturation effect. This current saturation (or the so-called quasi-saturation) is different from the channel current saturation, which is caused by the pinch-off of the channel. Quasi-saturation is instead caused by carrier velocity saturation in the drain-drift region.

A modified LDMOS (Figure 8.93c) is used for comparison purposes. It is the same LDMOS except that the N+ drain is moved to the gate side to reduce the drain-drift region length. With a very short drain-drift region, we see no quasi-saturation effect in the I-V curves, as shown in Figure 8.93d.

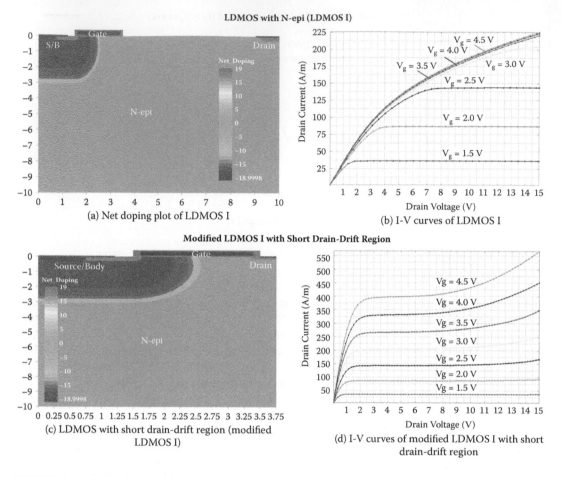

FIGURE 8.93 LDMOS net doping and I-V curves with (a) standard LDMOS I net doping plot, (b) I-V curves of standard LDMOS I, (c) modified LDMOS with short drain region, and (d) I-V curves of modified LDMOS.

For LDMOS to sustain a high breakdown voltage, the drain-drift region is usually lightly doped. The drain current density J_D can be expressed as

$$J_D = qn\mu_n E = qnv_n \tag{8.46}$$

$$v_n = \mu_n E \tag{8.47}$$

where n is the electron carrier density, μ_n is the mobility of electron, $q = 1.602 \times 10^{-19}$, C is the electron charge, v_n is the electron velocity, and E is the electric field in the drift region.

For a given current density J_D and field value E, if the carrier density/doping concentration is sufficiently high (i.e., $> 1E + 18$ cm^{-3}), then v_n does not set a constraint and the desired current value can be attained. However, if we have a lightly doped drain-drift region, then the maximum value of v_n, the saturation velocity, fixes a limit beyond that the current cannot increase any more: this causes the quasi-saturation of J_D.

8.3.7 LDMOS Self-Heating

When a large current passes through the transistor, the lattice temperature goes up with self-heating. This self-heating can be localized. Unlike bipolar devices, where localized self-heating causes a

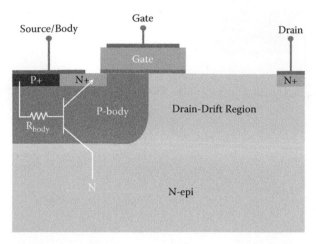

FIGURE 8.94 Parasitic BJT of an LDMOS.

positive feedback, allowing more current passes through and eventually hot spots to be created until the device fails, unipolar devices like LDMOS have better immunization. The carrier mobility is reduced when the device heats up, a negative feedback mechanism that stabilizes the device. This is especially beneficial for parallel configurations, where current flows can be more uniform across the cells connected in parallel. However, this does not mean that LDMOS will not suffer the same thermal failure as BJTs, since there is always a parasitic BJT within LDMOS, as shown in Figure 8.94.

For n-LDMOS, the drain-drift region acts as the collector, P-body as the base, and N+ source as the emitter. Normally, the P-body (base) is shorted to the source (emitter) with source contact. If a large body current exists and there is a large body resistance R_{body}, then the voltage drop within the P-body region may forward bias the base-emitter junction and turn on the parasitic BJT. When this happens, LDMOS can no longer support the high voltage drop between drain and source. It can even trigger thermal runaway by positive feedback mechanisms of current and lattice temperature.

A simulation example of LDMOS (LDMOS I with N-epi layer is used here) self-heating is shown in Figure 8.95. Snapbacks are observed with V_g = 1 V, 2 V, and 3 V. Due to the existence of multiple current solutions at the same voltage, it is necessary to use current as a control variable in TCAD simulations to observe the correct snapback effect. However, it is recommended for numerical stability reasons to use voltage in the low-bias range.

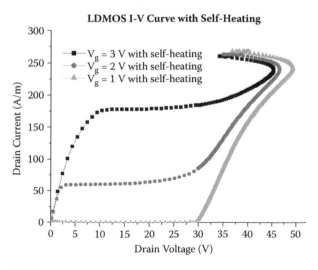

FIGURE 8.95 LDMOS I-V curves with self-heating.

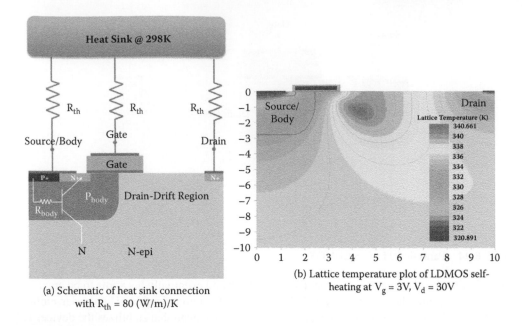

(a) Schematic of heat sink connection with R_{th} = 80 (W/m)/K

(b) Lattice temperature plot of LDMOS self-heating at V_g = 3V, V_d = 30V

FIGURE 8.96   (a) LDMOS heat sink connection; (b) Lattice temperature plot.

The heat sink is connected to all the contacts with a thermal conductance of 80 (W/m)/K, as illustrated in Figure 8.96a. Figure 8.96b is the lattice temperature plot of LDMOS (LDMOS I) with self-heating at V_g = 3 V, V_d = 30 V (before snapback).

8.3.8   LDMOS Parasitic Capacitances

Power LDMOS are switched on and off during different phases of operation, and the switching speed is limited by the parasitic capacitances. This section examines the simulation of four capacitances: gate to source (C_{GS}); gate to body (C_{GB}); gate to drain (C_{GD}); and drain to source (C_{DS}). C_{GD} is also known as reverse transfer capacitance. C_{GS} and C_{GB} are in parallel and are usually summed up to form the input capacitance (C_{in}). Figure 8.97a shows the various capacitances in an LDMOS, and a schematic plot of the parasitic capacitances is provided in Figure 8.97b.

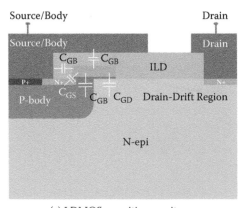

(a) LDMOS parasitic capacitors

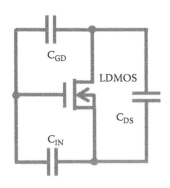

(b) Schematic of LDMOS and its parasitic capacitors

FIGURE 8.97   Capacitances in an LDMOS.

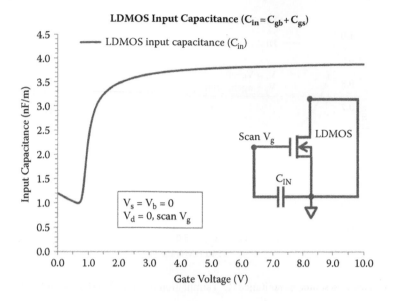

FIGURE 8.98 Input capacitance (C_{in}) simualtion result of LDMOS with N-epi (LDMOS II).

The LDMOS capacitance simulations are performed by superimposing the small AC signal to the DC drain bias voltage. In TCAD simulation, this is often a postprocessing step done after the main DC simulation is completed. The LDMOS II design with N-epi and field plate is simulated with source/body and drain contacts grounded. The voltage on the gate contact is scanned in the DC analysis and is superimposed with an AC signal to extract the capacitance. At frequency of 1 MHz, the simulated input capacitance and simulation setup schematic are shown in Figure 8.98.

The reverse transfer capacitance is simulated with source/body and gate contacts shorted to ground. Voltage on the drain is increased in DC with an AC signal superimposed to extract the capacitance and once again, the frequency is set at 1 MHz. The reverse transfer capacitor is drain-bias dependent, with lower values at high drain bias. Note that this reverse transfer capacitance is smaller, partly due to light doping concentration of the drain-drift region ($5E + 15$ cm^{-3}). Figure 8.99 shows the simulation result of reverse transfer capacitance.

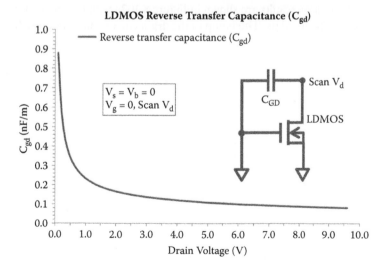

FIGURE 8.99 Reverse transfer capacitance (C_{gd}) simualtion result of LDMOS with N-epi (LDMOS II).

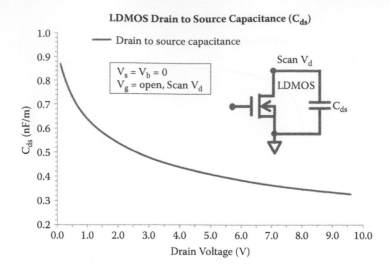

FIGURE 8.100   Drain to source capacitance (C_{ds}) simulation result of LDMOS with N-epi (LDMOS II).

Yet another important capacitance is the drain to source capacitance, C_{ds}. The C_{ds} is simulated with the gate open and source/body grounded. The voltage on the drain is increased in DC with a 1 MHz AC signal superimposed to extract the capacitance. Figure 8.100 is the simulation result of C_{ds}.

8.3.9   LDMOS GATE CHARGE

The gate charge plot is a standard industry practice to test the switching behavior of LDMOS. A constant current is applied to the gate terminal while turning on the LDMOS from the off-state. The basic simulation setup is illustrated in Figure 8.101, with parasitic capacitances shown. Since a load resistance is included, mixed-mode TCAD simulation is necessary. Here *mixed-mode* means resistors, capacitors, inductors, or even MOSFET and BJT compact models are added in addition to the physical (meshed) TCAD simulation.

The simulation steps are provided in Table 8.15 on how to use TCAD to simulate the LDMOS gate charge.

The gate charge simulation results are shown in Figure 8.102a with Figure 8.102b as a zoomed-in view for the period from 50 to 70 μs. The bump in the drain current plot in Figure 8.102a between 0

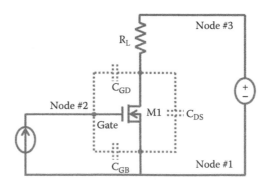

FIGURE 8.101   Simulation setup of LDMOS gate charge.

TABLE 8.15
Simulation Steps for LDMOS Gate Charge

Time	Simulation Steps for LDMOS Gate Charge
0 to 20 μs	Scan the voltage at node #3 (in Figure 8.101) from 0 V to 20 V over a period of 20 μs. During this simulation step, the parasitic reverse transfer capacitance (C_{GD}) and drain to source capacitance (C_{DS}) of the LDMOS are being charged and a charging current goes through the load resistor.
20 to 50 μs	The simulation time reference is increased from 20 μs to 50 μs to allow for complete charging of the parasitic capacitance; at this point, the drain current should decrease to almost zero.
50 to 70 μs	The gate current (at node #2) is increased from 0 to 500 μA/m over a 20 μs window (50 to 70 μs). During this period, the LDMOS gate capacitance (C_{GB}) is charged to about threshold voltage, and then the gate voltage is held at a plateau, due to the discharging of LDMOS reverse transfer capacitance (C_{GD}).
70 to 100 μs	The gate current (at node #2) is held constant at 500 μA/m for 30 μs. After the discharging of the LDMOS reverse transfer capacitance (C_{GD}) is completed, the gate voltage continues to increase to 6 V once the simulation time reference reaches 100 μs.

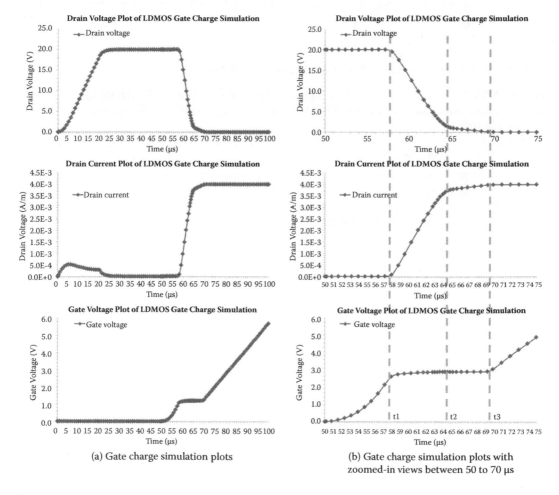

(a) Gate charge simulation plots

(b) Gate charge simulation plots with zoomed-in views between 50 to 70 μs

FIGURE 8.102   (a) Gate charge simulation plots; (b) Zoom-in view between 50 and 75 μs of the same plot.

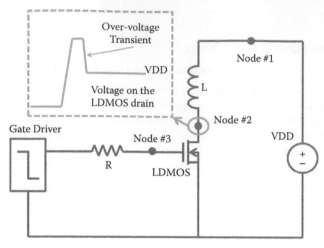

FIGURE 8.103   Test circuit for unclamped inductive switching.

and 25 μs is caused by charging and discharging of the LDMOS parasitic capacitance ($C_{GD} + C_{DS}$). There is also a plateau region in the gate voltage plot, which is caused by charging of the LDMOS reverse transfer capacitance C_{GD}.

8.3.10   LDMOS Unclamped Inductive Switching (UIS)

Power LDMOS are designed to work at high switching speeds. During the device turnoff period, small parasitic inductances or an inductive load can cause large transient voltage overshoots as seen in Figure 8.103.

A mixed mode (TCAD + lumped elements) simulation is performed based on the circuit configuration of Figure 8.103. LDMOS I with N-epi is chosen, with inductor L = 500 μH and gate resistor R = 0.05 Ω. The simulation is carried out with the following steps (Table 8.16).

The simulation result is illustrated in Figure 8.104. A large voltage overshoot is observed when the gate is turned off. If this is not properly clamped, then a much higher breakdown voltage will be required during the design of the LDMOS and a larger on-state resistance would be an unavoidable trade-off.

TABLE 8.16

LDMOS Unclamped Inductive Switching Simulation Steps

Time (ms)	TCAD Simulation Steps
0 to 1	Scan voltage at node #1 to 10 V within 1 ms (set 10 V to V_{DD})
1 to 2	Scan voltage at node #3 to 2.5 V within another 1 ms (turn on LDMOS by applying 2.5 V to the gate)
2 to 5	Wait for 3 ms
5 to 6	Scan voltage at node #3 to 0 V within 1 ms (turn off the LDMOS)
6 to 7	Wait for 1 ms

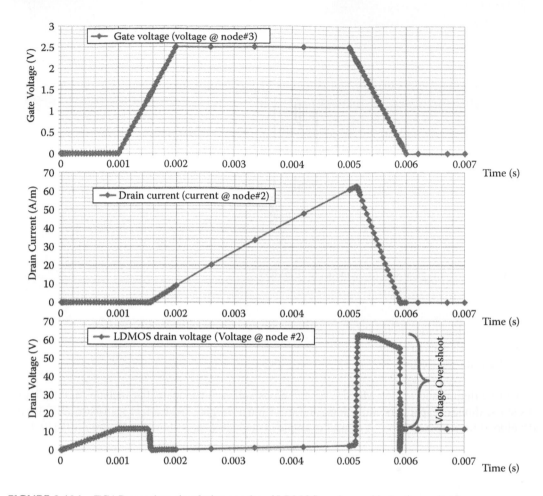

FIGURE 8.104 TCAD transient simulation results of LDMOS unclamped inductive switching.

8.3.11 COMPACT MODELS OF LDMOS

Compact models are basic building blocks for today's EDA industry. Most digital and analog IC simulations cannot be done with mixed-mode TCAD simulation, due to the rather complicated nature of the ICs. Compact models are built by physical and empirical mathematical equations. Typical examples for compact models include the BSIM and EKV models for MOSFETs [52] [188] and the VBIC model for BJTs [189]. For the user of a compact model, the model parameters need to be extracted either through experimental data or by TCAD simulation.

Compact modeling for LDMOS is more complicated. Standard MOSFET models are not capable of properly modeling LDMOS. A reliable LDMOS model should take into account the following physical effects [190]:

- Quasi-saturation effect
- Self-heating effect
- Geometry-related effects
- Graded channel effects

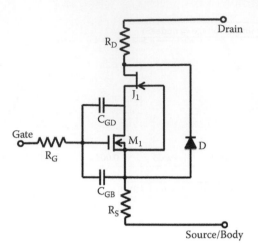

FIGURE 8.105   A typical subcircuit model for LDMOS.

- Bulk current
- Impact ionization in the drift region
- High-side switch effect
- Parasitic BJT effect

Unlike digital and analog MOSFETs, numerous subcircuit models that include MOSFET plus resistors, diodes, capacitors, inductors, and even junction gate field effect transistor (JFETs) are developed for LDMOS [190]. A typical example of such models is shown in Figure 8.105 [191].

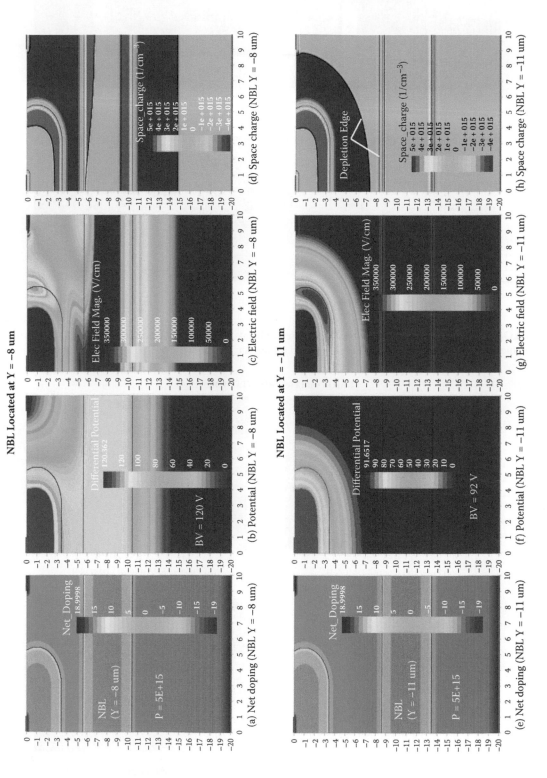

COLOR FIGURE 8.16 Comparison plots of different NBL locations with (a) to (d) located at $Y = -8$ μm and (e) to (h) located at $Y = -11$ μm.

LDMOS with N-epi (LDMOS I)

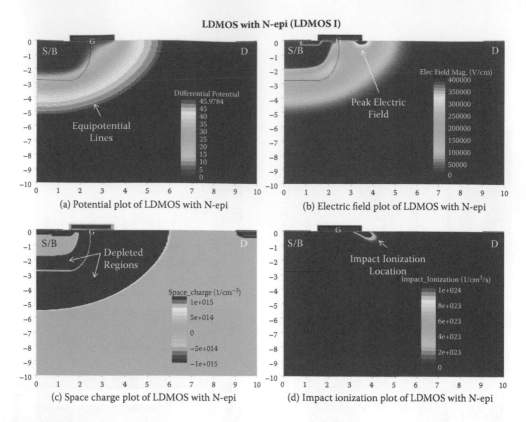

(a) Potential plot of LDMOS with N-epi

(b) Electric field plot of LDMOS with N-epi

(c) Space charge plot of LDMOS with N-epi

(d) Impact ionization plot of LDMOS with N-epi

COLOR FIGURE 8.54 Simulation results after device breakdown for LDMOS with N-epi (LDMOS I) with (a) potential, (b) electric field, (c) space charge, and (d) impact ionization.

LDMOS with Field Plate (LDMOS II)

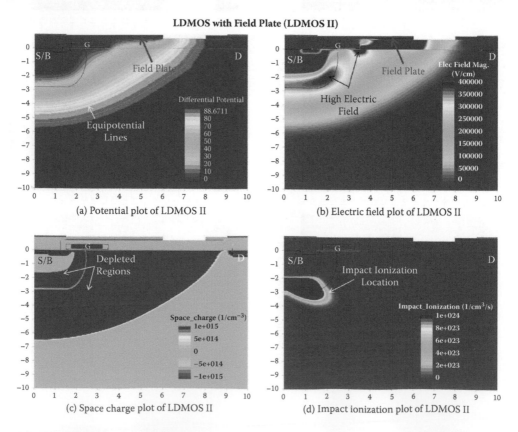

(a) Potential plot of LDMOS II

(b) Electric field plot of LDMOS II

(c) Space charge plot of LDMOS II

(d) Impact ionization plot of LDMOS II

COLOR FIGURE 8.58 Simulation results after device breakdown for LDMOS with N-epi and field plate (LDMOS II) with (a) potential, (b) electric field, (c) space charge, and (d) impact ionization.

LDMOS with STI (LDMOS III) at Breakdown

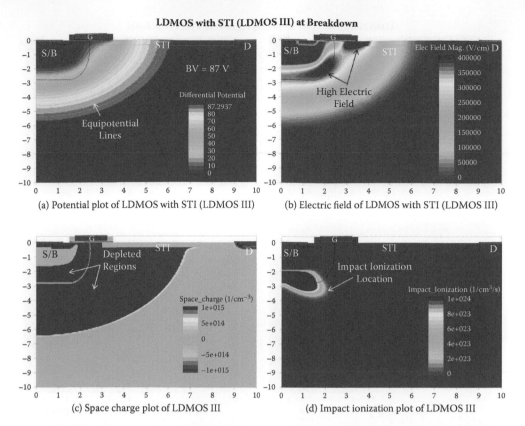

(a) Potential plot of LDMOS with STI (LDMOS III)

(b) Electric field of LDMOS with STI (LDMOS III)

(c) Space charge plot of LDMOS III

(d) Impact ionization plot of LDMOS III

COLOR FIGURE 8.64 Simulation results at device breakdown for LDMOS with N-epi and STI (LDMOS III) with (a) potential, (b) electric field, (c) space charge, and (d) impact ionization.

LDMOS with STI and Field Plate (LDMOS IV) at Breakdown

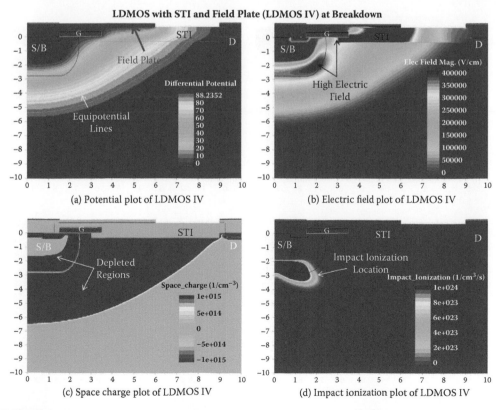

(a) Potential plot of LDMOS IV

(b) Electric field plot of LDMOS IV

(c) Space charge plot of LDMOS IV

(d) Impact ionization plot of LDMOS IV

COLOR FIGURE 8.67 Simulation results after device breakdown for LDMOS with STI and field plate (LDMOS IV) with (a) potential, (b) electric field, (c) space charge, and (d) impact ionization.

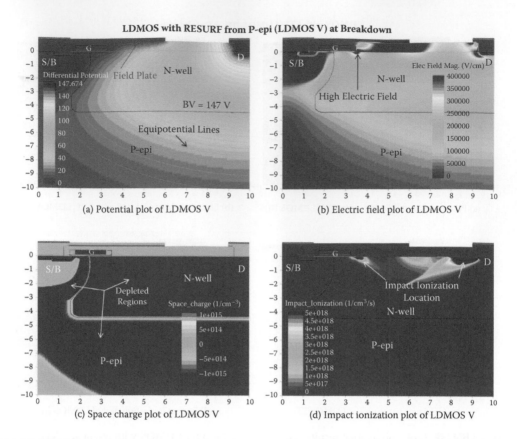

COLOR FIGURE 8.70 Simulation results at device breakdown for LDMOS with P-epi RESURF (LDMOS V) with (a) potential, (b) electric field, (c) space charge, and (d) impact ionization.

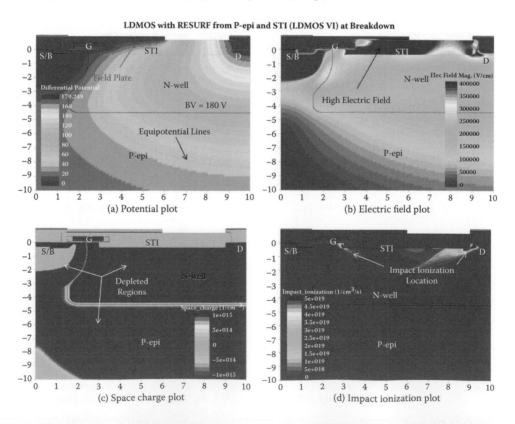

COLOR FIGURE 8.73 Simulation results at device breakdown for LDMOS with P-epi RESURF and STI (LDMOS VI) with (a) potential, (b) electric field, (c) space charge, and (d) impact ionization.

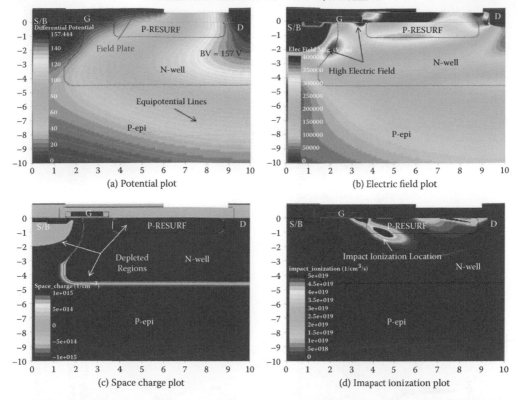

COLOR FIGURE 8.77 Simulation results after device breakdown for LDMOS VII with increased N-well doping (1E + 16 cm⁻³) and double RESURF: (a) Potential; (b) Electric field, (c) Space charge; (d) Impact ionization.

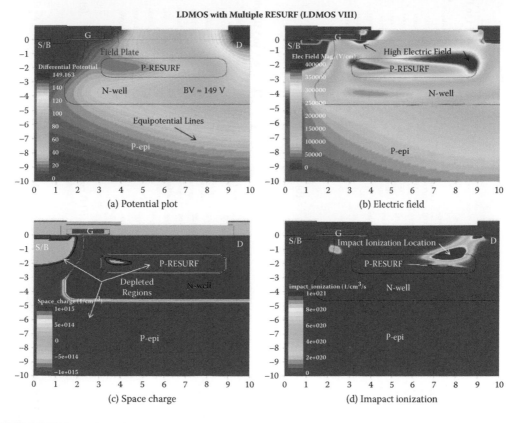

COLOR FIGURE 8.83 Simulation results after device breakdown for LDMOS VIII with multiple RESURF: (a) Potential; (b) Electric field; (c) Space charge; (d) Impact ionization.

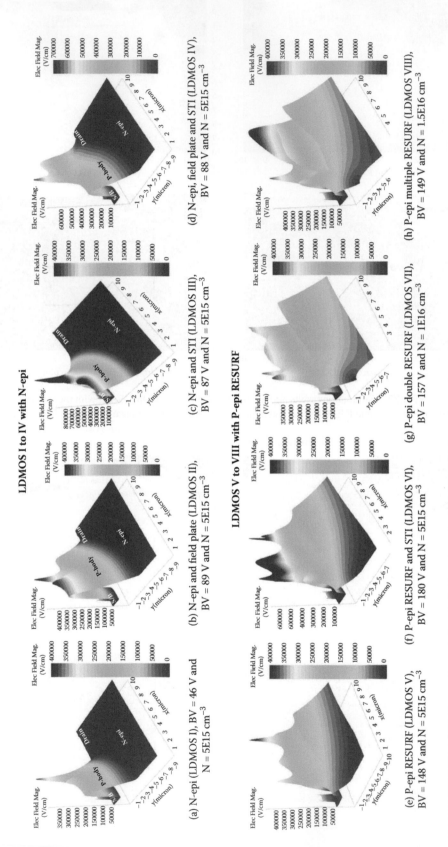

COLOR FIGURE 8.85 Comparison of electric field surface plots of all LDMOS structures discussed.

LIGBT Electron Concentrations at Various Turnoff Transient Time

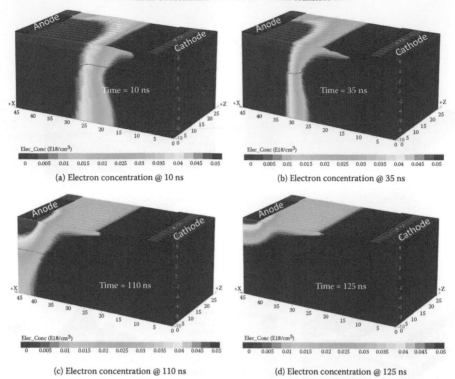

(a) Electron concentration @ 10 ns

(b) Electron concentration @ 35 ns

(c) Electron concentration @ 110 ns

(d) Electron concentration @ 125 ns

COLOR FIGURE 9.19   LIGBT turnoff transient plots of electron concentrations at (a) time = 10 ns, (b) time = 35 ns, (c) time = 110 ns, and (d) time = 125 ns.

LIGBT Hole Concentrations at Various Turnoff Transient Time

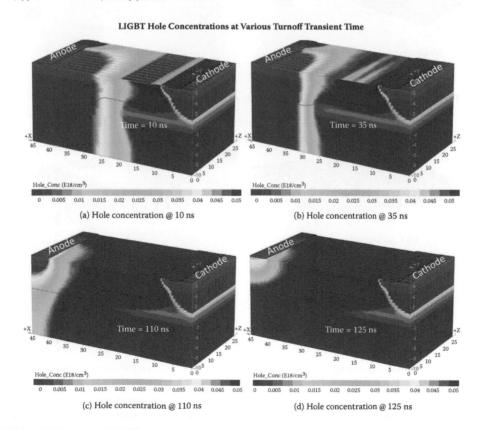

(a) Hole concentration @ 10 ns

(b) Hole concentration @ 35 ns

(c) Hole concentration @ 110 ns

(d) Hole concentration @ 125 ns

COLOR FIGURE 9.20   LIGBT turnoff transient plots of hole concentrations at (a) time = 10 ns, (b) time = 35 ns, (c) time = 110 ns, and (d) time = 125 ns.

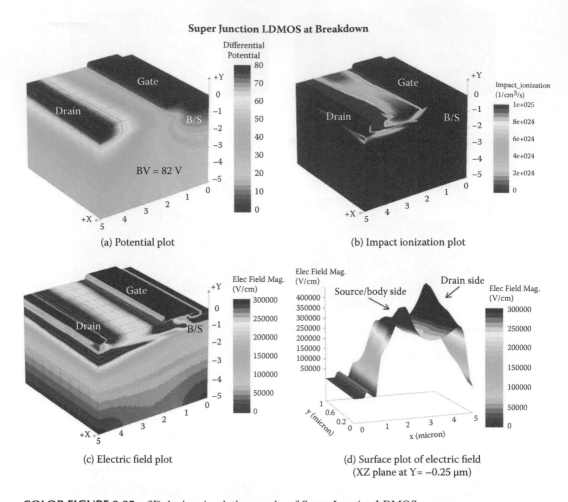

Super Junction LDMOS at Breakdown

(a) Potential plot

(b) Impact ionization plot

(c) Electric field plot

(d) Surface plot of electric field (XZ plane at Y = −0.25 μm)

COLOR FIGURE 9.35 3D device simulation results of Super Junction LDMOS.

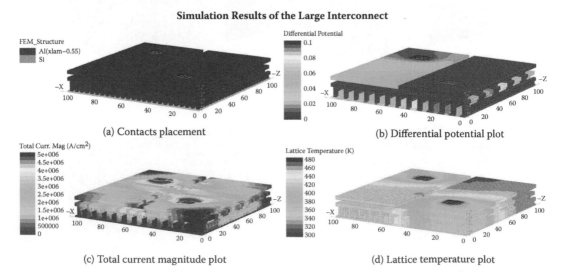

Simulation Results of the Large Interconnect

(a) Contacts placement

(b) Differential potential plot

(c) Total current magnitude plot

(d) Lattice temperature plot

COLOR FIGURE 9.63 3D device simulation results of the large interconnect with (a) contacts placement, (b) differential potential, (c) total current magnitude, and (d) lattice temperature.

9 Integrated Power Semiconductor Devices with 3D TCAD Simulations

This chapter provides a few 3D technology computer-aided design (TCAD) simulation examples for power semiconductor devices. Since the structure of power devices is getting more and more complex, traditional 2D simulations often fail to reveal important three-dimensional effects. Typical device examples with pronounced three-dimensional effects include Super Junction LDMOS, segmented anode lateral insulated gate bipolar transistor (LIGBT), and complicated interconnect structures.

9.1 3D DEVICE LAYOUT EFFECT

For power devices design, one of the most important considerations is the breakdown voltage. To enhance this value, we need to know how the electric field is distributed in the device: a uniformly distributed electric field is always preferred, but in a real design electric field crowding is unavoidable.

If we take a look at the power LDMOS layout, two of the most frequently used layout structures are given in Figure 9.1. Figure 9.1a has a racetrack layout structure, and Figure 9.1b has a rectangle shaped layout.

When a positive voltage is applied to the N+ drain contact while keeping gate and source/body terminals grounded, the device is reverse biased. For illustration purpose, if we plot the electric field lines, we can see a "flat" portion on both racetrack and square/rectangle shaped structures. The electric field in the flat regions (highlighted in dotted rectangles) resembles the ideal parallel-plane electric field in a power device. The curved regions in the racetrack structure and the corner points in the rectangle shaped structure will modify the electric field distribution and cause crowding of the electric field lines, as seen in Figure 9.2. This field crowding is a 3D effect and will degrade breakdown voltage.

To analyze this 3D field crowding effect, we can use 3D TCAD to simulate simplified structures and observe the distribution of the electric field. We first compare a square shaped structure and a circular shaped structure, with simulation parameters shown in Figure 9.3.

3D net doping plots for both structures are shown in Figures 9.4a and 9.4b; the differential potential is also illustrated in Figures 9.4c and 9.4d. For both devices, 40 V is applied to the cathode, and the anode is grounded.

Comparison plots of electric field magnitude are given in Figure 9.5. Figures 9.5a and 9.5b are 3D electric field plots for the square and circular structures; Figures 9.5c and 9.5d are 3D surface plots (XZ plane at Y = 0) to better visualize the same data. We can see from Figures 9.5c and 9.5d that the square-shaped structure has a higher average electric field than the circular structure. The electric field also peaks at the corners of the device, indicating high electric field crowding.

Further breakdown simulations using the same model parameters are performed for both structures, and the results are shown in Table 9.1. As expected, the higher field values in the square structure correlate with a much lower breakdown voltage than that of the circular structure. We also observe that while the 2D result is consistent with the ideal parallel-plane breakdown result given in (Table 6.6) both 3D structures show a significant deviation from this value. This clearly shows the importance of three-dimensional TCAD simulations.

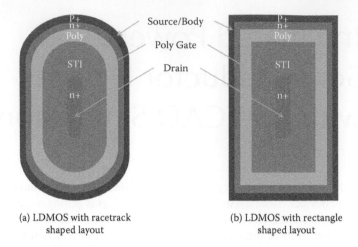

(a) LDMOS with racetrack
shaped layout

(b) LDMOS with rectangle
shaped layout

FIGURE 9.1   Two layout structures of LDMOS with (a) racetrack and (b) rectangle.

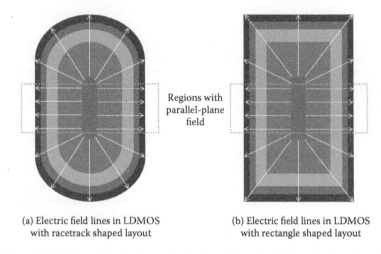

(a) Electric field lines in LDMOS
with racetrack shaped layout

(b) Electric field lines in LDMOS
with rectangle shaped layout

FIGURE 9.2   Electric field lines plots for (a) racetrack-shaped layout and (b) rectangle-shaped layout LDMOS.

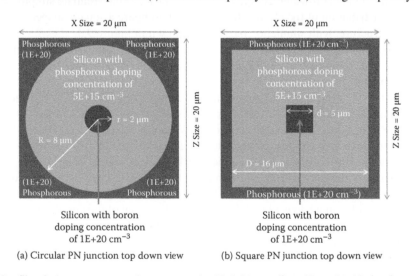

(a) Circular PN junction top down view

(b) Square PN junction top down view

FIGURE 9.3   Simulation structures and parameters for 3D field crowding effect with (a) circular and (b) square.

3D Net Doping and Potential (40 V) Plot: Circular and Square Shapes

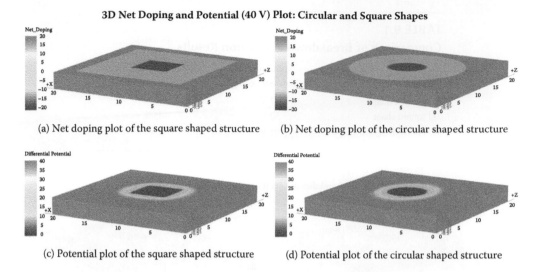

(a) Net doping plot of the square shaped structure   (b) Net doping plot of the circular shaped structure

(c) Potential plot of the square shaped structure   (d) Potential plot of the circular shaped structure

FIGURE 9.4   3D simulation results with net doping plots of (a) square and (b) circular structures and potential plots of (c) square and (d) circular structures.

3D Electric Field Comparison: Circular and Square Shapes

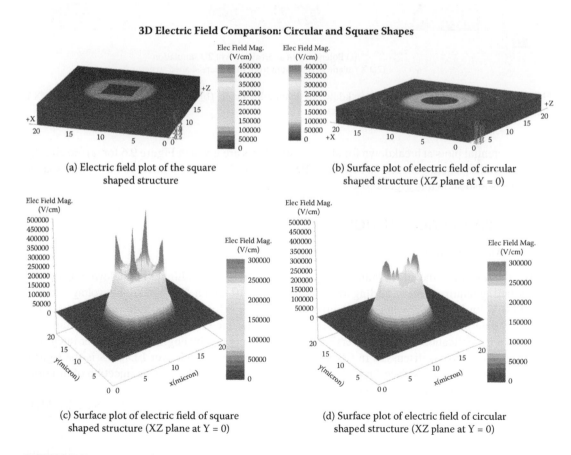

(a) Electric field plot of the square shaped structure   (b) Surface plot of electric field of circular shaped structure (XZ plane at Y = 0)

(c) Surface plot of electric field of square shaped structure (XZ plane at Y = 0)   (d) Surface plot of electric field of circular shaped structure (XZ plane at Y = 0)

FIGURE 9.5   3D simulation results with electric field plots of (a) square and (b) circular structures and surface plot of electric field of (c) square and (d) circular structures.

TABLE 9.1

Comparison of Breakdown Simulation Results

Simulated Structures	Simulated Breakdown Voltage	Deviation from 2D Breakdown Voltage
3D circular shape	67 V	−26%
3D square shape	50 V	−45%
2D cut from square shape	91 V	0%

Potential Plots at Breakdown: Circular, Square and 2D Structure

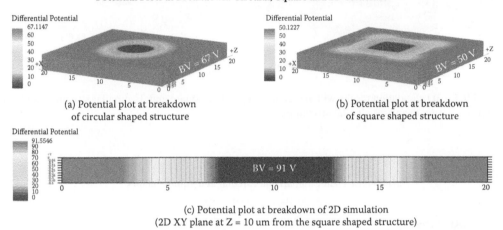

(a) Potential plot at breakdown of circular shaped structure

(b) Potential plot at breakdown of square shaped structure

(c) Potential plot at breakdown of 2D simulation
(2D XY plane at Z = 10 um from the square shaped structure)

FIGURE 9.6   The potential plots at breakdown of (a) circular shape, (b) square shape, and (c) 2D plane (XY plane at Z = 10 μm of the square-shaped structure).

The potential plots at breakdown for all three structures are given in Figure 9.6 for (a) the circular structure and (b) square-shaped structure. A 2D XY cut at Z = 10 μm (center) of the square-shaped structure is also shown in Figure 9.6c.

9.2 3D SIMULATION OF LIGBT

9.2.1 ABOUT LIGBT

As we already know, the on-resistance of an LDMOS increases with higher breakdown voltage due to conflicting requirements regarding low doping in the drift region. For breakdown voltages higher than 600 V, an LDMOS, even a Super Junction LDMOS, exhibits a very high on-resistance. The insulated gate bipolar transistor (IGBT) structure was invented in the 1980s for high-voltage and high-current applications; its lateral version, LIGBT, was created by engineers for power integrated circuit applications (as discussed in Chapter 4). The merit of LIGBT over LDMOS is the significantly lower on-state resistance for high voltage designs, thanks to the plasma injection of minority carriers when the LIGBT is turned on.

In terms of device structure, LIGBT is very similar to an LDMOS with a gate-controlled MOS structure and a long lightly doped drift region to support high voltages. However, LIGBT has an extra P+/N junction at the anode (or collector) side, which differentiates it from an LDMOS. Figure 9.7 illustrates a comparison between LDMOS (Figure 9.7a) and LIGBT (Figure 9.7b). In the

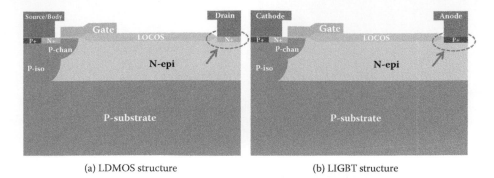

(a) LDMOS structure (b) LIGBT structure

FIGURE 9.7 Comparison between the LDMOS structure and LIGBT structure.

LIGBT structure, a P+ anode replaces the N+ drain of the LDMOS, and the identical source/body in the LDMOS is now called the cathode.

9.2.2 Segmented Anode LIGBT

LIGBT is promising for high-power applications in smart power ICs, thanks to the much lower on-resistance resulting from conductivity modulation of the high-resistivity drift region by injection of minority carriers. Unfortunately, this minority carrier injection also causes minority carrier storage effects, which substantially degrade the device switching speed. A shorted anode N+ device was proposed to alleviate this problem by providing an electron extraction path incorporated at the anode to extract electrons during device turnoff, the structure of which is illustrated in Figure 9.8. With access to the electrons, the electron–hole plasma in the drift region can be extracted, and a significant improvement in switching speed with only moderate decrease in current handling capability compared to the P+ anode [105]. This improvement is due to the incorporation of the N+ diffusion at the anode, which helps in removing the plasma during device turnoff. However, the continued forward bias of the anode during device turnoff, which causes unnecessary minority-carrier injection, retains the turnoff time of the

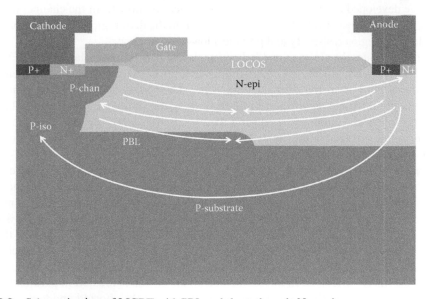

FIGURE 9.8 Schematic view of LIGBT with PBL and shorted anode N+ region.

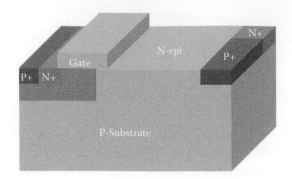

FIGURE 9.9   3D schematic view of segmented LIGBT.

device in the order of hundreds of nanoseconds [192]. Also, the additional area required for implementing the anode N+ diffusion makes the device less area-efficient.

For the shorted anode structure, the injection efficiency of the P+/N+ anode depends on the P+
length for a constant N+ length. For good injection efficiency, a large P+ length is needed; however,
this will slow down the switching speed due to the overinjection during on-state and excessive injection during device turnoff. Furthermore, the device size will be increased due to the large anode
size. On the other hand, a faster device can be achieved by reducing the P+ length. The maximum
speed of the device is limited by the practical minimum of the P+ length [192].

A smart structure called segmented anode LIGBT was proposed in [193]. A simplified 3D schematic view of this segmented anode LIGBT is shown in Figure 9.9. This design eliminates the
problems of long anode length and excessive hole injection during device turnoff. In this structure,
the electron conduction path at the anode is implemented along the device width by periodically
eliminating part of the P+ diffusion and replacing it with an N+ diffusion. Since the anode is implemented using segments of P+ and N+ diffusions, it is named as segmented anode LIGBT. At low
forward bias, electrons that flow from the channel through the drift region are collected by the
anode N+ diffusion in a way identical to that of the LDMOS. At high forward bias, the P+ portion
of the anode, which is situated in parallel with the N+ diffusion, is turned on. The turn-on is caused
by the voltage developed in the drift region due to the electron flow from the channel to the anode.
Turning on of the P+ injector injects minority carriers into the drift region. These minority carriers
modulate the drift region resistivity and provide a low on-resistance.

During the on-state, the electron and hole current flow paths are similar to that of the shorted
anode LIGBT. However, during the turnoff transient, most electrons in the segmented anode
structure flow directly from the drift region and the substrate to the N+ region without flowing
underneath the P+ diffusion as different from the case of the shorted anode device. Therefore, no
continued turn-on of the anode exists during the turnoff transient. The electrons present in this
structure are extracted through the anode N+ diffusion and the holes are extracted through the P+
cathode for fast device turnoff.

The segmented anode design provides the following advantages. First, this device requires a
smaller anode area compared to that of the shorted anode LIGBT. This is because the N+ diffusion, which is located behind the P+ anode diffusion in the shorted anode LIGBT, is not needed,
and the area needed by the N+ diffusion at the anode of the segmented anode LIGBT is relatively
small. The smaller anode area makes the device more area-efficient despite the fact that part of
the P+ anode area is used for implementing the N+ diffusion. Second, faster switching speed
can be achieved compared to the shorted anode device for the same current handling capability.
This is because the N+ diffusion in the segmented anode design is located in parallel with the P+
diffusion instead of behind the P+ diffusion as in the case of the shorted anode design. Most of

the electrons that are extracted from the drift region can flow to the anode N+ diffusion without flowing underneath the P+ diffusion. Only a small amount of electrons that flow near the two ends of the P+ injector will cause that part of the injector to turn on. This minimizes the forward bias of the P+ injector caused by electron extraction during device turnoff, resulting in faster switching speed compared with the shorted anode design. Finally, the trade-off between the on-resistance and the turnoff time can be achieved simply by adjusting the ratio of the P+ to N+ diffusion width along the device width. The wider the P+ diffusion, the lower the on-resistance and the longer the device turnoff time will be for a constant N+ diffusion width. Whereas in the case of the shorted anode LIGBT, increasing the P+ anode length for lower on-resistance may have to increase the device area as the P+ anode length is made longer [105].

9.2.3 3D PROCESS SIMULATION OF SEGMENTED ANODE LIGBT

Since the segmented anode LIGBT can be revealed only with 3D simulations, both 3D process simulation and device simulation are performed for better understanding of the device physics. Nine masks are used to create the LIGBT structure: P-type buried layer (PBL); local oxidation of silicon (LOCOS); P-iso; P-body; gate, N-plus; P-plus; contact; and Metal1. The device width is 25 µm: P+ anode occupies 20 µm, and N+ anode takes 5 µm. The total cell pitch of the device is 45 µm. The layout views of the nine masking steps are provided in Figure 9.10.

The simplified process steps of the segmented anode LIGBT are shown in Figure 9.11, with detailed information given in Table 9.2. In this process, LOCOS is used instead of STI. N-epi is grown on P-substrate, with a thickness of 6.5 µm. PBL mask and implant are utilized. The RESURF (Reduced Surface Electric Field) implant is a blanket implant to adjust the doping concentration in the N-epi layer. The P-ISO and drive-in step is applied so that a deep and heavily doped P+ region connects the P-body to the P-substrate. P+ and N+ sections coexist in the anode region and are self-aligned to the LOCOS. In this particular simulation example, P+ has a width of 20 µm, while N+

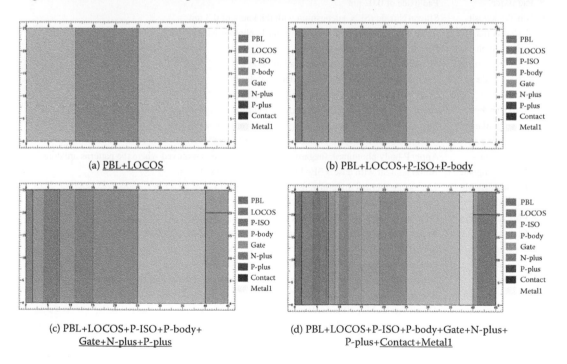

(a) PBL+LOCOS

(b) PBL+LOCOS+P-ISO+P-body

(c) PBL+LOCOS+P-ISO+P-body+
Gate+N-plus+P-plus

(d) PBL+LOCOS+P-ISO+P-body+Gate+N-plus+
P-plus+Contact+Metal1

FIGURE 9.10 Layout views of the segmented anode LIGBT.

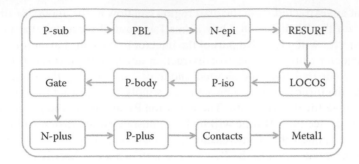

FIGURE 9.11 Simplified process steps of 3D TCAD simulation of a segmented anode LIGBT.

TABLE 9.2

Simplified Process Steps of 3D TCAD Simulation of a Segmented Anode LIGBT

Process Steps	Description
1. P-sub	Boron concentration of 3E+14 cm^{-3}, orientation = (100)
2. PBL mask	Apply PBL photomask (negative mask, photoresist is removed where the patterns are drawn)
3. PBL implant	Boron implant with dose = 2E+12 cm^{-2}, energy = 120 keV
4. PBL anneal	1 minute thermal anneal at temperature of 1000°C
5. N-epi layer	N-type epitaxial layer with thickness of 6.5 μm is deposited. The phosphorus doping concentration is 5E+14 cm^{-3}
6. RESURF implant	Phosphorus blanket implant with dose = 1.5E+12 cm^{-2}, energy = 150 keV
7. RESURF anneal	1 minute thermal anneal at temperature of 1000°C
8. Pad oxide	Pad oxide of 0.02 μm
9. Nitride deposit	Silicon Nitride (Si$_3$N$_4$) deposition with thickness of 0.3 μm
10. LOCOS mask	Apply LOCOS photomask (negative mask, photoresist is removed where the patterns are drawn)
11. Nitride etch	Nitride is etched where unmasked
12. LOCOS growth	60 minutes wet oxide at temperature of 1000°C
13. Etch nitride	Excessive nitride is etched away (thick = 0.3 μm)
14. Oxide etch	Excessive oxide is etched away (thick = 0.02 μm)
15. P-iso mask	Apply P-iso photomask (negative mask, photoresist is removed where pattern is drawn)
16. P-iso implant	Boron implant with dose = 1.5E+16 cm^{-2} and energy = 150 keV
17. P-iso drive-in	100 minute thermal drive-in at temperature of 1150°C
18. P-body mask	Apply P-body photomask (negative mask, photoresist is removed where the patterns are drawn)
19. P-body implant	Boron implant with dose = 3E+13 cm^{-2}, energy = 50 keV
20. P-body drive-in	120 minute thermal drive-in at temperature of 1150°C
21. Gate oxide	Gate oxide growth to yield oxide thickness of 600 Å
22. Poly deposition	Polysilicon deposition of 0.4 μm with phosphorus concentration of 1E+20 cm^{-3}
23. Gate mask	Apply gate photomask (positive mask, photoresist is preserved where the patterns are drawn)
24. Gate etch	Polysilicon and oxide are subsequently dry-etched
25. Gate anneal	Poly gate anneal at 1150°C for 1 minute
26. N-plus mask	Apply N-plus photomask (negative mask, photoresist is removed where the patterns are drawn)
27. N-plus implant	Phosphorus implant with dose = 2E+16 cm^{-3} and energy = 20 keV
28. N-plus anneal	Thermal anneal for 10 seconds at 900°C
29. P-plus mask	Apply P-plus photomask (negative mask, photoresist is removed where the patterns are drawn)
30. P-plus implant	Boron implant with dose = 1E+16 cm^{-3} and energy = 15 keV

TABLE 9.2 (CONTINUED)

Simplified Process Steps of 3D TCAD Simulation of a Segmented Anode LIGBT

Process Steps	Description
31. P-plus anneal	Thermal anneal for 10 seconds at 900°C
32. Oxide deposition	Deposit oxide with thickness of 0.5 μm
33. Contact mask	Apply contact photomask (negative mask, photoresist is removed where the patterns are drawn)
34. Contact etch	Etch oxide to expose contact locations
35. Metal deposition	Deposit metal with thickness of 0.7 μm
36. Metal mask	Apply metal photomask (positive mask, photoresist is preserved where the patterns are drawn)
37. Metal etch	Etch metal to leave cathode, gate, and anode contacts

has a width of 5 μm. We will make a comparison between different P+ to N+ ratios later in terms of forward and turnoff transit characteristics.

The process simulation results are given in Figure 9.12 with material plot and net doping plots. For a better view of net doping in silicon, all other materials except silicon are ripped off in Figure 9.12c to expose silicon only. A 2D cut of XY plane at Z = 0 (with P+ anode) is also shown in Figure 9.12d.

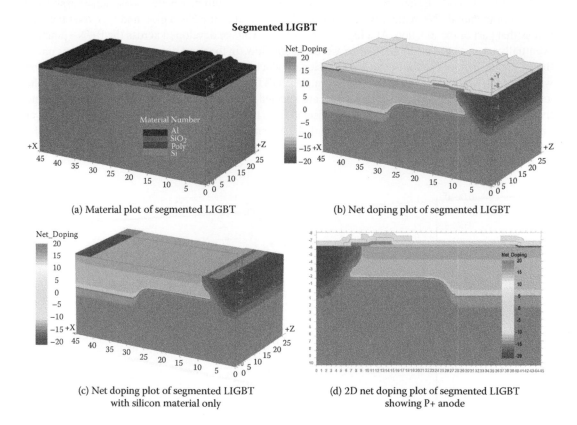

(a) Material plot of segmented LIGBT

(b) Net doping plot of segmented LIGBT

(c) Net doping plot of segmented LIGBT with silicon material only

(d) 2D net doping plot of segmented LIGBT showing P+ anode

FIGURE 9.12 Simulation results of segmented anode LIGBT with (a) material plot, (b) net doping plot, (c) net doping plot with silicon material only, and (d) 2D net doping plot showing a P+ anode.

9.2.4   3D Device Simulation of Segmented Anode LIGBT

9.2.4.1   Forward Characteristic Simulation

Since the breakdown characteristic of the segmented anode LIGBT is similar to LDMOS, the break-down analysis will not be repeated here. Forward characteristics of segmented anode LIGBT are analyzed at different P+ anode to N+ anode section ratios, with the following simulation conditions:

1. Gate voltage is first biased to 5 V to ensure the device is fully turned on.
2. Anode voltage is scanned to 1.5 V followed by a current scan till the anode current reaches 0.004 A. (With 45 μm by 25 μm device size, this current density is about 360 A/cm².)

The simulation results of different P+ anode to N+ anode ratios are compared in Figure 9.13. N+ only structure (LDMOS) has the highest on-resistance and no minority carrier injection effect. With higher ratio of P+ section over N+ section in the anode, higher current handling capability is obtained with the same anode voltage. P+ only (pure LIGBT) has the highest current handling capability and the lowest on-resistance.

In comparison with pure P+ anode LIGBT, which has an onset voltage (P+/N− junction turn-on voltage) of less than 1 V, for segmented anode LIGBT there is a voltage clamping effect at the P+/N+ anode. The N+ region clamps the voltage near the anode such that the P+/N− diode does not turn on at a low forward bias; it behaves like a resistor with electrons flowing from the cathode through the N− region to the N+ anode. Before the injector turns on, electrons accumulate at the P+/N− junction situated away from the anode N+ region, which reduces the potential at the N− region and increases the bias across that part of the junction. At a higher forward bias, voltage developed across the P+/N− junction eventually turns on the junction. In other words, lateral flow (in the third dimension) of electrons under the P+ anode diffusion to the N+ anode diffusion, producing lateral Ohmic drops, is the mechanism by which the P+/N− junction in the segmented anode LIGBT structure becomes forward biased. After the junction is forward biased, minority carriers are injected into the N− region to conductivity-modulate the drift region resistance. Thus, holes are injected from the P+ anode into the N− region. Many fewer holes are observed in the N− region underneath the N+ anode due to the voltage clamping effect at the P+ injector near the N+ diffusion. Since the conductivity modulation effect reduces the drift region

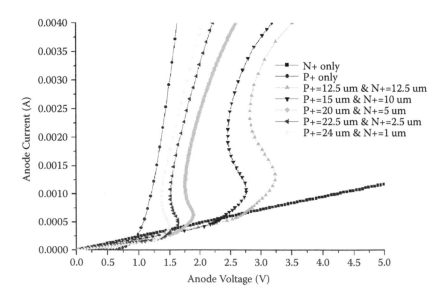

FIGURE 9.13   Forward I-V curves of different anode P+/N+ ratios.

resistance, forward voltage of the device after the injector turns on is actually lower than the forward voltage before the injector turns on and at a much higher forward current.

The simulated I-V characteristics of the segmented anode LIGBT with different anode P+ to N+ ratio are illustrated in Figure 9.13. Except anode cases with P+ only (pure LIGBT) and N+ only (LDMOS), the segmented anode LIGBTs exhibit bistable I-V characteristics with two conduction regimes. The bistable characteristic, as will be described shortly, is due to the conductivity modulation effect coupled with an Ohmic voltage drop developed at the electron conduction path implemented by the anode N+ diffusion. The first conduction regime represents the LDMOS conduction regime, in which the forward bias is so low that the P+/N– diode at the anode is not active, and the current flow is taking place at the anode N+ diffusion. At a high forward bias, the voltage developed across the diode causes the diode to turn on and conductivity-modulates the drift region shunting resistance, which shunts the current flowing into the P+/N– junction. Since the shunting resistance drops after the P+/N– diode turns on, the forward voltage drops correspondingly, even at a higher current density. This causes the I-V characteristic to snap back. As the conductivity modulation effect is saturated, further increase in current level causes the forward voltage to increase, and the I-V characteristic reaches the bipolar regime. The voltage at which the snapback starts to occur is defined as the onset voltage, V_{onset}. The V_{onset} is adjustable and is a function of the shunting resistance. The lower the shunting resistance, the higher the V_{onset}. Since the shunting resistance is partially determined by the anode P+ to N+ ratio, the onset voltage and the level of conductivity modulation can then be determined by adjusting the P+ to N+ ratio. The smaller P+ to N+ ratio represents a smaller shunting resistance, which allows more current to flow through before the diode turns on and therefore results in a higher onset voltage. This higher onset voltage translates into a higher forward voltage and gives rise to a lower current handling capability compared with a larger P+ to N+ ratio. Thus, by adjusting the P+ to N+ ratio, a different injection efficiency can be achieved and segmented anode LIGBTs can be designed to have different current handling capability [193].

Except the N+ structure (LDMOS), with anode voltage passes some onset voltage (V_{onset}), electron–hole plasma is formed in the drift region, which modulates the conductivity of the drift region and lowers the on-resistance of the transistor, as mentioned before. Figure 9.14 is the 3D TCAD simulation results of a segmented anode LIGBT with P+ (20 µm) and N+ (5 µm). We can see the electron–hole plasma in 3D from Figure 9.14 and in 2D cut planes from Figure 9.15. At a forward voltage of 1.75 V and a forward current of 0.002 A (after snapback), it is evident that very high electron and hole concentrations are located at the anode side of the drift region and that relatively

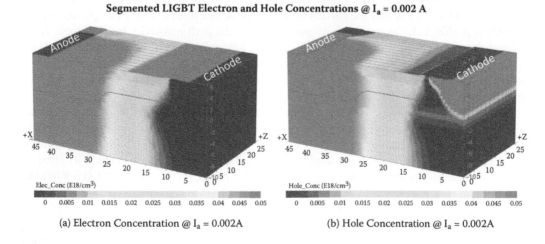

(a) Electron Concentration @ I_a = 0.002A   (b) Hole Concentration @ I_a = 0.002A

FIGURE 9.14 3D simulation results of electron and hole plasma injection of segmented anode LIGBT at I_a = 0.002 A (P+ = 20 µm and N+ = 5 µm).

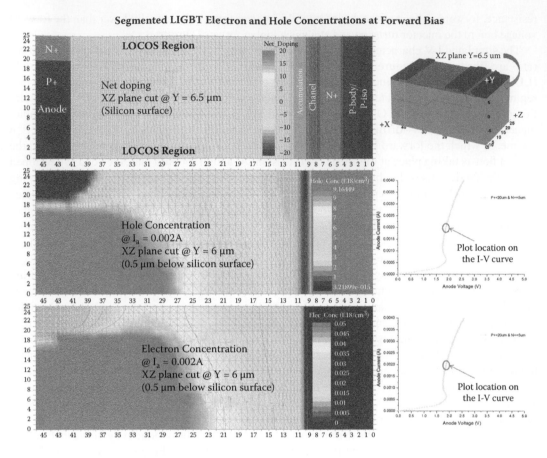

FIGURE 9.15 XZ plane at Y = 6 μm plots with net doping, hole concentration, and electron concentration of segmented anode LIGBT at I_a = 0.002 A (P+ = 20 μm and N+ = 5 μm).

lower concentrations are located at the cathode side of the drift region. Ideally, we would like to see a more uniform plasma concentration in the entire drift region for a low on-state resistance. Figures 9.14 and 9.15 show that the current design of the device does not have an optimized plasma distribution profile as far as conductivity modulation of the drift resistivity is concerned. A more optimum design of the plasma profile should have the amount of holes injected by the anode to be reduced (for faster switching as will be discussed in the next section) and the amount of electrons injected by the cathode to be increased (for attracting more holes to flow to the cathode side of the drift region for further conductivity modulation).

As described already, hole injection can be reduced by adjusting the P+/N+ ratio at the anode or by implementing the anode using a Schottky contact, which was first introduced to minimize the anode hole injection in the LIGBT. This device was called the Schottky injection field effect transistor (SINFET) [194]. A lightly doped P anode was also used later on to reduce the hole injection in the IGBT [195]. This structure also allows a lighter N-buffer region to be used to produce a field stop IGBT (FS IGBT) for a more optimum switching loss performance. Electron injection enhancement can be done by increasing the resistive path for holes flow to the cathode using a narrow mesa trench gate structure [196]. With a pair of deep trenches and a narrow N-mesa, accumulation of holes at the narrow mesa raises the electrostatic potential at the P-body/N-drift junction for enhancement of electron injection from the channel. This much increased electron–hole plasma concentration at the cathode side of the N-drift is responsible

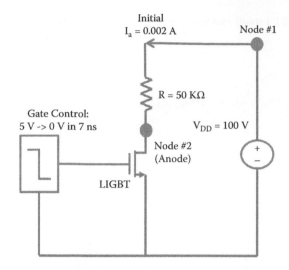

FIGURE 9.16   Circuit setup of LIGBT turnoff transient simulation.

for the further reduction of the on-resistance in the structure. With this increase in plasma, the entire drift region is now conductivity modulated for the optimum on-state performance.

9.2.4.2   LIGBT Turnoff Transient Simulation

The turnoff transient of the segmented anode LIGBT can be simulated with a resistive load; the simulation setup is illustrated in Figure 9.16. A resistor of 50 KΩ is connected in series to the anode of the LIGBT. The simulation procedure is as following:

1. Equilibrium is first established.
2. Gate voltage is scanned to 5 V in 10 μs.
3. Anode voltage is scanned to 1 V in 10 μs.
4. Anode current is scanned to 0.002 A in 10 μs (178 A/cm² in anode current density, the voltage on Node #1 of Figure 9.16 should reach approximately 100 V).
5. Turn off the LIGBT by rapidly ramping down the gate voltage from 5 V to 0 V in 7 ns.
6. Scan the time till the anode current decreases to a negligible level.

The transient simulation results of the LIGBT turnoff are illustrated in Figure 9.17. Note that compared with the N+ only and other segmented anode LIGBTs, the P+ only configuration takes too long for the current to drop to a negligible level; therefore, it is drawn in a separate curve in Figure 9.18. The long turnoff time of the P+ only LIGBT is expected, since the minority carriers injected when the anode is forward biased is not possible to be extracted with a P+ only anode. Removal of the electron–hole plasma for device turnoff can be relied only on electron–hole recombination, which is mainly a function of the minority carrier lifetime. The stored minority carriers prevent the transistor to switch fast. With even a small N+ region, the switching time decreases significantly. The switching time can be adjusted by modifying the P+/N+ ratio: the smaller the ratio, the shorter the turnoff time, with the N+ only (LDMOS) being the extreme case.

Unfortunately, as we have observed in Figure 9.13, under the forward bias condition, a smaller P+/N+ ratio means lower current handling capability (higher on-resistance). A trade-off has to be made between fast switching and current handling capability (on-resistance).

The electron and hole plasma distributions in the segmented anode LIGBT structure (P+ = 20 μm and N+ = 5 μm) during turnoff transient are shown in Figures 9.19 and 9.20, respectively. In

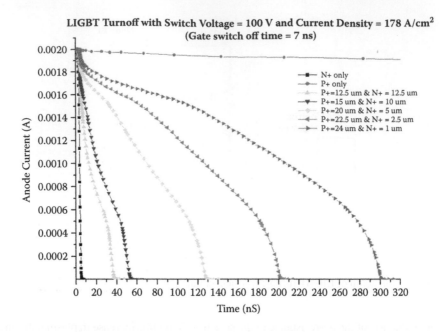

FIGURE 9.17 Turnoff transient simulation results of segmented anode LIGBTs.

Figure 9.19, one can see that electrons at the cathode region are removed almost instantly after the removal of the MOS channel in approximately the first 10 ns into turnoff. This causes the sudden drop of the anode current in the first 10 ns, as shown in the I-V curve in Figure 9.17. As the device gets deeper into turnoff, the electron plasma was removed through electron–hole recombination in the drift region as well as electron extraction through the anode (at time = ~35 ns), in which the

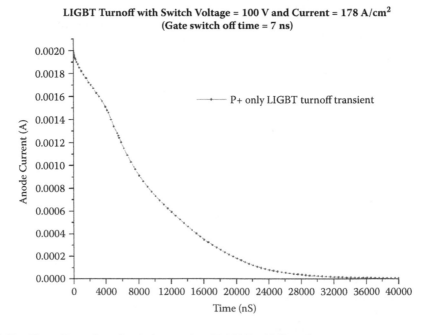

FIGURE 9.18 Turnoff transient simulation results of LIGBT with P+ only anode.

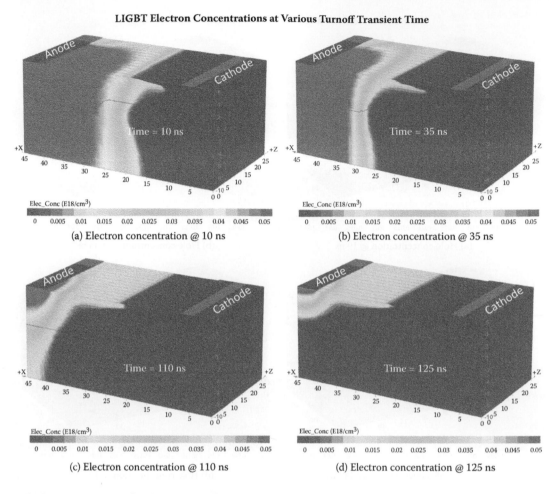

LIGBT Electron Concentrations at Various Turnoff Transient Time

(a) Electron concentration @ 10 ns

(b) Electron concentration @ 35 ns

(c) Electron concentration @ 110 ns

(d) Electron concentration @ 125 ns

FIGURE 9.19 **(See color insert)** LIGBT turnoff transient plots of electron concentrations at (a) time = 10 ns, (b) time = 35 ns, (c) time = 110 ns, and (d) time = 125 ns.

latter process dominates since the electron–hole recombination is a much more lengthy process. The remaining electrons were removed quickly through electron extraction at the anode (at time = ~110 ns), and eventually very few electrons remained at the anode at time = 125 ns.

A slightly different distribution profile is observed for the holes, as shown in Figure 9.20. Even with the removal of the MOS channel and the sudden stop supplying of electrons, a considerable amount of holes are still situated near the cathode region at the first few tens of nanoseconds after turnoff. While electrons were extracted through the anode, holes were extracted through the cathode at the same time during the entire turnoff. The extraction was first concentrated closer to the surface and eventually shifted to the bulk of the device via the P-body at the end of turnoff because holes are more difficult to be extracted deeper at the substrate.

With the mechanism of plasma extraction having taken place in the segmented anode LIGBT, minority carrier storage effect in conventional LIGBT can be significantly reduced, and LIGBT with fast switching characteristics can be obtained.

9.2.4.3 Forward I-V Family of Curves Comparison with LDMOS

If we plot the forward I-V family of curves for LIGBT (P+ only anode) and compare them to LDMOS (the same structure but with N+ only anode or drain), we will see a much better performance is

LIGBT Hole Concentrations at Various Turnoff Transient Time

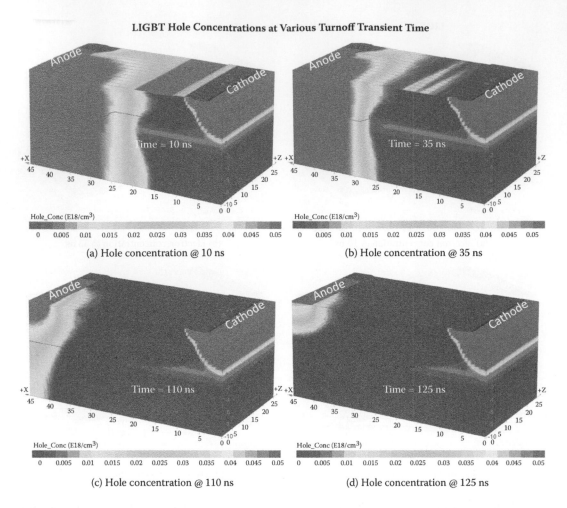

(a) Hole concentration @ 10 ns   (b) Hole concentration @ 35 ns

(c) Hole concentration @ 110 ns   (d) Hole concentration @ 125 ns

FIGURE 9.20   (See color insert) LIGBT turnoff transient plots of hole concentrations at (a) time = 10 ns, (b) time = 35 ns, (c) time = 110 ns, and (d) time = 125 ns.

obtained with the LIGBT structure, except for a tiny region before the onset of injection at low anode (or drain) voltage, as shown in Figure 9.21.

9.3   SUPER JUNCTION LDMOS

9.3.1   BASIC CONCEPT

Like RESURF, which employs a charge coupling effect, a Super Junction LDMOS uses alternating N- and P-regions to mutually deplete each other so that a higher doping concentration and lower on-resistance can be achieved. It also means that for the same doping concentration in the drain-drift region, a higher breakdown voltage can be obtained.

To explain the Super Junction concept, we start with a simple 2D simulation. In Figure 9.22, a simple 2D structure with alternating N- and P-pillars is shown. These are used to mimic the N- and P-stripes in the drain-drift region of a Super Junction LDMOS. (In this chapter, we do not distinguish between N/P pillars, N/P stripes, and N/P wells as long as they have alternating N- and P-regions to mutually deplete each other to support high breakdown voltages). Assuming the width of N-pillars equals P-pillars ($W_N = W_P$), L is the length of N/P pillars. This 2D structure has a

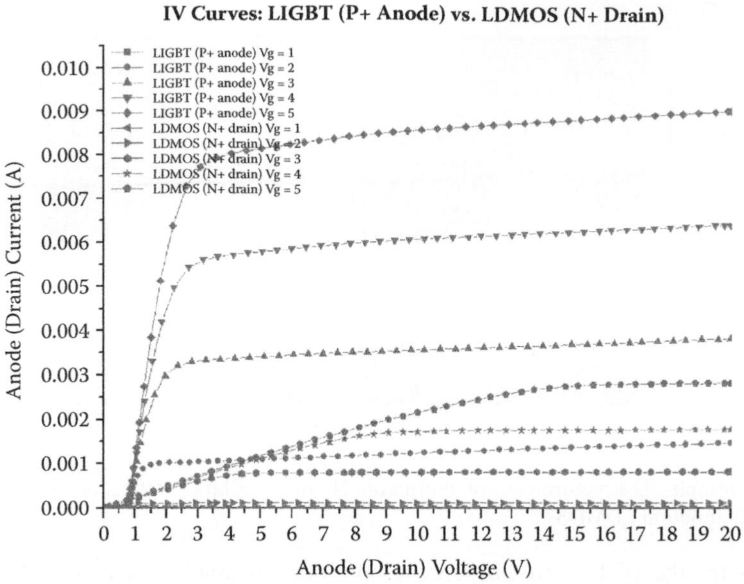

IV Curves: LIGBT (P+ Anode) vs. LDMOS (N+ Drain)

FIGURE 9.21 The I-V curve comparison between LIGBT (P+ anode) and LDMOS (N+ drain) with the same structure except anode (drain) dopant type.

heavily doped N+ cathode (to mimic the N+ drain contact in Super Junction LDMOS) and a heavily doped P+ anode (to mimic the P-body and the source/body contact).

For this kind of alternating N/P structure, it has been found that the maximum breakdown voltage takes place with maximum (critical) electric field E_C at the optimum charge Q_{opt} in the N- and P-regions [158]:

$$Q_{opt} = qN_D W_N = qN_P W_P = \varepsilon_s E_C \tag{9.1}$$

where $\varepsilon_s = 11.68$ is the dielectric constant of silicon, $q = 1.602\text{E-}19$ C is the electron charge, and N_D and N_P are the doping concentrations of the N- and P-regions, respectively. If we assume $E_C = 3\text{E} + 5$ V/cm, the optimum charge Q_{opt} is found to be $3.1\text{E-}7$ C/cm². It is often more convenient to use optimum dopant density per unit area ($dose = N_D W_N$, which is calculated to be $1.94\text{E} + 12$ cm⁻²) than an optimum charge, but this should not be confused with the device implant dose. Following the axis convention used throughout this book, the unit area for the dopant density is defined on the XY plane (along the N/P boundary in this case), while the implant dose is defined on the XZ plane (the wafer surface, perpendicular to XY plane). This optimum charge/dose is often oversimplified and not suitable for cases where depletion is from multiple sources (like double RESURF or multiple RESURF).

9.3.1.1 How to Choose the Pillar Width W_N

The N/P pillars widths W_N and W_P are often designed to be equal to each other, with their values critical for a successful design of Super Junction LDMOS. 2D simulations with different doping concentrations and pillar widths for the simple structure of Figure 9.22 were performed. Different pillar widths were chosen ($W_N = 0.1$, 0.2, 0.3, and 0.4 μm), coupled with different doping concentrations. Note that during every simulation, $W_N = W_P$ and doping concentration in N-pillars always equals P-pillars to keep dose balanced. Figure 9.23 shows the simulation results.

For each pillar width, there is an optimum doping concentration, beyond which the breakdown voltage drops dramatically. In general, narrower pillars yield a higher optimum doping concentration,

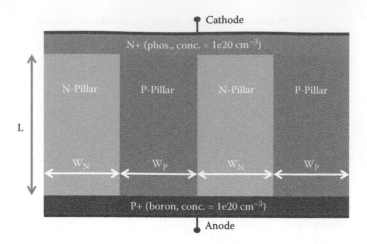

FIGURE 9.22 Simple 2D structure view of alternating N- and P-pillars to reflect charge coupled drain-drift region of a Super Junction LDMOS.

as predicted by the theory. However, depending on process technology, narrow pillars, like the one with a width of only 0.1 μm, is difficult to obtain with current smart power IC technology. In the near future, with smart power IC feature sizes below 90 nm node, this layout width should be realized.

To better understand how the Super Junction concept works, two simulations are compared. Both structures have N/P pillars doping concentration of 1E + 17 cm^{-3} but have different pillar widths: as illustrated in Figure 9.24, one device has $W_N = W_P = 0.1$ μm, while the other has $W_N = W_P = 0.4$ μm. Voltage is applied to the cathode terminals while keeping anode terminals grounded.

The breakdown simulation results for the 0.1 μm device are shown in Figure 9.25. The breakdown voltage for this structure reaches 104 V (Figure 9.25a) and both N- and P-pillars are fully depleted as can be seen from the space charge plot in Figure 9.25b. The electric field is very uniform across the device as shown in Figures 9.25c and 9.25d. The heavily doped N+ cathode and P+ anode regions are excluded in the 3D surface electric field plot of Figure 9.25d.

The breakdown simulation results for the 0.4 μm structure are shown in Figure 9.26. The breakdown voltage is limited to 36 V (Figure 9.26a), and the N/P pillars are not fully depleted, as can be

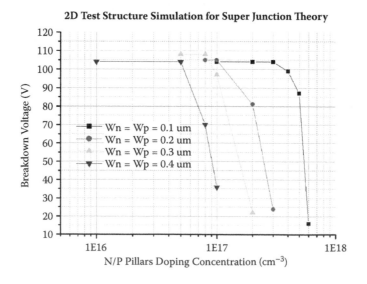

FIGURE 9.23 2D test structure simulation results for Super Junction theory.

Net Doping Plot of 2D Structures with $W_N = 0.1$ µm and $W_N = 0.4$ µm

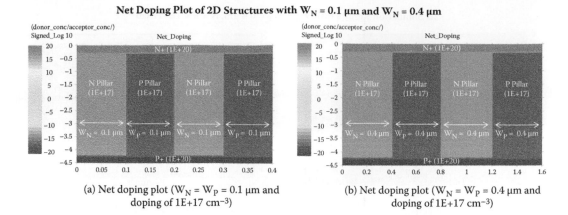

(a) Net doping plot ($W_N = W_P = 0.1$ µm and doping of 1E+17 cm^{-3})

(b) Net doping plot ($W_N = W_P = 0.4$ µm and doping of 1E+17 cm^{-3})

FIGURE 9.24 Net doping plots of the two simulation structures.

2D Simulation Results with $W_N = W_P = 0.1$ µm, $N = P = 1E+17$ cm^{-3}

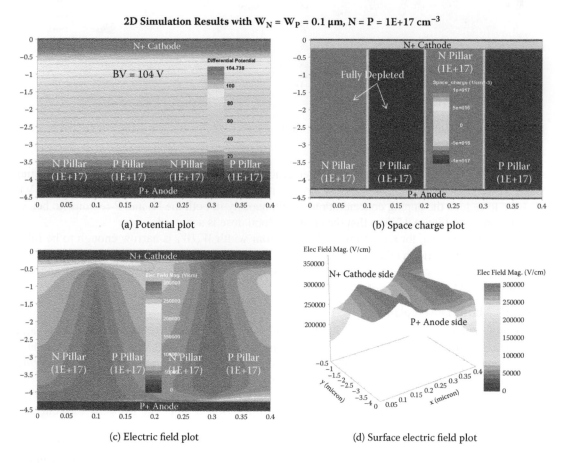

(a) Potential plot

(b) Space charge plot

(c) Electric field plot

(d) Surface electric field plot

FIGURE 9.25 Simulation results of structure with $W_n = W_p = 0.1$ µm and doping concentration of 1E+17 cm^{-3}.

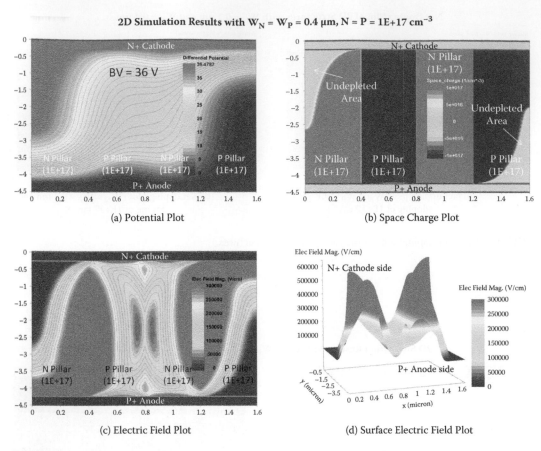

FIGURE 9.26 Simulation results of structure with $W_n = W_p = 0.4$ µm and doping concentration of N = P = 1E + 17 cm^{-3}.

seen in Figure 9.26b. We also note that the electric field is not very uniform in Figures 9.26c and 9.26d. If we want to keep $W_N = W_P = 0.4$ µm and achieve a high breakdown voltage (over 100 V), we have to reduce the doping concentration simultaneously in both the N- and P-regions (to below 5E + 16 cm^{-3}, see Figure 9.23) so that the optimum total dose is achieved.

Generally speaking, for an optimized design (pillar width W_N/W_P is narrow enough to be fully depleted); we will notice a relatively flat surface from the 3D surface plot (cf. Figure 9.25d). For an improperly designed case, such as that shown in Figure 9.26, we will get distinctive peaks and valleys from the 3D surface plot (refer to Figure 9.26d).

9.3.1.2 About Dose Balance

A key design consideration of a Super Junction device is to keep a dose balanced so that the N- and P-regions mutually deplete each other. Even a slight dose imbalance will lead to significant reduction of breakdown voltage. This places a great challenge on Super Junction device manufacturing and may lower the yield due to fabrication variations.

To illustrate the dose imbalance problem, we modify the previous simulation case with $W_N = W_P = 0.1$ µm, N = P = 1E + 17 cm^{-3} to $W_N = W_P = 0.1$ µm, N = 1E + 17 cm^{-3}, P = 5E + 16 cm^{-3}, as shown in Figure 9.27.

The breakdown simulation results are given in Figure 9.28. The breakdown voltage is dramatically reduced for the unbalanced case (only 34 V vs. 104 V), see Figure 9.28a. With a higher dose in the N-region compared with the P-region, part of the N-region is not depleted, as seen in Figure 9.28b. We also note crowding of the electric field close to the anode (P+) region as shown in Figures 9.28c and 9.28d.

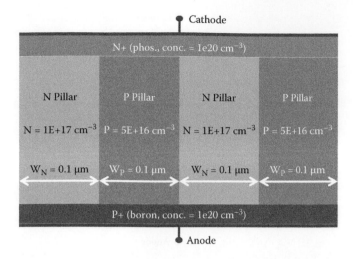

FIGURE 9.27 Schematic net doping plot of a dose imbalanced structure ($W_N = W_P = 0.1$ μm, N = 1E + 17 cm^{-3}, P = 5E + 16 cm^{-3}).

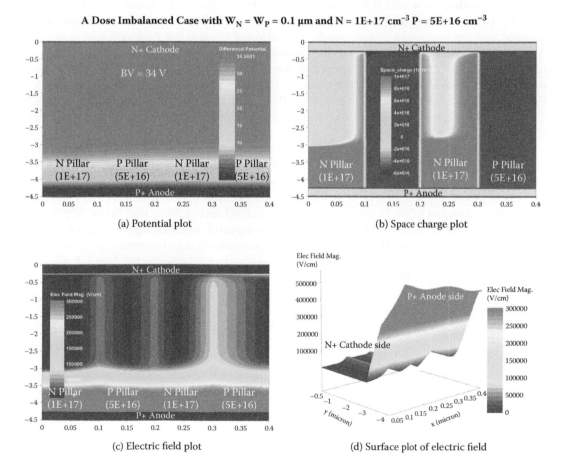

FIGURE 9.28 Simulation results of a dose imbalanced structure ($W_N = W_P = 0.1$ μm, N = 1E + 17 cm^{-3}, P = 5E + 16 cm^{-3}).

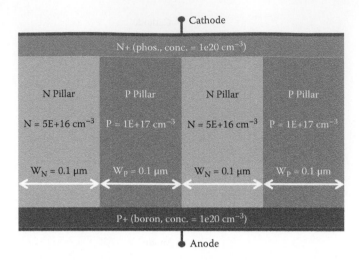

FIGURE 9.29 Schematic net doping plot of a dose imbalanced structure ($W_N = W_P = 0.1$ μm, N = 5E + 16 cm^{-3}, P = 1E + 17 cm^{-3}).

As we have demonstrated, a higher dose in the N-region results in a degraded breakdown voltage, so it is only natural to wonder if the opposite is true. At first glance, we would guess that since the N-region (where the drain-drift region in a Super Junction LDMOS is located) would now be fully depleted and the breakdown voltage would be unaffected because the P-region's role is to "help" deplete the N-region. However, more detailed simulations with a second unbalanced structure shown in Figure 9.29 show that is not true. The simulation results for this case are given in Figure 9.30; similar to the previous unbalanced case, the breakdown voltage is again reduced to 36 V (Figure 9.30a). From the space charge plot, we see it is now the P-regions that are partially depleted (Figure 9.30b). The higher dose in the P-region now causes electric field crowding close to the cathode (N+) region (Figures 9.30c and 9.30d).

We learned that it is not the depletion of a particular pillar that matters; it is the uniformity of the electric field that governs the optimized design. To get a uniformed electric field, all N/P pillars must be fully depleted. Only by balancing the dose and making pillar width smaller than a critical value can we achieve this goal. The critical pillar width can be obtained from experiments or TCAD simulations.

In a Super Junction LDMOS, the dose balance is much harder to achieve than in our simplified 2D case with constant N/P regions. Many different regions are involved including the depletion from the P-epi layer at the bottom and P-body region from the source/body side; sometimes an N buffer layer close to the drain is also used in the design. Fortunately, we can tell whether the dose is balanced by looking at the electric field peaks: a well-designed Super Junction device should have a nearly uniform electric field throughout the device. A significantly higher peak at the P+ (or body/source side for N-LDMOS) side indicates that the N-region dose is too high (as in Figure 9.28d), while a significant higher peak at the N+ (or drain) side indicates that the P-region dose is too high (as in Figure 9.30d).

9.3.1.3 On-Resistance

The primary reason for designing a Super Junction LDMOS is to reduce the on-resistance without jeopardizing the breakdown voltage. According to references [158] [186], it has been found that for a standard LDMOS, on-resistance is proportional to the square of breakdown voltage:

$$R_D \propto BV^2$$

(9.2)

while for Super Junction the relationship is linear:

$$R_{D,sj} \propto BV$$

(9.3)

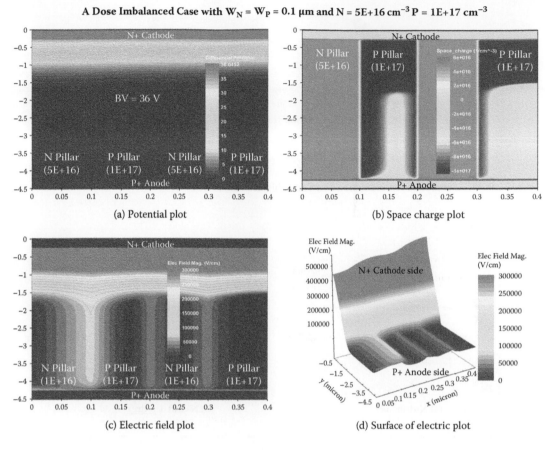

FIGURE 9.30 Simulation results of a dose imbalanced structure ($W_N = W_P = 0.1$ μm, N = 5E + 16 cm^{-3}, P = 1E + 17 cm^{-3}).

Qualitatively, the square-power increase of breakdown voltage for a standard LDMOS results in a much higher on-resistance than for a Super Junction LDMOS. Please note that this relationship is valid only for resistance coming from the drain-drift region; the channel resistance is not included in this relationship.

For lower voltage ratings (< 80 V), a Super Junction does not have too much advantage since channel resistance is the same for both devices and accounts for a large percentage of the total resistance. The costs of a more complex design (more mask layers and lower yield) versus the benefits it provides must also be considered. Super Junction LDMOS are therefore mostly used in applications where breakdown voltages greater than 300 V are required.

9.3.2 SUPER JUNCTION **LDMOS** STRUCTURE

With the basic understanding given in the previous section, we now introduce a 3D TCAD simulation of a Super Junction LDMOS, a schematic of which is illustrated in Figure 9.31. Instead of the uniform N-drift region found in a standard LDMOS, alternating N-regions and P-regions are introduced along the z-axis (depth) to create a Super Junction LDMOS structure. If we take a top-down view of the drain-drift region, the alternating N and P-regions structure is just like the 2D structure we previously discussed.

The simulation includes both process and device simulations. For the process simulation, the mask design is shown in Figure 9.32. Two N-wells (N-drifts) and two P-wells (P-drifts) with a width of 0.25 μm are used; we recall from previous sections and 2D simulations that this width is critical

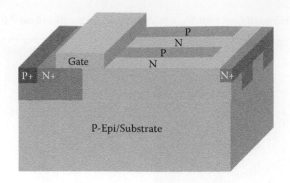

FIGURE 9.31   Schematic representation of a Super Junction LDMOS.

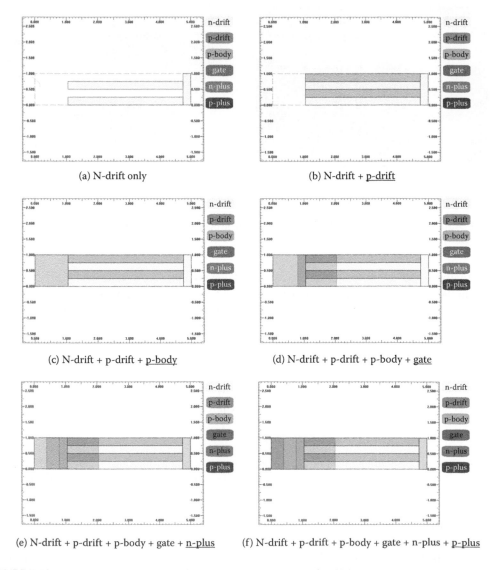

FIGURE 9.32   Layout of a Super Junction LDMOS for 3D TCAD simulation.

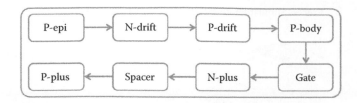

FIGURE 9.33 Simplified process flow of 3D TCAD simulation of a Super Junction LDMOS.

for the breakdown voltage and on-resistance improvements. The device pitch size is chosen to be 5 µm, with a gate length of 1.25 µm.

9.3.3 3D Process Simulation of Super Junction LDMOS

Figure 9.32 shows the layout of the Super Junction LDMOS for 3D TCAD simulation. Mask files can be created by various software tools and saved in GDSII format. In this simplified 3D simulation, six masks are used. The gate mask is for the polysilicon gate; the P-plus mask is used to create P+ contact for P-body; and the N-plus mask is for the source/drain N+ implant. N-drift and P-drift masks are used for the alternating N/P wells with an additional N-region (which shares the mask with N-drift for simplicity) placed close to the drain to balance the depletion effect from substrate [197]. The final P-body mask is for the P-type body implant.

The simplified process simulation steps of this Super Junction LDMOS structure are given in Figure 9.33, while Table 9.3 gives a more complete description of the process steps. All the process

TABLE 9.3

Simplified Process Steps of 3D TCAD Simulation of a Super Junction LDMOS

Process Steps	Description
1. P-epi	Boron concentration of 1E + 14 cm^{-3}, orientation = 100
2. N-drift mask	Apply N-drift photomask (negative mask, photoresist is removed where the patterns are drawn)
3. N-drift implant	Phosphorus implant with dose = 8E + 12 cm^{-2}, energy = 450 keV
4. N-drift anneal	10 seconds thermal anneal at temperature of 1000°C
5. P-drift mask	Apply P-drift photomask (negative mask, photoresist is removed where the patterns are drawn)
6. P-drift implant	Boron implant with dose = 8E + 12 cm^{-2}, energy = 300 keV
7. P-drift anneal	10 seconds thermal anneal at temperature of 1000°C
8. P-body mask	Apply P-body photomask (negative mask, photoresist is removed where the patterns are drawn)
9. P-body implant	Chain implant of boron with dose = 9E + 12 cm^{-2} and energy = 20, 50, 120 and 200 keV
10. P-body anneal	10 seconds thermal anneal at temperature of 1000°C
11. Gate oxide	Gate oxide of 150 Å
12. Poly deposit	In-situ phosphorus doped poly is deposited (thickness = 0.25 µm, doping concentration = 1E + 20 cm^{-3})
13. Gate mask	Apply gate photomask (positive mask, photoresist is preserved where pattern is drawn)
14. Gate etch	Polysilicon and oxide are subsequently etched away
15. N-plus mask	Apply N-plus photomask (negative mask, photoresist is removed where the patterns are drawn)
16. N-plus implant	Arsenic implant with dose = 3E + 15 cm^{-3} and energy = 20 keV
17. N-plus anneal	Thermal anneal for 30 seconds at 1050°C
18. Spacer	Silicon Nitride (Si$_3$N$_4$) is deposited (0.15 µm) and then etched away leaving spacers at gate corners
19. P-plus mask	Apply P-plus photomask (negative mask, photoresist is removed where the patterns are drawn)
20. P-plus implant	Boron implant with dose = 2E + 15 cm^{-3} and energy = 5 keV
21. P-plus anneal	Thermal anneal for 10 seconds at 1100°C
22. Export	Export to device simulator

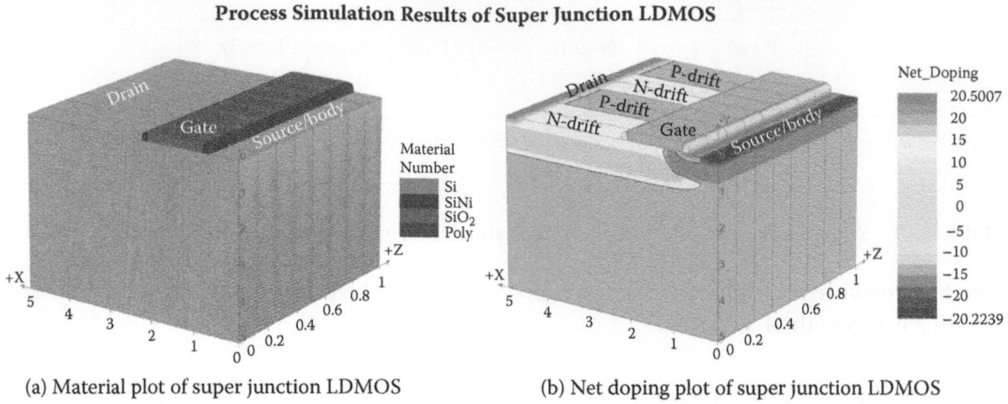

Process Simulation Results of Super Junction LDMOS

(a) Material plot of super junction LDMOS   (b) Net doping plot of super junction LDMOS

FIGURE 9.34 3D process simulation results of Super Junction LDMOS with (a) material plot and (b) net doping plot.

steps and parameters used here are arbitrary since real process steps are usually proprietary know-how of foundries and IDMS. The purpose of this simulation is to give the reader a template for designing and simulating a Super Junction LDMOS.

The 3D process simulation results (material and net doping plots) of this Super Junction LDMOS are given in Figure 9.34.

9.3.4 3D DEVICE SIMULATION OF SUPER JUNCTION LDMOS

The next step is to use the exported files from 3D process simulation to perform a 3D device simulation and estimate the breakdown voltage of this device. We stress again that the simulation examples listed in this book are not optimized, nor are they calibrated to any specific process, foundry, or fab. Readers should use their own sense when designing a device and choosing relevant material parameters.

Three contacts are defined for this device: drain, source, and gate. The device simulation shows a high breakdown voltage of 82 V, as shown in Figure 9.35. The device pitch size is only 5 μm. From Figure 9.35c, we notice that the electric field is fairly flat in the z direction, and relatively uniform in the x direction, as further revealed in Figure 9.35d, which is a 3D surface plot of electric field on the XZ plane at Y = −0.25 μm (just below N+/P+ source/drain/body diffusions). However, there is still room for further optimization: as we can see from Figure 9.35d, there are two electric field peaks: one close to the drain side; and the other close to the source/body side. The drain side peak is higher than the source/body side peak, which indicates a slight dose imbalance in the N/P pillars. From our 2D simulations discussed previously, in the 3D electric field surface plot, if the electric field peak at the N+ drain side is higher than the peak at the source/body (P+) side, it suggests that we might have a higher P-drift dose than N-drift dose. In this particular case, we are using the same dose for both N-drift and P-drift. The higher P-dose may stem from the P-epi (or P-substrate in some device structures) under-neath the N/P drift regions that increases the total P-dose. Although we have placed an additional N-region at the drain side, the dose and energy of this N-type implant are the same as N-drift, since we are sharing the same mask with N-drift. An N-well at the drain side with a dedicated mask layer and optimized implant recipe may help further improve the breakdown voltage and create an even more uniform electric field throughout the N/P drift regions; the exercise is left to interested readers.

9.3.5 3D SIMULATION OF A STANDARD LDMOS WITH THE SAME N-DRIFT DOPING

If we do a slight modification on the device structure, by removing all P-drifts and replace them with N-drifts to make a uniform N-drift region in the Z direction, the device becomes a standard LDMOS as shown in Figure 9.36.

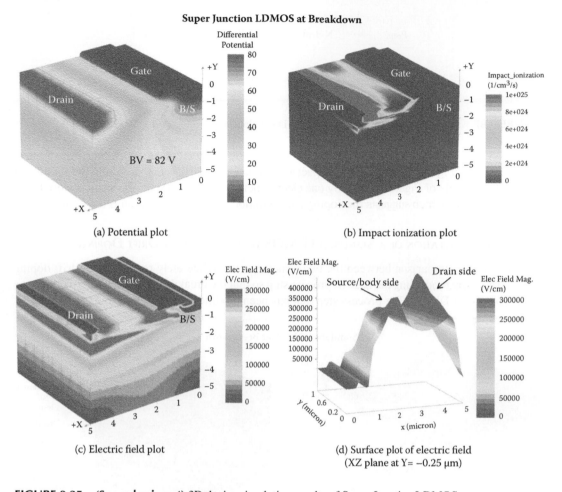

Super Junction LDMOS at Breakdown

(a) Potential plot

(b) Impact ionization plot

(c) Electric field plot

(d) Surface plot of electric field
(XZ plane at Y= −0.25 μm)

FIGURE 9.35 **(See color insert)** 3D device simulation results of Super Junction LDMOS.

Now we perform the same process and device simulation, except that the P-drift process step is removed. The process steps are given in Figure 9.37, and the detailed process steps are the same as Super Junction LDMOS in Table 9.3 except that the P-drift mask and implant steps are skipped.

After process simulation, the device breakdown simulation is carried out using the exported files from the process simulation. The simulation results for the standard LDMOS with the same N-drift

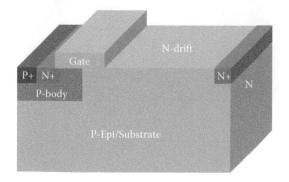

FIGURE 9.36 Schematic representation of a standard LDMOS.

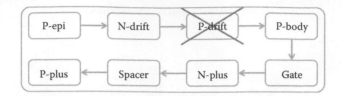

FIGURE 9.37 Simplified process steps of 3D TCAD simulation of a standard LDMOS.

doping are shown in Figure 9.38. We show in Figure 9.38a that the breakdown voltage has dropped from 82 to 26 V. Figures 9.38b and c are impact ionization and electric field plots. From the 3D surface plot of electric field of Figure 9.38d, only one electric field peak is observed. This peak is close to the source/body side, which suggests the doping concentration in the N-drift region is too high.

9.3.6   3D Simulation of a Standard LDMOS with Reduced N-Drift Doping

To make a fair comparison between the two structures, we deliberately reduce the N-drift doping concentration by modifying the N-drift process step so that the breakdown voltage is on par with the Super Junction LDMOS. The process step that has been modified is shown in Table 9.4.

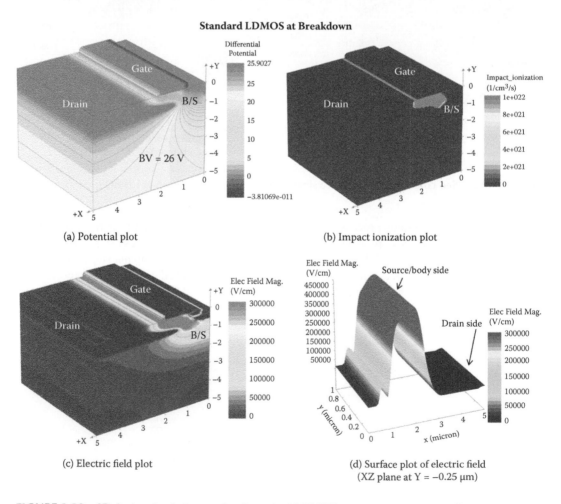

FIGURE 9.38   3D device simulation results of standard LDMOS.

TABLE 9.4

Modified Process Step of 3D TCAD Simulation of a Standard LDMOS

Process Step	Description
3. N-drift implant	implant dose = 1e11 energy = 150 phosphorus
	implant dose = 3e11 energy = 400 phosphorus

Figure 9.39 illustrates the device breakdown simulation results after modifications. The breakdown voltage is now at 80 V (Figure 9.39a), slightly lower than the Super Junction device of 82 V (Figures 9.39b and c are impact ionization and electric field plots). If we take a look at the electric field surface plot (Figure 9.39d) of the XZ plane at the same location (Y = –0.25 µm), we will see that the electric field distribution is much improved compared with the standard LDMOS (Figure 9.38) with an unmodified N-well by introducing another peak near the drain end.

9.3.7 Comparison of Super Junction LDMOS and Standard LDMOS

The LDMOS with a modified N-drift doping condition is compared with the Super Junction LDMOS in terms of breakdown voltage and on-resistance in Figure 9.40. The Super Junction LDMOS is clearly superior in terms of breakdown voltage and on-resistance.

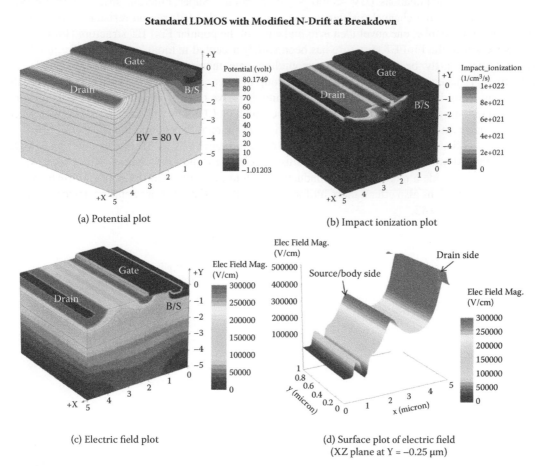

Standard LDMOS with Modified N-Drift at Breakdown

(a) Potential plot

(b) Impact ionization plot

(c) Electric field plot

(d) Surface plot of electric field (XZ plane at Y = –0.25 µm)

FIGURE 9.39 3D device simulation results of standard LDMOS with modified N-drift.

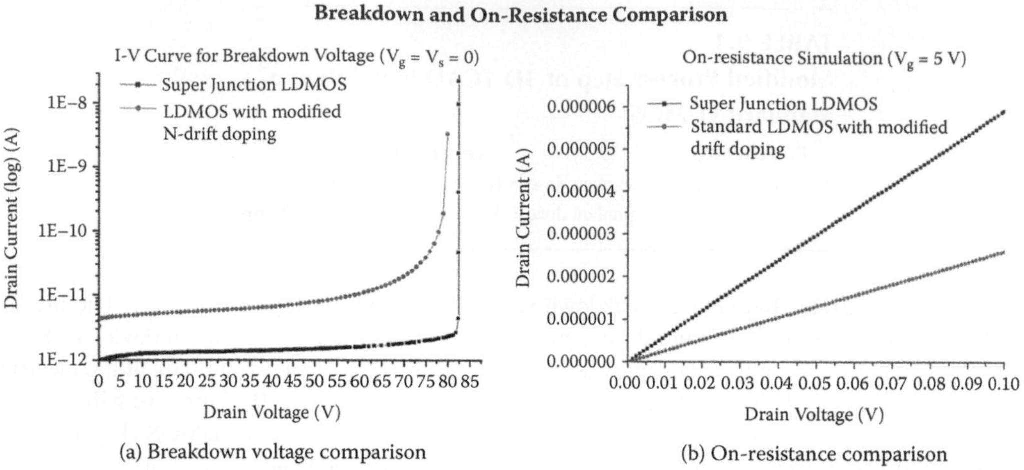

Breakdown and On-Resistance Comparison

I-V Curve for Breakdown Voltage ($V_g = V_s = 0$)

On-resistance Simulation ($V_g = 5$ V)

(a) Breakdown voltage comparison

(b) On-resistance comparison

FIGURE 9.40 Comparison of breakdown voltage and on-resistance between Super Junction LDMOS and standard LDMOS with modified N-drift doping concentration.

9.4 SUPER JUNCTION POWER FinFET

For low-voltage applications (BV < 100 V), conventional Super Junction structure is not very attractive because the channel resistance is comparable to the drift region resistance. To improve the channel resistance, one novel idea is to make use of the popular FinFET structure [198].

As we know, the FinFET structure has been widely accepted in today's deep submicron CMOS technologies, partly because the 3D surrounding gate increases the total channel width and effectively reduces the channel resistance. A group of researchers from the University of Toronto and Hong Kong University of Science and Technology have recently demonstrated the interesting idea of integrating a FinFET structure to the Super Junction LDMOS and have successfully reduced the total on-resistance by 30% compared with the conventional Super Junction LDMOS [198].

The basic idea of FinFET + LDMOS is illustrated in Figure 9.41. A 3D corrugated trench MOS gate on the sidewalls and the top surface is used to increase the total channel width (i.e., $W_{top} + W_{side}$) and to provide a more effective conduction path to the drift region. The deep trench source/drain contact regions also offer a more uniform current flow [198]. A more detailed device structure is given in Figure 9.42.

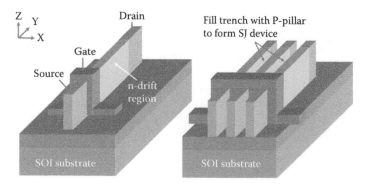

FIGURE 9.41 Basic idea schematics of Super Junction FinFET. (From Abraham Yoo, Jacky C. W. Ng, Johnny K. O. Sin, and Wai Tung Ng, University of Toronto and Hong Kong University of Science and Technology. With permission.)

Detailed Device Structure and Cross-Sectional Views

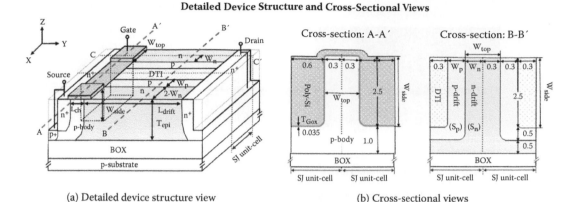

(a) Detailed device structure view   (b) Cross-sectional views

FIGURE 9.42 A more detailed view of the device structure with cross sections. (From Abraham Yoo, Jacky C. W. Ng, Johnny K. O. Sin, and Wai Tung Ng, University of Toronto and Hong Kong University of Science and Technology. With permission.)

9.4.1 Process Flow of the Super Junction Power FinFET

The device is fabricated on a silicon on insulator (SOI) substrate with CMOS compatible technology. Nine masks are used in fabricating this device. The starting wafer has an N-epi layer on BOX (Buried OXide). A deep trench is etched first to expose the sidewall for the P-body implant. The P-body is formed by ion implantation of the sidewall with an implant tilt of 45 degrees (Figure 9.43a). A deep trench isolation (DTI) process is then carried out to form a thick layer of oxide followed by active lithography and nitride hard-mask etching to define the device active area. The sidewall

Breakdown and On-Resistance Comparison

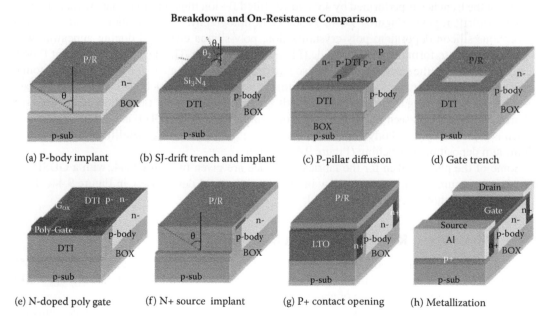

(a) P-body implant   (b) SJ-drift trench and implant   (c) P-pillar diffusion   (d) Gate trench

(e) N-doped poly gate   (f) N+ source implant   (g) P+ contact opening   (h) Metallization

FIGURE 9.43 Simplified process steps for Super Junction FinFET. (From Abraham Yoo, Jacky C. W. Ng, Johnny K. O. Sin, and Wai Tung Ng, University of Toronto and Hong Kong University of Science and Technology. With permission.)

Layout and Fabricated Device Views of SJ Power FinFET

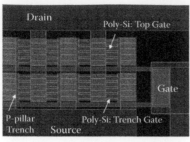

(a) Layout view of the SJ power FinFET

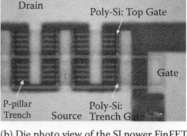

(b) Die photo view of the SJ power FinFET

(c) SEM view of the SJ power FinFET

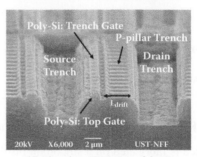

(d) Closed up SEM view of the SJ power FinFET

FIGURE 9.44 Some photos of fabricated Super Junction power FinFET. (From Abraham Yoo, Jacky C. W. Ng, Johnny K. O. Sin, and Wai Tung Ng, University of Toronto and Hong Kong University of Science and Technology. With permission.)

doping of the trenches is performed by 45° and 12° tilted B+ ion implantations that are then diffused to form P-drift regions (Figures 9.43b, 9.43c). Gate lithography, gate oxidation, *in situ* N-doped amorphous silicon deposition, poly-crystallization, poly-silicon etch, and doping annealing were then carried out to form the gate electrode (Figures 9.43d, 9.43e). The deep trench adjacent P-body is etched to prepare the sidewall for the N+ source/drain implant. Some oxide is left at the bottom to prevent N+ implant penetration to the bottom that is reserved for the P+ body contact implant later on. The sidewall doping at the source/drain ends is carried out by a 45° tilted dual implant of arsenic and phosphorus, as seen in Figure 9.43f. Low temperature oxide (LTO) is used for passivation, which is followed by a P+ body contact opening (Figure 9.43g). Finally, metallization is done with aluminum deposition and etching (Figure 9.43h) [198].

Some of the photos taken for the fabricated device are given in Figure 9.44, with a GDSII layout view in Figure 9.44a, die photo view in Figure 9.44b, SEM device view in Figure 9.44c, and a closed-up SEM photo view in Figure 9.44d.

9.4.2 MEASUREMENT RESULTS OF SUPER JUNCTION POWER FINFET

The device was fabricated and tested with I_d-V_d and I_d-V_g curves shown in Figures 9.45a and 9.45b. The saturation drain current of the SJ power FinFET over 380 mA/mm was attained at $V_{GS} = 10$ V with a measured threshold voltage of about 1.7 V.

The fabricated device is compared with a planar Super Junction LDMOS. It has been found that the SJ power FinFET devices with smaller total channel width W exhibit 33% lower R_{on} when compared with the SJ-LDMOSFETs, as seen in Figure 9.46.

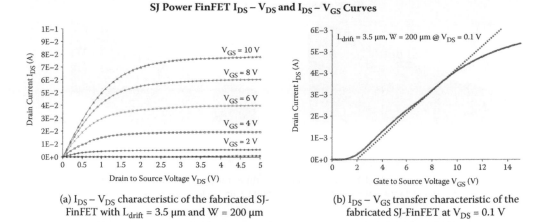

(a) $I_{DS}-V_{DS}$ characteristic of the fabricated SJ-FinFET with L_{drift} = 3.5 μm and W = 200 μm

(b) $I_{DS}-V_{GS}$ transfer characteristic of the fabricated SJ-FinFET at V_{DS} = 0.1 V

FIGURE 9.45 Measurement results of I_d-V_d and I_d-V_g curves of Super Junction (SJ) power FinFET. (From Abraham Yoo, Jacky C. W. Ng, Johnny K. O. Sin, and Wai Tung Ng, University of Toronto and Hong Kong University of Science and Technology. With permission.)

9.4.3 3D Simulation of Super Junction Power FinFET

The 3D TCAD simulation of Super Junction power FinFET is performed with the net doping plot shown in Figure 9.47. The device unit cell with gate oxide and DTI is shown in Figure 9.47a, and the same structure with gate oxide and DTI removed to reveal the interior is given in Figure 9.47b. Constant doping is used in this simulation.

The simulated curves for on-state (a) and off-state (b) are shown in Figure 9.48. We see the trade-off between on-resistance and breakdown voltage with different drift region lengths.

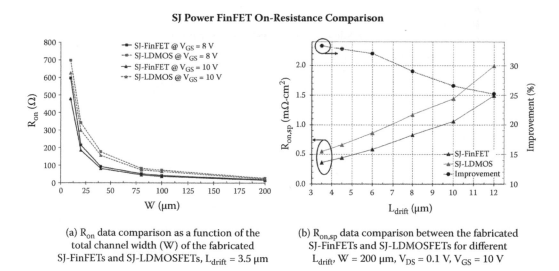

(a) R_{on} data comparison as a function of the total channel width (W) of the fabricated SJ-FinFETs and SJ-LDMOSFETs, L_{drift} = 3.5 μm

(b) $R_{on,sp}$ data comparison between the fabricated SJ-FinFETs and SJ-LDMOSFETs for different L_{drift}, W = 200 μm, V_{DS} = 0.1 V, V_{GS} = 10 V

FIGURE 9.46 On-resistance comparison between Super Junction power FinFET and Super Junction LDMOS. (From Abraham Yoo, Jacky C. W. Ng, Johnny K. O. Sin, and Wai Tung Ng, University of Toronto and Hong Kong University of Science and Technology. With permission.)

3D TCAD Simulation: Net Doping Plots

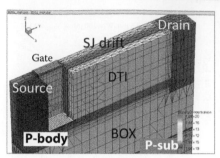
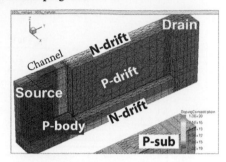

(a) Net doping plot of the 3D Super Junction
FinFET unit cell with gate oxide and DTI

(b) Net doping plot of the 3D Super Junction
FinFET unit cell with gate oxide and DTI removed

FIGURE 9.47 3D TCAD simulation results of Super Junction FinFET unit cell. (From Abraham Yoo, Jacky C. W. Ng, Johnny K. O. Sin, and Wai Tung Ng, University of Toronto and Hong Kong University of Science and Technology. With permission.)

From Figure 9.49a, we see the simulated impact of gate sidewall width and drain-drift region length on the breakdown voltage. This kind of information is normally not easy to obtain from experiments since a lot of design variations need to be performed and one cannot always tell whether the difference comes from fabrication or real device physics. A benchmark comparison among published data for SJ-LDMOS and SJ power FinFET is shown in Figure 9.49b: the SJ power FinFET is approaching or exceeding the silicon limit.

More simulation results are given in Figure 9.50. On-resistance with respect to distance (a) and electric field with respect to distance (b) are plotted. One of the important merits of TCAD simulation is that it can give the engineers an idea of the important physics inside the device, such as electric field distribution, which are difficult to access using pure measurements.

Breakdown and On-Resistance Comparison

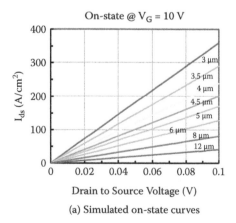
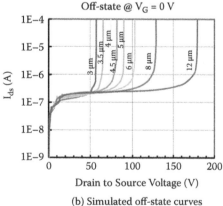

(a) Simulated on-state curves

(b) Simulated off-state curves

FIGURE 9.48 Simulated on-state (a) and off-state (b) curves. (From Abraham Yoo, Jacky C. W. Ng, Johnny K. O. Sin, and Wai Tung Ng, University of Toronto and Hong Kong University of Science and Technology. With permission.)

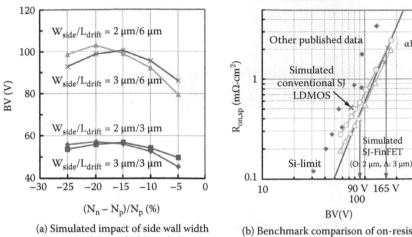

(a) Simulated impact of side wall width (W$_{side}$) and drain-drift length (L$_{drift}$) on breakdown voltage

(b) Benchmark comparison of on-resistance vs. breakdown voltage for various devices

FIGURE 9.49 Simulated impact of sidewall width (W$_{side}$) and drain-drift length (L$_{drift}$) on (a) breakdown voltage and (b) benchmark comparison of on-resistance versus breakdown voltage for various devices. (From Abraham Yoo, Jacky C. W. Ng, Johnny K. O. Sin, and Wai Tung Ng, University of Toronto and Hong Kong University of Science and Technology. With permission.)

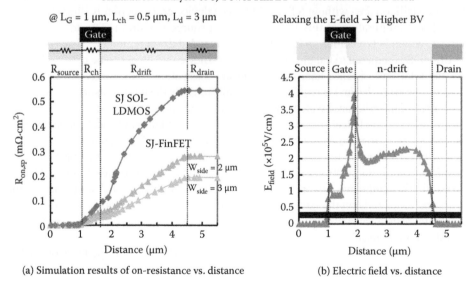

(a) Simulation results of on-resistance vs. distance

(b) Electric field vs. distance

FIGURE 9.50 Simulation results of (a) on-resistance versus distance and (b) electric field versus distance. (From Abraham Yoo, Jacky C. W. Ng, Johnny K. O. Sin, and Wai Tung Ng, University of Toronto and Hong Kong University of Science and Technology. With permission.)

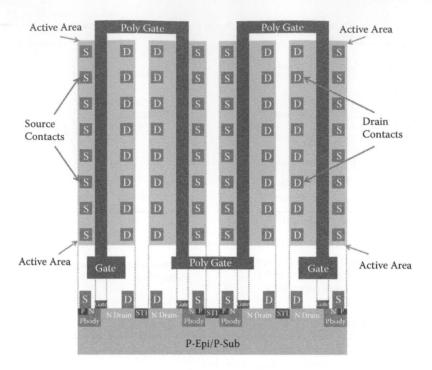

FIGURE 9.51   Layout example of a large LDMOS.

9.5   LARGE INTERCONNECT SIMULATION

For power devices, the device size is usually fairly large to handle high current and high power. Figure 9.51 is an example of an LDMOS layout showing the source/body and drain contacts, gate, and active area (silicon area without STI or DTI). The device cross section is illustrated underneath the layout. The configuration of this layout features a source (S)–drain (D)–drain (D)–source (S) arrangement, which is typical for an LDMOS with extended drain region.

If the LDMOS is required to handle high current, the device width has to be designed large enough to reduce the total on-resistance and, thus, the power dissipation. With larger device sizes, the on-resistance coming from the device itself drops, but unfortunately the total measured resistance does not drop accordingly. This is because with high current levels and large device sizes the voltage drop on the metal interconnects is no longer negligible: the resistance contribution of metal interconnects can be even greater than that from the device itself, a phenomenon frequently referred to as metal debiasing.

TCAD is used to simulate the large interconnect resistance and see the current/temperature distributions. Our goal in this section is to build an interconnect structure with three metal layers that resides on a highly conductive silicon substrate. To make simulation easier, we neglected the active device part and focus only on the interconnect structure. The highly doped silicon substrate is used to eliminate the resistance contribution from the device so that the majority of the resistance comes from the interconnect itself.

Figure 9.52 shows the layout view of contacts that touch the highly doped silicon surface, with a cross-section view. We see a symmetrical (S–D–S–D) configuration with a small voltage (0.1 V) applied to all the drain contacts while keeping source contacts grounded. Current flows in silicon substrate are depicted in the cross-section view of Figure 9.52. The resistance between source and drain is made very small by laying out all the S–D pairs in parallel; the width is large enough to neglect the small resistance from source to drain in the highly doped substrate. In the following sections, we will present a detailed simulation demonstration of a complicated interconnect structure.

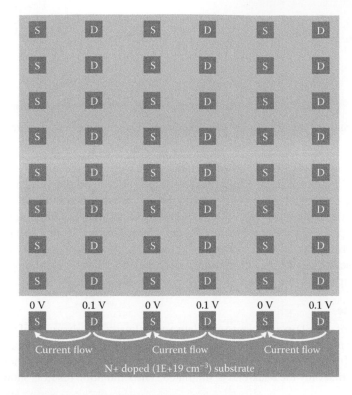

FIGURE 9.52 The simplified layout for large interconnect simulation.

9.5.1 3D Process Simulation of the Large Interconnect

The simulation starts with a heavily doped N-type silicon substrate (with phosphorus) to mimic the active device when the gate is fully on. With a doping concentration of $1E + 19$ cm^{-3}, the resistance from the substrate is kept at a minimum. A total of three metal layers are simulated with simulation size of 100 μm by 100 μm. The metal layers are arranged so that metal1 corresponds to the stripes, metal2 the buses (wider than metal1), and metal3 the blocks (wider than metal2); a schematic view of the metal scheme is given in Figure 9.53. Metal1 stripes are connected to source or drain by the contacts, and they are patterned in a S–D–S–D configuration. Via1 connects metal1 and metal2, and they are placed so that metal1 for the source is connected to metal2 for the source. Metal2 regions are typically thicker and wider than metal1, but for the purpose of a simulation demonstration the thicknesses of all the metals are set to be 0.5 μm. This, of course, is not the case in real

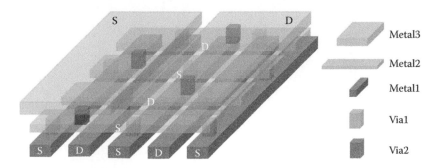

FIGURE 9.53 Metal layout scheme with metal1 in stripes, metal2 in buses, and metal3 in blocks.

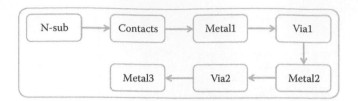

FIGURE 9.54 Simplified process steps of 3D TCAD simulation of a complicated interconnect.

manufacturing where metal3 is substantially thicker than metal1 and metal2. From Figure 9.53, we see that metal2 buses are placed horizontally, with S–D–S–D in the vertical direction. Metal3 regions are designed to be in large blocks: two blocks are used in this simulation, with the source on the left and the drain on the right. Metal3 blocks are connected by via2 from metal2.

The simplified process steps are shown in Figure 9.54, whereas the complete layout of the interconnect structure is illustrated in Figure 9.55. Six mask layers (contact, metal1, via1, metal2, via2, and metal3) are used in this design, and aluminum is used for all layers (including contacts and vias) for the sake of simplicity.

9.5.1.1 Substrate

Since the purpose of this 3D example is to simulate a complicated interconnect, if we use the standard procedure to build an LDMOS or an analog MOS, the simulation will be much more complicated and time-consuming. Instead, we just use an N+ doped silicon substrate with phosphorus concentration of 1E + 19 cm^{-3} so that only the back-end is simulated. The simulation result is shown in Figure 9.56.

9.5.1.2 Contacts

The ILD1 layer is deposited, and contact holes are etched. Aluminum is used in the simulation to fill those contact holes. After filling, a CMP step is performed to prepare the surface for the metal1 layer.

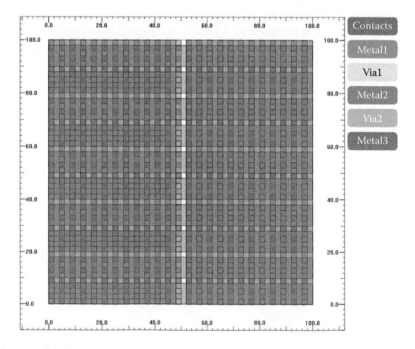

FIGURE 9.55 Layout of the large interconnect structure.

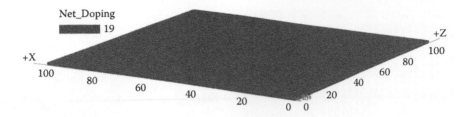

FIGURE 9.56   Substrate net doping plot.

The height of the contacts is 0.5 μm with a size of 2 μm by 2 μm. There are totally $25 \times 25 = 625$ contacts in this simulation, as shown in Figure 9.57.

9.5.1.3   Metal1

Aluminum metal stripes are used to connect the contacts in the vertical direction. After deposition, aluminum is etched with the metal1 photomask. The thickness of the metal1 layer is 0.5 μm. The metal stripe has a width of 2 μm, length of 100 μm and runs in the z direction, as shown in Figure 9.58. A total of 25 metal1 stripes are simulated. Please note that in real practice design rules may require the dimension of a metal1 width larger than the width of square contacts and vias.

9.5.1.4   Via1

The next step is to place via1 on top of metal1. Since metal1 connects the underlying contacts in S–D–S–D fashion, via1 are used to connect metal1 (source) to metal2 (source) and metal1 (drain) to metal2 (drain). Each via1 row is placed onto the metal1 stripes on either an odd number or even number of the metal1 columns, as shown in Figure 9.59. Like the contacts, oxide is first deposited to create ILD2 before via1 holes are etched and filled. Chemical mechanical polishing is performed to flatten the surface and prepare for the metal2 step.

9.5.1.5   Metal2

The process step of metal2 is similar to metal1, with a thickness of 0.5 μm. Metal2 lines are drawn in wider buses and are placed in a horizontal direction as shown in Figure 9.60. 10 metal2 buses are simulated, each with a width of 8 μm and a length of 100 μm.

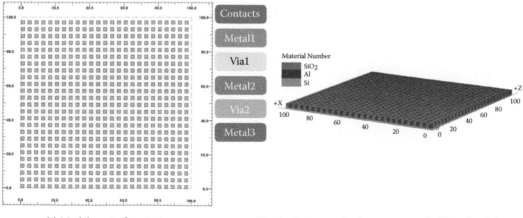

(a) Mask layout of contacts   (b) Simulation result after contacts etch, fill, and polish

FIGURE 9.57   3D process simulation of large interconnect: (a) Mask; (b) After contacts CMP.

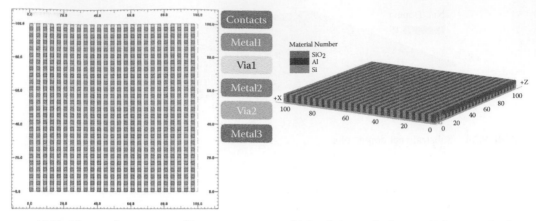

(a) Mask layout of contact + metal1   (b) Simulation result after metal1 deposit and etch

FIGURE 9.58   3D process simulation of large interconnect: (a) Mask; (b) After metal1 etch.

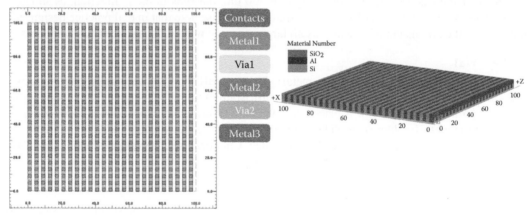

(a) Mask layout of contact + metal1 + via1   (b) Simulation result after via1 etch, fill and polish

FIGURE 9.59   3D process simulation of large interconnect: (a) Mask; (b) After via1 step.

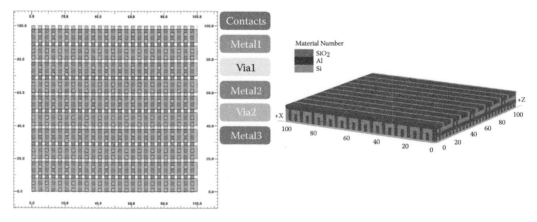

(a) Mask layout of contacts + metal1 + via1 + metal2   (b) Simulation result after metal2 deposit and etch

FIGURE 9.60   3D process simulation of large interconnect: (a) Mask; (b) After metal2 step.

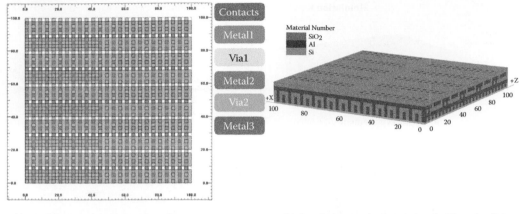

(a) Mask layout of contacts + metal1 + via1 + metal2 + via2

(b) Simulation result after via2 etch, fill, and polish

FIGURE 9.61   3D process simulation of large interconnect: (a) Mask; (b) After via2 step.

9.5.1.6   Via2

Via2 are placed on top of metal2 to connect metal2 and metal3. Like via1 and contacts, each via2 region is a 2 μm by 2 μm square. They are placed so that metal2 buses for the source are connected to the metal3 block for the source, and metal2 buses for drain are connected to the metal3 block for the drain, as shown in Figure 9.61.

9.5.1.7   Metal3

The final step is to deposit and etch metal3. Two metal3 blocks are drawn for the source and drain, respectively. Each metal3 block is 48 μm wide and 100 μm long, as shown in Figure 9.62. The oxide ILDs are deliberately removed to show the inside.

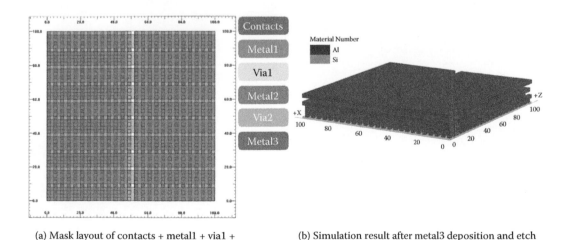

(a) Mask layout of contacts + metal1 + via1 + metal2 + via2 + metal3

(b) Simulation result after metal3 deposition and etch

FIGURE 9.62   3D process simulation of large interconnect: (a) Mask; (b) After metal3 step.

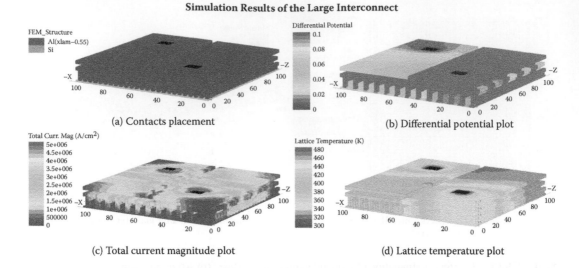

FIGURE 9.63   (See color insert) 3D device simulation results of the large interconnect with (a) contacts placement, (b) differential potential, (c) total current magnitude, and (d) lattice temperature.

9.5.2   3D DEVICE SIMULATION OF THE LARGE INTERCONNECT

Before we can do device simulation, we need to place contacts (i.e., equipotential boundary conditions) on the metal3 blocks. In this simulation, the contacts have a square shape with a size of 10 μm by 10 μm. Since the metal3 blocks are very large, the location of contacts will influence the device simulation result. We arbitrarily placed the contacts at the locations shown in Figure 9.63a. A small voltage bias of 0.1 V is applied to the drain (right contact on the metal3 block) with the self-heating model turned on. We can view the differential potential, total current magnitude, and lattice temperature plots in Figure 9.63b–d. The thermal boundary for the contacts consists of a heat sink with a constant temperature of 300 K connected via a thermal conductance of 100 W/K.

10 GaN Devices, an Introduction

Strictly speaking, GaN power devices are not considered integrated power devices even though high electron mobility transistors (HEMTs) have a lateral current flow path similar to that of LDMOS. However, researchers [199] [200] have already started exploring the possibility of using gallium nitride (GaN) for smart power IC. It is therefore useful to give readers an introduction to this topic, especially since many people in the field are familiar only with silicon.

10.1 COMPOUND MATERIALS VERSUS SILICON

Silicon has been the dominant material for power semiconductor devices for many years. Given the economies of scale due to existing infrastructure, mature technology, and innovative structures and devices like RESURF, Super Junction, and IGBT, silicon devices are still predicted to be the dominant material in the foreseeable future. However, ever-increasing demands for higher efficiency and better performance are beginning to approach the silicon limit, and this technology is expected to saturate in a few years' time. Recently, compound materials such as 4H silicon carbine (4H-SiC) and GaN are drawing the attention of semiconductor companies. In fact, if we define figures of merits (FOMs) such as R_{dson}, $R_{dson} \times Q_{SW}$ or efficiency $\times$ density/cost [201], as shown in Figure 10.1 we see that as silicon-based power device technology reaches maturation GaN-based power devices appear ready to continue to provide the rapid and dramatic improvement in performance that the electronics industry enjoyed over the past 30 years.

A comparison is made between silicon, 4H-SiC, and GaN [202] [203] in Table 10.1. Both 4H-SiC and GaN have large bandgaps and can support a 10x higher critical electric field. SiC also has a better thermal conductivity, which makes thermal design of power devices more convenient. Unfortunately, SiC is too expensive a material for most power electronics system providers. As we mentioned earlier, power electronics and especially power supplies are very price-sensitive. On the other hand, GaN is less expensive, since there is already an installed base for high-brightness light-emitting diode (HB LED) applications.

The higher FOMs in Table 10.1 indicate the capability of these materials of operating at higher switching frequencies and at higher voltages with faster reverse recovery [203]. GaN also benefits from a particularity of its crystal structure (polarization), which can be harnessed to create a 2DEG (2-Dimensional Electron Gas); this will be discussed later in this chapter. The remainder of this chapter focuses solely on GaN devices; interested readers can refer to books such as [158] for a detailed analysis of SiC power devices.

This chapter gives the reader a step-by-step introduction to GaN devices by TCAD simulation. Starting from a single layer of GaN, we will explore the unique characteristics of the device that are associated with the polarization effect. This is followed by a stack of AlGaN layers to examine the band diagram around the 2DEG channel. A gate contact is then introduced to see how the 2DEG channel is pinched off by applying a gate potential. Finally, the source and drain are added to make a complete HEMT device. More complicated devices are introduced at the end of this chapter.

TABLE 10.1

Material Property Comparison between Silicon, 4H-SiC and GaN

Material Property	Silicon	4H-SiC	GaN
Bandgap (eV)	1.1	3.2	3.4
Critical field (1E+6V/cm)	0.3	3	3.5
Electron mobility (cm²/V-Sec.)	1450	900	2000
Electron saturation velocity (1E+6 cm/Sec.)	10	22	25
Thermal conductivity (W/cm²K)	1.5	5	1.3
Baliga FOM = $\varepsilon_s\,\mu E_c^3$	1	675	3000

10.2 SUBSTRATE MATERIALS FOR GaN DEVICES

GaN devices can be grown on various substrates. The most commonly used substrates include silicon, silicon carbine (SiC), and GaN. It has been found that, in terms of defect density, lattice mismatch, reliability and yield, as well as off-state leakage, GaN substrate outperforms Si or SiC counterparts [204], [205], as listed in Table 10.2. (The Al_2O_3 substrate is omitted here due to very low thermal conductivity and a relatively high cost that limits its usefulness for power applications.)

Unfortunately, a relatively modest 2-inch GaN substrate can cost as much as $1900 in 2012 [206], compared with as little as $25 for a 6-inch silicon wafer. Since power electronics and power management systems are often cost-sensitive, the reasons for the push toward GaN-on-silicon are obvious. Besides much lower wafer cost, GaN-on-silicon can also leverage existing infrastructure, despite the fact that it has a large lattice mismatch and difference in thermal expansion coefficients. In the meantime, it is expected that cheaper GaN and SiC wafers will become available as the technology matures. This chapter focuses on simulations of GaN devices with Si or SiC substrates.

TABLE 10.2

GaN Device Properties on Different Substrates [204] [205]

Attributes	Si Substrate	SiC Substrate	GaN Substrate
Defect density (cm⁻²)	1E+9	5E+8	1E+3 to 1E+5
Lattice mismatch (%)	17	3.5	0
Thermal conductivity (W/cm-k at 25 °C)	1.5	3.0~3.8	1.5
Coefficients of thermal expansions (%)	54	25	0
Breakdown voltage (V)	200	600	1700
Off-state leakage	High	High	Low
Reliability and yield	Low	Low	High
Lateral or vertical device	Lateral	Lateral	Lateral or vertical
Integration possibility	Very high	Moderate	—
Substrate size (mm) (as of 2012)	400	150	50
Substrate cost (relative)	Low	Very high	Very high

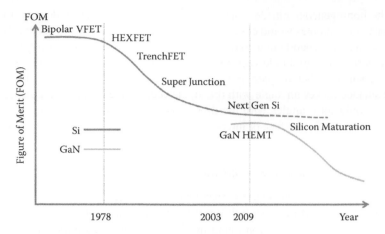

FIGURE 10.1   Trend of FOM for silicon and GaN devices over the years.

10.3   POLARIZATION PROPERTIES OF III-NITRIDE WURTZITE

10.3.1   Microscopic Dipoles and Polarization Vector

It is well-known that III-Nitride materials (GaN, InN, AlN, and their alloys) exhibit polarization charge behavior that must be accounted for in designing GaN power devices. Without getting into theoretical details, we use a simplistic view to explain the origin and underlying physics of polarization in this section.

By polarization behavior, we mean that microscopic dipoles form within the bulk of the material (see Figure 10.2a). A microscopic dipole can be denoted by the charge (e) and the separation between the negative and positive charges (d): $e\vec{d}$ (unit m-Coulomb). If the density of dipoles is defined as ρ_d, we can express the quantity of interest as the polarization vector $\vec{p}$:

$$\vec{p} = \rho_d e \vec{d} \tag{10.1}$$

where the polarization is given in units of Coulomb/m^2.

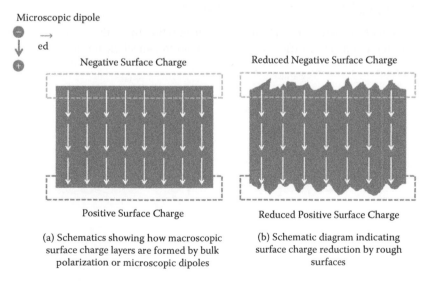

(a) Schematics showing how macroscopic surface charge layers are formed by bulk polarization or microscopic dipoles

(b) Schematic diagram indicating surface charge reduction by rough surfaces

FIGURE 10.2   (a) Schematics showing how macroscopic surface charge layers are formed by bulk polarization or microscopic dipoles; (b) Schematic diagram indicating surface charge reduction by rough surfaces.

In an infinite homogeneous nitride material with uniform polarization density, the positive and negative charges of the dipoles would cancel each other within the bulk, and net charges appear only on a flat surface that is perpendicular to the vector of the polarization (pointing from the negative charge to the positive charge) dipole vector (see Figure 10.2a). Please note that net surface charges do not appear on surfaces that are parallel to the dipole vector, and the amount of charges would be reduced if the surface makes an angle with the dipole vector. In a rough surface or rough interface, nonflat regions would form angles with the dipole vector; thus, the net surface charges would be reduced compared with a flat surface or interface (see Figure 10.2b). Such a geometric charge reduction effect explains why surface and interface quality affects the amount of net surface charges.

10.3.2 CRYSTAL STRUCTURE AND POLARIZATION

Similar to how silicon atoms connect to four neighbors in a lattice, in GaN each Ga atom connects to four N-atoms; alloys in the III-nitride family are the same with In or Al substituting for Ga. However, unlike the diamond cubic configuration found in silicon, the most common and stable crystal configuration for III-nitrides is a hexagonal close packed (hcp) Wurtzite structure (Figure 10.3).

We note that N has a high electronegativity (3.04) compared with Ga (1.81), so the GaN bonds are strongly ionic. This means that the hexagonal crystal structure is responsible for the microscopic polarization observed in GaN: if the dipoles do not cancel each other out because of how the atoms are placed in the lattice, then a net polarization vector forms in the direction in which the atoms stack (c-axis).

Two arrangements are possible depending on how the bonds are aligned with respect to the substrate surface. By convention, the Ga-to-N bond along the c-axis is defined as the [0001] direction, and substrates with this bond on the outer surface are called Ga-face. Substrates grown in the other direction [000$\bar{1}$] are called N-face and exhibit a polarization vector direction opposite to that of Ga-face substrates.

The formation of Ga-face or N-face structures is almost entirely the result of the process and growth conditions; if one is not careful, different N-face and Ga-face regions can even form on the same substrate (domain reversal). In general, MOCVD processes tend to produce Ga-faced substrates, whereas N-faced substrates are favored in MBE. With the technology currently available, Ga-face crystals also tend to be of a higher quality and are cheaper and faster to grow. Since MOCVD is also the favored growth method for the industry, Ga-faced crystals are commonly used, and we will therefore limit ourselves to this configuration.

10.3.3 IDEAL c_0/a_0 RATIO FOR ZERO NET POLARIZATION

Close examination of the structure in Figure 10.3 indicates that the GaN microscopic dipoles vector projected along the c-axis have different signs and tend to cancel each other. Defining the lattice

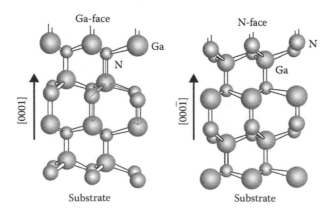

FIGURE 10.3 Atomic arrangement in Ga-face (a) and N-face (b) GaN crystals. The arrows show the direction of microscopic dipoles.

constant of Wurtzite in the stacking direction as c and in the hexagonal plane as a, we will show that for a specific c/a ratio the net dipole will be zero and the crystal will have no net polarization. A zero subscript is added to the lattice constant of a relaxed and bulk material since strained lattice constants are often used.

Projecting the microscopic dipoles along the c-axis is the same as projecting the bond length (d) onto the c-axis (d_z). The dipoles will therefore cancel each other out if

$$d = 3d_z \tag{10.2}$$

Examining the structure, we find the lattice constants can be defined using the bond length and its projections:

$$c = 2d + 2d_z \tag{10.3}$$

$$a = 2d_x \tag{10.4}$$

Some basic vector analysis using the previous equations leads to the ideal c/a ratio for zero polarization:

$$\frac{c}{a} = \sqrt{\frac{8}{3}} = 1.6330 \tag{10.5}$$

Since all of the III-N compounds (Ga-faced) have c_0/a_0 ratios (1.6010 for AlN; 1.6259 for GaN; 1.6116 for InN) smaller than the aforementioned ideal ratio, the three Ga-N bonds have shorter d_z and thus all exhibit a negative spontaneous polarization with the polarization vector pointing toward the substrate. This kind of polarization due to crystal asymmetry is termed spontaneous polarization.

10.3.3.1   Strain-Induced Polarization

If the c_0/a_0 ratio of the III-N lattice is forced to deviate from the above ideal ratio of $\sqrt{8/3}$, the polarization of the crystal is affected. One way to change the c_0/a_0 ratio is through strain, such as in pseudomorphic epitaxial growth on a different III-N material. During pseudomorphic epitaxy, the material expands or contracts to match the in-plane lattice constants of the underlying layer. When in-plane lattice constant a changes in one way, the c must change in the other way to preserve the total bulk volume.

As a result of the in-plane lattice constant change, the c_0/a_0 ratio changes, which in turn changes the polarization strength. The additional polarization in strained III-N layers is termed piezoelectric polarization.

Let us consider the case of InGaN and AlGaN growing on GaN and the overall lattice spacing is that of the GaN substrate. The lattice constants (a_0) are 3.189, 3.544, and 3.111 Å for GaN, InN, and AlN, respectively. Therefore, InGaN would experience biaxial compressive strain in the plane parallel to the substrate and the c_0/a_0 ratio would be increased toward the ideal ratio, which in turn reduces the polarization strength. In another word, the sign of piezoelectric polarization for InGaN is positive.

On the other hand, AlGaN would experience biaxial tensile strain in the plane parallel to the substrate and the c_0/a_0 ratio would be reduced from the ideal ratio, which in turn increases the polarization strength. In other words, the sign of piezoelectric polarization for AlGaN is negative. Please note that AlGaN enhancing the polarization charges is especially significant since a strong polarization induces the triangle quantum well for formation of 2D electron gas in a power GaN HEMT structure.

We therefore note the importance for TCAD modeling of GaN devices to know the reference lattice of the crystal and any external stress terms to properly calculate the strain and polarization in the device. This is especially challenging for GaN-on-silicon where the difference in thermal expansion coefficients leads to external stress or cracks when the structure cools down to room temperature. In practice, one or more thick layers of GaN is usually grown and allowed to relax to ensure that all subsequent epi layers will follow the GaN lattice.

10.3.3.2 Empirical Approach to Modeling Polarization

The previous subsections use a simplistic explanation to account for the origin of spontaneous and piezoelectric polarizations in III-N materials. A rigorous calculation is rather involved, and a common approach is to express the total polarization as an empirical function of material composition (see, e.g., [207]).

Usually, the polarization is given for binaries of GaN, InN, and AlN and for arbitrary composition; a linear combination is taken with a second-order correction term (bowing). We take one example of $Al_xGa_{1-x}N$ from reference [207] here. The spontaneous polarization p_{sp} in units of e/m^2 can be expressed as

$$P_{sp} = -0.09x - 0.034(1-x) + 0.019x(1-x) \tag{10.6}$$

For computation of piezoelectric polarization, we need to calculate the strain ε as

$$\varepsilon = (a_{0,\ GaN} - a_{0,\ AlGaN})/a_{0,\ AlGaN} \tag{10.7}$$

Then, the piezoelectric polarization $P_{pz,AlN}$ for AlN can be written as

$$P_{pz,AlN} = -1.808\varepsilon + 5.624\varepsilon^2; \quad for\ (\varepsilon < 0) \tag{10.8}$$

$$P_{pz,AlN} = -1.808\varepsilon - 7.888\varepsilon^2; \quad for\ (\varepsilon > 0) \tag{10.9}$$

Similarly, for GaN

$$P_{pz,GaN} = -0.918\varepsilon + 9.541\varepsilon^2 \tag{10.10}$$

Piezoelectric polarization for the compound is obtained by linear combination:

$$P_{pz} = P_{pz,AlN}x + P_{pz,GaN}(1-x) \tag{10.11}$$

The total polarization for AlGaN can be finally expressed as

$$P_{AlGaN} = P_{sp} + P_{pz} \tag{10.12}$$

Please note that the empirical formula (10.12) is the full amount of polarization charges that can be measured. In real devices, rough surfaces and interface defects may introduce a screening factor, and usually some fraction (say 30 to 70 percent) of the previous full polarization is used in TCAD simulation.

As an example, the polarization is simulated with a TCAD device simulator (APSYS) according to the formula found in [208] with a screen factor of 0.7 (70 percent); a factor of unity means 100 percent of the theoretical interface charge is used. This screening factor is used to account for deviations from the theory such as screening by charged fixed defects and partial relaxation. As seen Figure 10.4a, polarization charges in the TCAD simulator are represented by fixed charges. The top has negative fixed charges, while the bottom has positive charges. Unlike silicon, which has no polarization effect, a tilted band diagram is obtained if we draw a 1D cutline from top to bottom, as shown in Figure 10.4b. Note that for the moment we ignore the compensation and passivation of charges on the exposed surface that would occur in a real device.

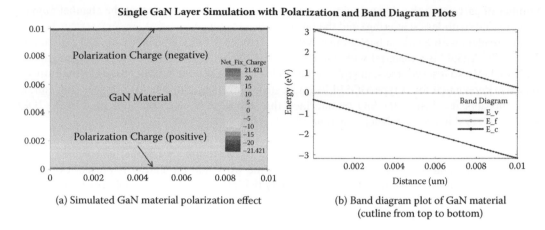

(a) Simulated GaN material polarization effect

(b) Band diagram plot of GaN material (cutline from top to bottom)

FIGURE 10.4 Simulated GaN material polarization effect.

10.4 AlGaN/GaN HETEROJUNCTION

By adding an AlGaN layer on top of a GaN layer, we create a heterojunction: the $Al_xGa_{(1-x)}N/GaN$ heterojunction is widely used in devices such as HEMTs. Here, x is the mole fraction of aluminum in the AlGaN material. With AlGaN grown on top of GaN, the band diagram near the interface is altered due to the polarization mismatch of the two materials and the resulting interface charges. As an example, the band diagram for x = 0.3 ($Al_{0.3}Ga_{0.7}N/GaN$) is plotted side by side with x = 1 (AlN/GaN) on Figure 10.5. The $Al_xGa_{(1-x)}N$ layer is 30 nm thick in both cases, but only a 10 nm window near the material boundary is shown.

From Figure 10.5, we note the formation of a triangular well in the conduction band at the interface. This region is very thin and will serve to confine carriers on a two-dimensional plane (i.e., a quantum well) leading to the formation of a two-dimensional electron gas (2DEG). We note that as the Al fraction in AlGaN increases, the bandgap of AlGaN increases as does the polarization mismatch; as a result, the triangular electron well becomes deeper and deeper. As the bottom of the well dips below the Fermi level, a high carrier sheet density is formed that provides for a large

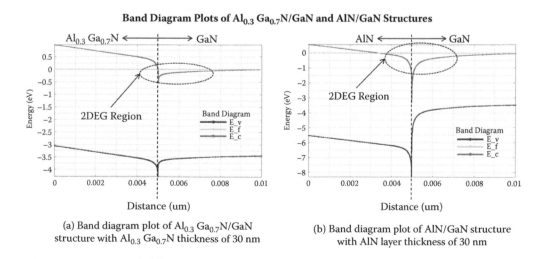

(a) Band diagram plot of $Al_{0.3}Ga_{0.7}N/GaN$ structure with $Al_{0.3}Ga_{0.7}N$ thickness of 30 nm

(b) Band diagram plot of AlN/GaN structure with AlN layer thickness of 30 nm

FIGURE 10.5 Band diagram plots of $Al_xGa_{(1-x)}N/GaN$ heterojunction with (a) $Al_{0.3}Ga_{0.7}N/GaN$ and (b) AlN/GaN.

number of carriers in the channel without relying on doping. This means that the channel mobility is increased (with less scattering) and the on-resistance is reduced.

For readers with a silicon background, we can correlate the AlGaN/GaN with SiO_2/Silicon in a MOSFET. AlGaN is a material with a much larger (4 eV for 30% Al, 6 eV for AlN) bandgap than GaN (3.4 eV), so it serves the same role as the oxide (which has a much larger bandgap than silicon). It is also known that for N-type MOSFETs under strong inversion a thin layer of highly concentrated electrons appears close to the SiO_2/Si interface: this thin layer of electrons analogous to the 2DEG formed in our AlGaN/GaN heterojunction.

The most important difference compared with a silicon MOSFET is that it is not necessary to create a strong inversion or use doping to form a 2DEG in a GaN HEMT; instead, the built-in polarization mismatch at the interface does all of the work and the 2DEG is present even under equilibrium conditions. We therefore need to study the influence of the Al mole fraction x and the thickness of the AlGaN layer on our 2DEG.

10.4.1   BAND DIAGRAM PLOTS WITH A FIXED AL MOLE FRACTION

If we arbitrarily select a fixed x value, for example, x = 0.3, we can vary the thickness of the $Al_xGa_{(1-x)}N$ layer to see the how the layer thickness affects the 2DEG. In Figure 10.6, simulated band diagrams at equilibrium for $Al_{0.3}Ga_{0.7}N$ layer thicknesses of 15 nm, 20 nm, 35 nm, and 45 nm are plotted. No metal contacts or surface passivation are used in these simulations. We note that

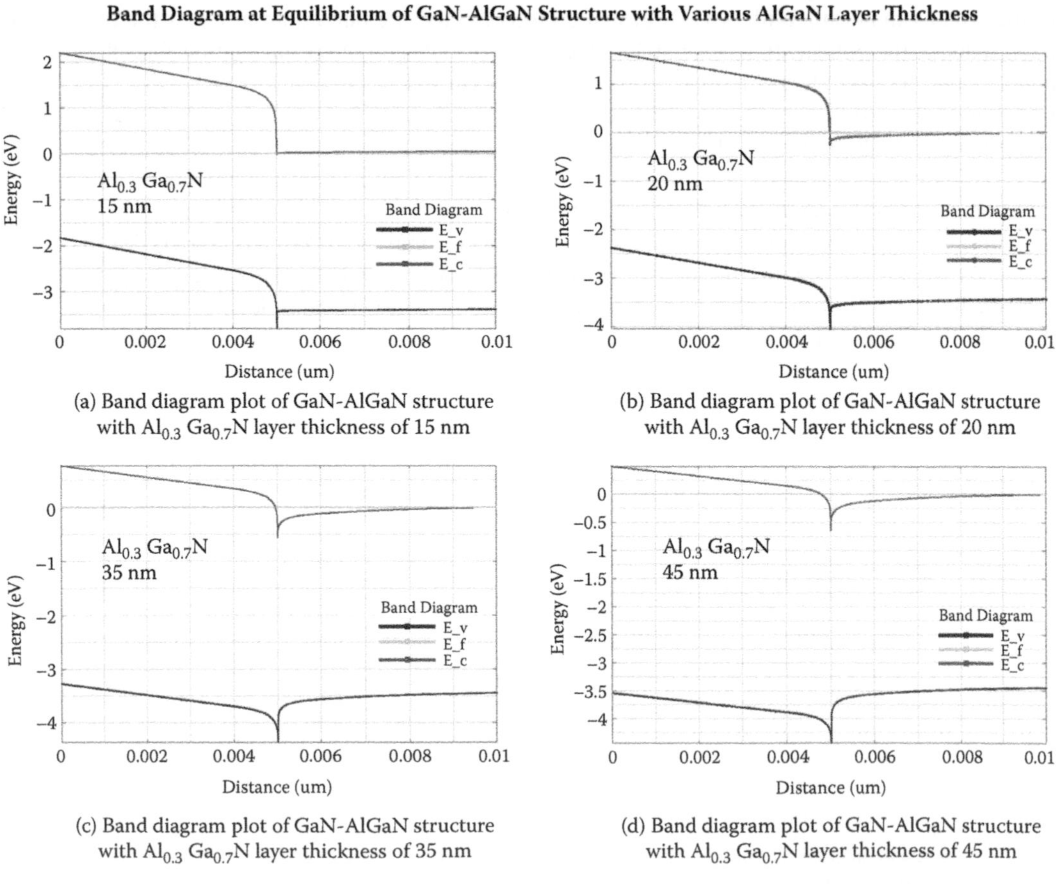

Band Diagram at Equilibrium of GaN-AlGaN Structure with Various AlGaN Layer Thickness

(a) Band diagram plot of GaN-AlGaN structure with $Al_{0.3}Ga_{0.7}N$ layer thickness of 15 nm

(b) Band diagram plot of GaN-AlGaN structure with $Al_{0.3}Ga_{0.7}N$ layer thickness of 20 nm

(c) Band diagram plot of GaN-AlGaN structure with $Al_{0.3}Ga_{0.7}N$ layer thickness of 35 nm

(d) Band diagram plot of GaN-AlGaN structure with $Al_{0.3}Ga_{0.7}N$ layer thickness of 45 nm

FIGURE 10.6   Band diagram plots of $Al_{0.3}Ga_{0.7}N$/GaN heterojunction with different layer thickness of (a) 15 nm, (b) 20 nm, (c) 35 nm, and (d) 45 nm.

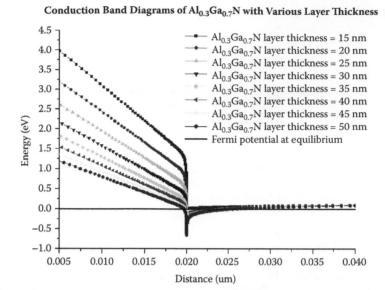

FIGURE 10.7   Conduction bands plot of $Al_{0.3}Ga_{0.7}N/GaN$ with different $Al_{0.3}Ga_{0.7}N$ layer thickness.

the 2DEG region gets larger with a thicker $Al_{0.3}Ga_{0.7}N$ layer thickness; at 15 nm, the triangular well is so shallow that no 2DEG region can be found. The band diagrams give us a clue on how high the electron concentration can be near the interface: thicker AlGaN layers result in a higher sheet charge and thus a lower resistivity during the on-state. This layer thickness-2DEG relationship applies to all x values.

This band diagram dependence on the $Al_{0.3}Ga_{0.7}N$ layer thickness can be clarified by superposing the conduction bands in Figure 10.7. Different layer thickness yields different 2DEG sizes, and for any given Al fraction there is a critical thickness below which the 2DEG ceases to exist.

In addition to this minimum thickness, in real manufacturing processes there is also a maximum thickness that can be used. Readers familiar with compound semiconductor growth will surely know about the concept of critical film thickness [209]: beyond this thickness, thin films under stress start to relax and take on their natural bulk lattice constant rather than that of the substrate. In addition to the cracking such relaxation would induce, the 2DEG would also be affected since relaxation reduces strain in the layer and therefore affects the polarization mismatch [210] [211]. According to published data [205], the most commonly used AlGaN layer thicknesses are around 15 nm to 30 nm.

10.4.2   BAND DIAGRAM PLOTS WITH A FIXED AlGaN THICKNESS

For a fixed AlGaN layer thickness to 30 nm, we can observe the effects of the Al mole fraction in Figure 10.8. For small Al content, the 2DEG is also not found for this thickness, which suggests that a thicker layer would be needed. At larger concentrations, the well is deeper, which leads to a larger 2DEG concentration and lower on-resistance in the channel. A more detailed comparison is also made in Figure 10.9.

Again, real device processes limit what can be accomplished, and it is not always possible to simply use a larger Al fraction. Cracks may appear if the Al mole fraction is too large [211]. According to published data [205], the most commonly used Al mole fractions are from x = 0.15 to x = 0.3.

10.4.3   AlGaN/GaN STRUCTURE WITH DOPED AlGaN OR GaN LAYER.

It has been found that adding dopants in the AlGaN layer can modify the 2DEG sheet charge. For example, for the case of $Al_{0.25}Ga_{0.75}N/GaN$ with an AlGaN layer thickness of 25 nm, a heavier

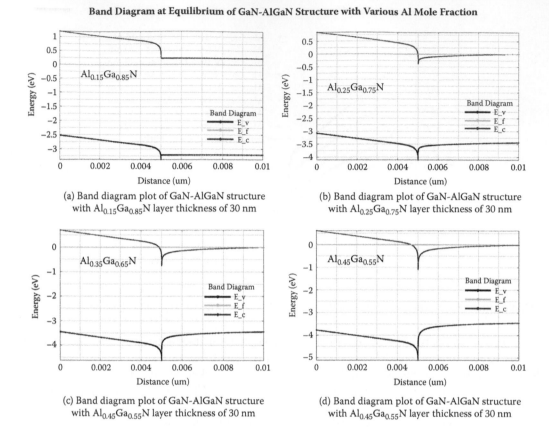

Band Diagram at Equilibrium of GaN-AlGaN Structure with Various Al Mole Fraction

(a) Band diagram plot of GaN-AlGaN structure with $Al_{0.15}Ga_{0.85}N$ layer thickness of 30 nm

(b) Band diagram plot of GaN-AlGaN structure with $Al_{0.25}Ga_{0.75}N$ layer thickness of 30 nm

(c) Band diagram plot of GaN-AlGaN structure with $Al_{0.45}Ga_{0.55}N$ layer thickness of 30 nm

(d) Band diagram plot of GaN-AlGaN structure with $Al_{0.45}Ga_{0.55}N$ layer thickness of 30 nm

FIGURE 10.8 Band diagram plots of AlxGa(1−x)N/GaN structure with fixed layer thicknesses and different Al mole fractions: (a) $Al_{0.15}Ga_{0.85}N$; (b) $Al_{0.25}Ga_{0.75}N$; (c) $Al_{0.35}Ga_{0.65}N$; (d) $Al_{0.45}Ga_{0.55}N$.

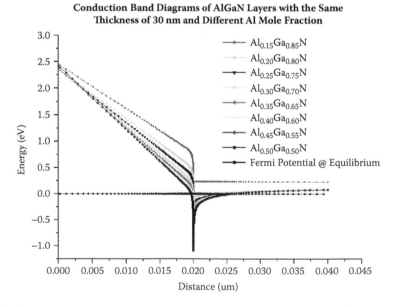

Conduction Band Diagrams of AlGaN Layers with the Same Thickness of 30 nm and Different Al Mole Fraction

FIGURE 10.9 Conduction bands plot for $Al_xGa_{(1-x)}N/GaN$ structure with fixed layer thickness (of 30 nm) and different Al mole fractions.

Band Diagram at Equilibrium with Different AlGaN Doping Conditions

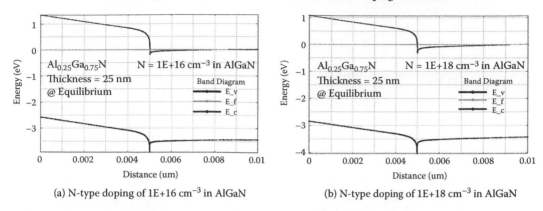

(a) N-type doping of 1E+16 cm^{-3} in AlGaN   (b) N-type doping of 1E+18 cm^{-3} in AlGaN

FIGURE 10.10   Band diagram plot of doping effect in AlGaN layer with (a) N-type doping of 1E+16 cm^{-3} and (b) N-type doping of 1E+18 cm^{-3}.

N-type doping in the AlGaN layer creates a larger 2DEG sheet charge than a lightly doped AlGaN layer, as shown in Figure 10.10.

N-type doping in the GaN layer, on the other hand, has much less impact on the 2DEG sheet charge density, as seen in the band diagram plots in Figure 10.11.

10.4.4 AlGaN/GaN Structure with Metal Contacts

The 2DEG channel formed by the AlGaN/GaN heterojunction is filled with a high concentration of electrons (>1E+19 cm^{-3}). When no external vertical field exists, it is a good conducting channel, and with a vertical electric field it is possible to fully deplete this conducting channel to achieve gate control.

We now add a metal contact on top of an AlGaN layer and another metal contact to the GaN. A Schottky contact is applied to the AlGaN with a metal work function of 5.1 eV to represent Au; for the GaN contact, we use an Ohmic contact. In this simulation, x = 0.2 (Al$_{0.2}$Ga$_{0.8}$N), and Al$_{0.2}$Ga$_{0.8}$N layer thickness is 30 nm.

Band Diagram at Equilibrium with Different GaN Doping Conditions

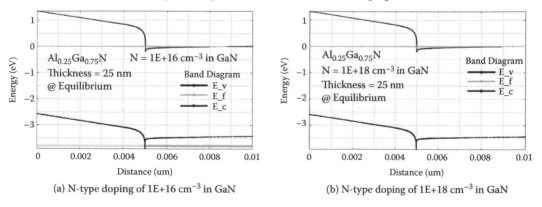

(a) N-type doping of 1E+16 cm^{-3} in GaN   (b) N-type doping of 1E+18 cm^{-3} in GaN

FIGURE 10.11   Band diagram plot of doping effect in GaN layer with (a) N-type doping of 1E+16 cm^{-3} and (b) N-type doping of 1E+18 cm^{-3}.

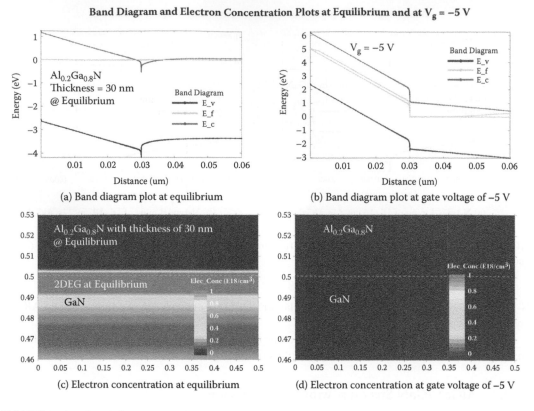

(a) Band diagram plot at equilibrium

(b) Band diagram plot at gate voltage of −5 V

(c) Electron concentration at equilibrium

(d) Electron concentration at gate voltage of −5 V

FIGURE 10.12   Band diagram and electron concentration plots at equilibrium ($V_g = 0$ V) and at $V_g = -5$ V.

If we keep the bottom metal contact (GaN) grounded and apply a negative voltage bias (−5 V) on the top metal contact (AlGaN contact, called a gate contact in this section), we see a diminishing 2DEG and electron concentration at the AlGaN/GaN material boundary.

Figure 10.12a shows the band diagram plot at equilibrium ($V_g = 0$ V). A 2DEG region exists at equilibrium, and this is also verified by the electron concentration plot of Figure 10.12c. For a gate contact voltage of −5 V, the band diagram is plotted again in Figure 10.12b; the 2DEG region no longer exists, and the high electron concentration at the material boundary has vanished, as shown in Figure 10.12d. This phenomenon is caused by the electric field exerted from the gate contact that pinches off the 2DEG channel.

In fact, HEMT devices use this gate control effect to deplete or enhance the 2DEG channel so that the device can be controlled like a field-effect transistor (FET).

10.5   TRAPS IN AlGaN/GaN STRUCTURE

Traps are recombination centers associated with crystal defects and primarily form near the substrate during crystal growth; there is a period during the growth when the GaN's lattices adapt to that of the substrate (e.g., Si, SiC, GaN). Traps can also be introduced at interfaces depending on surface passivation technologies. We will discuss how traps can affect device performance in this section. We will simulate a 30 nm thick $Al_{0.3}Ga_{0.7}N$ layer on top of a GaN layer. The GaN layer has a thickness of 500 nm and will be simulated with and without traps. Please refer to Section 6.5 in Chapter 6 for detailed information on the trap models.

As seen in Figure 10.13a, if no traps are placed (intrinsic material), then the Fermi level in GaN is very close to the conduction band, which means the GaN has a very low resistance that can cause problems during simulations. For example, for HEMTs, we expect that during the on-state only the 2DEG channel is conducting. During the off-state when the 2DEG channel is pinched-off, no current is conducting except a very small leakage current. If the GaN material below 2DEG has low resistance, then current will flow in this region even if we apply a bias on the gate to fully deplete the 2DEG. We know that in real fabricated devices the GaN-substrate (not the 2DEG layer) is a semi-insulating material with high resistance. This discrepancy between the simulated case and a real device is because of the presence of traps, which are often ignored in simulation.

To solve this problem, traps need to be placed within the GaN for the TCAD simulation to turn it into a semi-insulating material. Since in our case, the GaN is acting like an N-doped material (as seen in our previous band diagrams), we need to add P-type traps to move the Fermi level away from the conduction band.

In this simulation, we will place acceptor-type traps so that we can create a semi-insulating GaN material. Figure 10.13a is the simulation result without traps. The GaN material behaves like an N-doped semiconductor with a conduction band very close to the Fermi level. Figures 10.13b–d illustrate simulation results with different trap concentrations in the GaN layer, with band diagrams showing the deviation of the conduction band from the Fermi level. The traps are located at 1.5 eV below the conduction band and have a capture cross section of 1E-13 cm^2.

Band Diagram at Equilibrium of GaN-AlGaN Structure with Different Trap Placement

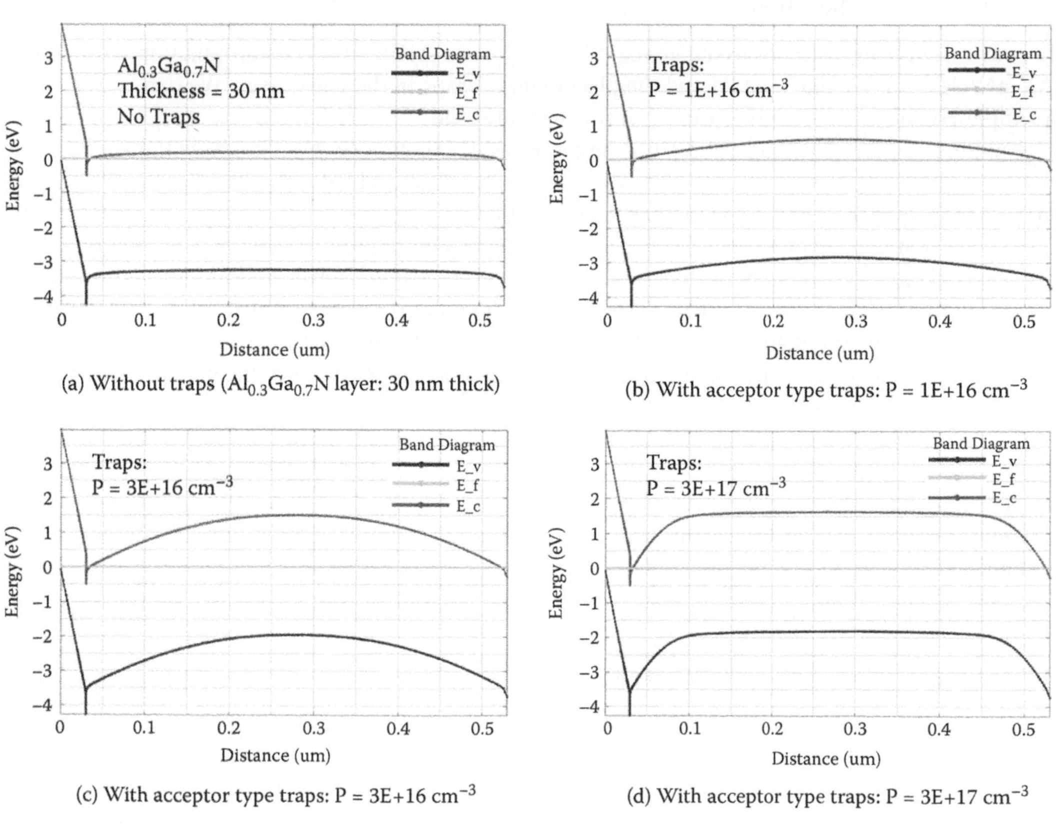

(a) Without traps (Al$_{0.3}$Ga$_{0.7}$N layer: 30 nm thick)

(b) With acceptor type traps: P = 1E+16 cm^{-3}

(c) With acceptor type traps: P = 3E+16 cm^{-3}

(d) With acceptor type traps: P = 3E+17 cm^{-3}

FIGURE 10.13 Band diagram plots of different trap placement in the GaN layer.

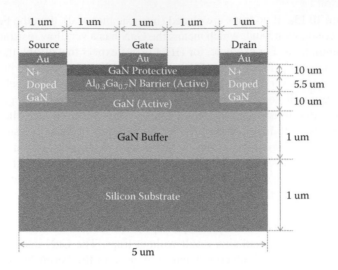

FIGURE 10.14   HEMT device structure illustration.

10.6   A SIMPLE AlGaN/GaN HEMT

10.6.1   DEVICE STRUCTURE

In previous sections, we discussed a single GaN layer, an AlGaN/GaN heterojunction with 2DEG, AlGaN/GaN with a gate contact and the placement of traps in the GaN layer. We can now move to a simple HEMT by adding the source, drain, and substrate.

Figure 10.14 is a schematic view of a simple HEMT. We choose $Al_{0.3}Ga_{0.7}N$ as the barrier layer with thickness of 30 nm. Besides the AlGaN/GaN layer, we added the source/drain, a GaN cap layer, a GaN buffer layer, and a silicon substrate. The GaN cap layer is used to fix the surface inhomogeneity that may be present in the AlGaN layer to get a more uniform device performance across the wafer [205]. A thick GaN buffer layer is grown on the silicon substrate to relieve stress caused by lattice mismatch. Layers marked as *active* are where quantum mechanical models are applied, and the Schrodinger equation is solved in addition to the classical drift-diffusion model [156]. To model the confinement of the 2DEG, quantum well models are used, with the channel region as the "well" and surrounding layers as the "barriers."

The work function for the gate metal's Schottky contact is chosen to be 5.1 eV (Au). N+ doped GaN are used in the source and drain to create Ohmic contacts. Acceptor-like traps are placed in GaN buffer and GaN active regions to suppress the drain to source leakage current beneath the 2DEG channel and make the substrate semi-insulating. Polarization is set at all interfaces in this case, with a screening factor of 0.7 to reflect the polarization level in a real device. The simulation parameters for GaN HEMT are listed in Table 10.3, and trap placement parameters are listed in Table 10.4.

10.6.2 I_D-V_G CURVES FOR GaN HEMT

For the HEMT device we are simulating, at equilibrium a 2DEG channel is formed at the $Al_{0.3}Ga_{0.7}N$/GaN heterojunction. The channel is not intentionally doped, and a high sheet charge density of electrons is generated from both piezoelectric and spontaneous polarization. If a small bias is applied to the drain, current will flow so the channel is always on. We call this type of device a depletion mode device, because to turn off the device the gate has to be biased at a negative potential to deplete the channel. This is in contrast with enhancement mode devices, which are normally off, and a gate bias is used to turn on and modulate the channel conductivity.

TABLE 10.3

Simulation Parameters for GaN HEMT

Simulation Parameters	Value
Gate length	1 μm
Gate to source distance	1 μm
Gate to drain distance	1 μm
GaN cap layer thickness	10 nm
$Al_xGa_{(1-x)}N$ composition	0.3
$Al_xGa_{(1-x)}N$ layer thickness	30 nm
GaN active layer thickness	10 nm
GaN buffer active layer thickness	1 μm
Substrate material	Silicon
Substrate thickness	1 μm
Source/drain doping	1E+19 cm^{-3}
Polarization screening	0.7

For depletion mode devices, the threshold voltage is defined as the gate voltage that turns off the 2DEG channel. (The device is on when the gate is floating or grounded.) To simulate the threshold voltage, the drain terminal is first biased at +5 V, and then the gate voltage is scanned from 0 to −5 V to completely pinch off the channel, with the source grounded. Unlike enhancement mode MOSFET, whose threshold voltage is usually defined at the linear region (i.e., a small positive voltage on the drain), a depletion mode HEMT needs to be biased in the saturation region because we need to see when the channel is completely pinched off from normal operation under a high drain bias and high current level. However for MOSFET, we need to know when the channel can be turned on with the lowest possible drain voltage and drain current.

A comparison of I_d-V_g curves is made at a different AlGaN layer thickness with the same aluminum mole fraction of 0.3 ($Al_{0.3}Ga_{0.7}N$), as seen in Figure 10.15. The simulations are performed with a drain bias to +5 V and gate scanned from 0 to −5 V.

From the I_d-V_g curves, we noticed that a thicker AlGaN layer brings higher on-state current level (with $V_d = 5$ V, $V_g = V_s = 0$ V) and higher threshold voltage. This is because a thicker AlGaN layer will generate a higher 2DEG sheet charge and thus lower on-resistance and higher on-state current levels. However, the thicker AlGaN layer also increases threshold voltage and makes it harder to turn off the device.

10.6.3 SUMMARY

So far in this chapter, we presented the GaN material in a step-by-step approach. We started with an explanation of substrate materials, polarization effect, and 2DEG of a single GaN layer and then moved on to add a AlGaN layer on top of GaN so that a heterojunction is formed.

TABLE 10.4

Trap Parameters for GaN HEMT

Traps Placement in GaN Layers	Value
Acceptor (P)-type traps concentration	3E+17 cm^{-3}
Cross section for traps	1E-13 cm^2
Trap location in the band diagram	1.5 eV below conduction band

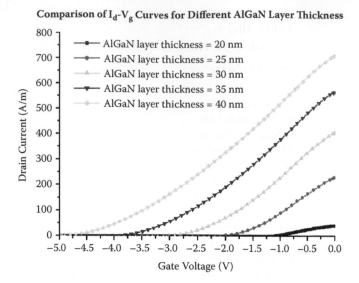

FIGURE 10.15   A comparison between of I_d-V_g curves between different AlGaN layer thickness with the same Al mole fraction of 0.3 (Al$_{0.3}$Ga$_{0.7}$N).

We then studied several parameters that affect the 2DEG density, including the AlGaN layer thickness, Al composition, and doping effect in AlGaN. From previous simulations, we have observed that higher Al composition and thicker AlGaN layers yield higher 2DEG density. However, there are limits for these parameters, as shown in Table 10.5, according to reference [205]. In practice, the most commonly used Al mole fraction is in the range of x = 0.15 to x = 0.3, and the AlGaN layer thickness will be in the 15 nm to 30 nm range. With proper design, the 2DEG will have a sheet charge density on the order of 0.8 ~ 1.1 E+13 cm^{-2} with a mobility of 1300 to 2000 cm^2/Vs [205].

Traps are placed in the GaN buffer layer under the 2DEG channel to create a semi-insulating layer and suppress the leakage current. We also demonstrated the field effect from the gate to pinch off the 2DEG channel so that the device can be controlled like other FETs. Finally, a simple HEMT device was simulated with added source/drain regions. I_d-V_g curves were simulated under a different AlGaN thickness. It has been found that a thicker AlGaN makes it harder to pinch off the 2DEG channel and causes higher threshold voltage (in absolute value) for HEMT (see Figure 10.15). This is similar to the influence of gate oxide thickness to MOSFET threshold voltage where a thicker gate oxide yields higher threshold voltage. The next section introduces some more complicated examples.

TABLE 10.5

Limits of Al Mole Fraction and AlGaN Thickness Engineering

Design Parameters	Benefit	Issues
High Al composition (x>0.3)	Higher 2DEG density	Device leakage and crystal quality degradation
Low Al composition (x<0.1)	—	No 2DEG
Thick AlGaN layer (thickness > 30 nm)	Higher 2DEG density	The lattice mismatch between AlGaN/GaN exceeds the maximum yield strength will cause the AlGaN to relax through the formation of cracks Harder to pinch off the 2DEG channel
Thin AlGaN layer (thickness < 10 nm)	—	No 2DEG

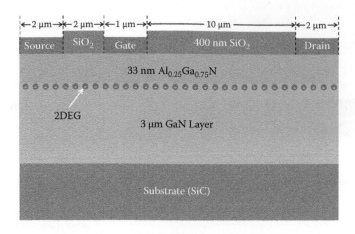

FIGURE 10.16 Schematic plot of the GaN power HEMT.

10.7 GaN POWER HEMT EXAMPLE I

This section gives a simulation example of a GaN power HEMT with a device structure based on the research paper (Gang Xie et al. ISPSD'10) [212] with some modifications. The simulated HEMT device achieved a breakdown voltage of about 575 V.

10.7.1 DEVICE STRUCTURE

The device structure is given in Figure 10.16. The source and drain have Ohmic contacts; each has a width of 2 μm. The distance between gate and source is 2 μm, while gate-to-drain distance is 10 μm. Like silicon LDMOS, the gate-to-drain distance is important because breakdown voltage needs to be supported along the region between gate and drain. The source, gate, and drain contacts are separated with 400 nm thick SiO_2. The simulated structure and enlarged views are given in Figure 10.17.

A fixed charge density of 1.1E+13 cm^{-2} caused by the piezoelectric and polarization dipole was modeled along the upper side of the AlGaN/GaN interface to determine the 2DEG sheet carrier concentration. Donor traps with a maximum concentration of 1E+15 cm^{-3} and acceptor traps with a concentration of 3E+16 cm^{-3} are used to make the substrate semi-insulating; these traps are located 1.5 eV away from the conduction band and use a capture cross section of 1E-13 cm^2. The semi-insulating traps are effective in suppressing parasitic conduction underneath the 2DEG channel [212]. The simulation parameters are listed in Table 10.6.

10.7.2 IMPACT IONIZATION COEFFICIENT OF GaN MATERIAL

Before we move on to simulate the breakdown characteristics of the GaN power HEMT, we need to have a discussion about the impact ionization. In silicon, impact ionization has been extensively studied, and the relevant models have been described in Section 6.3.5. GaN is a more recent material that is not as well understood, and published parameters vary from source to source [213], [214], [215]. In general, the Chynoweth model is widely used to define impact ionization in GaN. In this section, we will briefly survey some of the suggested GaN impact ionization coefficients for the Chynoweth model.

Chynoweth's Law [159] is repeated here. For electrons, the impact ionization coefficient $a_n(E)$ can be expressed as

$$\alpha_n(E) = a_n \cdot e^{\left(-\frac{b_n}{E}\right)}$$

(10.13)

Simulation Structure Views of AlGaN/GaN HEMT

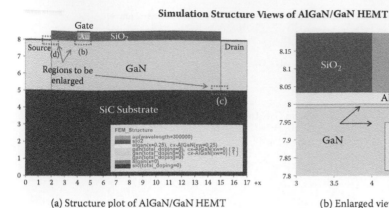

(a) Structure plot of AlGaN/GaN HEMT

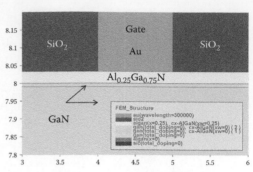

(b) Enlarged view of the HEMT structure around the gate region

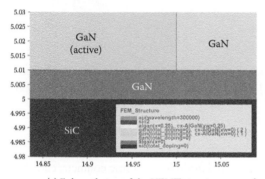

(c) Enlarged view of the HEMT structure around the substrate/GaN layer boundary at drain side

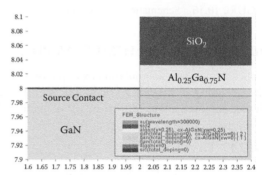

(d) Enlarged view of the HEMT structure around the AlGaN/GaN layer boundary at source side

FIGURE 10.17   Enlarged views of the GaN HEMT structure.

where a_n and b_n are fitting parameters for electrons, and E is the electric field strength. Similarly, for holes, the impact ionization coefficient $\alpha_p(E)$ can be expressed as

$$\alpha_p(E) = a_p \cdot e^{\left(-\frac{b_p}{E}\right)} \tag{10.14}$$

where a_p and b_p are fitting parameters for holes. Table 10.7 shows the Chynoweth impact ionization coefficients from various sources.

Note that a sapphire substrate is assumed for the coefficients value from Ozbek [215].

TABLE 10.6

Simulation Parameters for GaN Power HEMT

Simulation Parameters	Value
Gate length	1 μm
Gate-to-source distance	2 μm
Gate-to-drain distance	10 μm
Substrate	SiC
GaN layer thickness	3 μm
$Al_{0.25}Ga_{0.75}N$ barrier layer	33 nm

TABLE 10.7

Chynoweth Impact Ionization Coefficients from Various Sources

Parameters	Turin and Balandin [213]	Xie et al. [214]	Ozbek [215]
a_n for electrons	2.90E+8 cm^{-1}	2.6E+8 cm^{-1}	9.17E+5 cm^{-1}
b_n for electrons	3.40E+7 V/cm	3.40E+7 V/cm	1.72E+7 V/cm
a_p for holes	1.34E+8 cm^{-1}	4.98E+6 cm^{-1}	8.7E+5 cm^{-1}
b_p for holes	2.03E+7 V/cm	2.03E+7 V/cm	1.46E+7 V/cm

10.7.3 BREAKDOWN SIMULATION OF GaN HEMT

Unlike normally off devices such as LDMOS and LIGBT, to simulate the reverse characteristics of power HEMT, one needs to apply a negative voltage on the gate so that the 2DEG channel region is pinched off by the electric field from the metal gate. Figure 10.18a shows the potential plot. The simulated device has a breakdown voltage of 575 V at $V_g = -5$ V. Figure 10.18b is the electric field plot, which indicates the electric field peaks are located at gate and drain edges; note that a much

Simulation Results Views of Power GaN HEMT

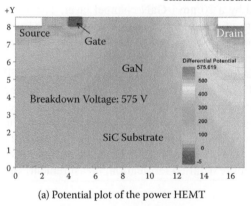

(a) Potential plot of the power HEMT

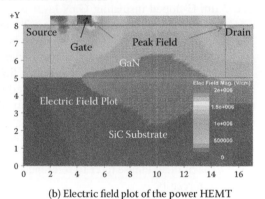

(b) Electric field plot of the power HEMT

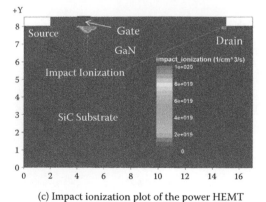

(c) Impact ionization plot of the power HEMT

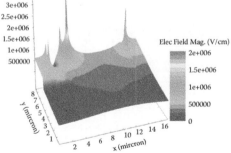

(d) Electric field 3D surface plot of the power HEMT

FIGURE 10.18 Simulation results of the power GaN HEMT at breakdown with (a) potential plot, (b) electric field plot, (c) impact ionization plot, and (d) 3D surface plot.

higher critical field is observed for GaN than silicon. Figure 10.18c illustrates the impact ionization plot, with impact ionization locations corresponding to electric field peaks. Finally, a 3D surface plot of the electric field is shown in Figure 10.18d. In this simulation, the impact ionization coefficients are taken from reference [213].

10.8   GaN POWER HEMT EXAMPLE II

This section briefly discusses a slightly more complicated GaN power HEMT example according to reference (Gang Xie et al. EDL 2012) [214]. A GaN power HEMT was fabricated on a 3-inch RF GaN-on-SiC process with a novel air bridge field plate (AFP). Traditionally, field plate (FP) structures have been effective in achieving high breakdown voltages and can also improve device reliability and suppress current collapse from occurring in GaN HEMTs. However, in most existing GaN processes, only limited breakdown voltage improvements can be obtained with conventional FP structures due to the undesirably thin Si_3N_4 passivation layer (e.g., 150 nm). To deal with this issue, a novel structure called AFP was proposed by Gang Xie et al. [214]. The AFP is a metal field plate that jumps from the source over the gate region and lands between the gate and drain. The proposed structure yields a high breakdown voltage even with a very thin Si_3N_4 passivation layer.

The novel AFP structure is compared side by side with a conventional field plate, as shown in Figure 10.19. The AFP structure has an air gap between the gate and the field plate. The device has a SiC substrate with a 20 nm AlN nucleation layer to alleviate the stress from the lattice mismatch. The 2 µm C-doped insulating GaN layer is followed by a 20 nm i-GaN channel layer. The 1 nm i-AlN spacing layer is deposited before a 20 nm unintentionally doped $Al_{0.28}Ga_{0.72}N$ barrier layer. The thickness of the SiN passivation layer is 150 nm. The settings of polarization-induced fixed charges, traps, and impact ionization coefficients are the same as in Section 10.7.

Results show that thanks to a reduced field near the gate corner, the device with the AFP has a breakdown voltage of 375 V compared with 125 V for a conventional FP and 37 V for the structure without FP. The leakage current for the AFP device is also an order of magnitude lower than that of FP, indicating a great advantage of using the AFP structure under reverse bias in the off-state. In terms of input capacitance, the air-bridge HEMT exhibits a much lower C_{gs} compared with the conventional FP case, but the device structure without AFP or FP has an even lower input capacitance. Please refer to reference [214] for more detailed fabrication, measurement, and simulation results.

Field Plate and Air Bridge Plate GaN HEMT Structures

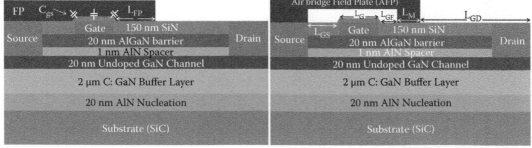

(a) Schematic illustration of conventional Field Plate (FP) GaN power HEMT

(b) Simulation material plot of the novel Air bridge Field Plate (AFP) GaN power HEMT

FIGURE 10.19   Schematic plots of GaN HEMT with (a) conventional FP and (b) AFP.

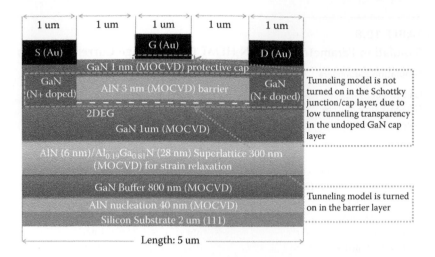

FIGURE 10.20   A more complicated GaN HEMT structure for gate leakage analysis.

10.9   GATE LEAKAGE SIMULATION OF GaN HEMT

One design headache that often plagues GaN HEMT design engineers is the gate leakage current that was thought to be associated with the traps found in the insulating materials. This section will use TCAD simulation to reveal the physics related to the gate leakage current in GaN HEMT [216].

10.9.1   DEVICE STRUCTURE

A slightly more complicated GaN HEMT structure is built based on the literature (Tongde Huang et. al, EDL 2012) [217] and shown in Figure 10.20; a silicon substrate with orientation of (111) is used. The epi-layers include a thin layer of AlN of 40 nm, an 800 nm GaN buffer layer, and a 300 nm AlN (6 nm)/$Al_{0.19}Ga_{0.81}N$ (28 nm) superlattice. The GaN (1 nm)/AlN (3 nm)/GaN heterojunction is grown by MOCVD, and a 1 nm GaN layer is used as a passivation cap. Ti/Al/Ni/Au metal system was deposited for gate source and drain contact. Here, the nucleation layer acts like a template to accommodate the lattice mismatch between the substrate and GaN [205]. For detailed process information, please refer to [217]. The simulation parameters are listed in Table 10.8. Note that the polarization model is based on the reference [207].

10.9.2   MODELS AND SIMULATION SETUP

To better focus on the gate leakage mechanism, we use the traditional drift-diffusion model for general carrier transport with some modification based on quantum mechanical considerations. First, the Schrodinger equation is solved in the 2DEG region to account for quantum confinement effects, and the resulting carrier density is used in the drift-diffusion equation solver. Quantum mechanical tunneling theory is also applied to the AlN barrier to enhance the drift-diffusion current.

Polarization charges were defined on most interfaces except for the top surface of the cap layer, where passivation was done and polarization effects are assumed to be zero. The tunneling probability is calculated only between two points, so a tunneling region (shown in the highlighted region on Figure 10.20) is defined. The dotted square on this figure around the G(Au)/GaN layer is the Schottky contact region, which uses standard thermionic emission theory: quantum tunneling was not enabled in the Schottky junction model due to the small tunneling transparency associated with the undoped barrier. Please refer to Section 6.6 for detailed explanations on how to set up a quantum tunneling simulation.

TABLE 10.8

Simulation Parameters for GaN HEMT Gate Leakage Current Analysis

Simulation Parameters	Value
Gate length	1 μm
Gate-to-source distance	1 μm
Gate-to-drain distance	1 μm
Substrate	Silicon
AlN nucleation layer thickness	40 nm
GaN buffer layer thickness	800 nm
AlN (6 nm)/$Al_{0.19}Ga_{0.81}N$ (28 nm) superlattice thickness	300 nm
GaN layer thickness	1 μm
AlN layer thickness	3 nm
GaN cap layer	1 nm
Metal work function	4.37 eV
Polarization screening	0.35

Both donor and acceptor types of traps are placed within the GaN buffer layer and AlN (6 nm)/ $Al_{0.19}Ga_{0.81}N$ (28 nm) superlattice. The trap concentration is set to be 1E+16 cm^{-3} in the GaN buffer layer with a capture cross-section area of 1E-13 cm^2 for both the acceptor and donor traps. For the superlattice layer, the trap concentration is set to be 1E+16 cm^{-3} for the donors, whereas 3E+17 cm^{-3} is used for the acceptors; the capture cross section is the same as the GaN layer. A deep trap level of 1.5 eV is chosen for both layers [217]. The simulation parameters for deep level traps within the superlattice layer are listed in Table 10.9, and those within the GaN buffer layer are listed in Table 10.10.

The simulation structure is illustrated in Figure 10.21. Layers marked as active are where the quantum mechanical models are applied in addition to the classical drift-diffusion. Layers with a star sign have polarization charges considered; charges on other layers are assumed to be either passivated and canceled by various effects or simply neglected for simplicity.

10.9.3   GATE LEAKAGE SIMULATION

10.9.3.1   Pure TCAD Simulation

The I_g-V_g gate leakage current is simulated and compared with experimental data found in reference [217]. It has been found that, with pure TCAD simulation, gate leakage current is well below the experimental leakage current level, no matter how many traps are placed within the device or where they are used, as shown in Figure 10.22 line(a). The reason behind the much lower leakage

TABLE 10.9

Simulation Parameters for Traps Placed within the Superlattice Layer

Simulation Parameters (Superlattice)	Value
Donor type trap concentration	1E+16 cm^{-3}
Acceptor type trap concentration	3E+17 cm^{-3}
Capture cross section for both types of traps	1E-13 cm^2
Traps location (from conduction band)	1.5 eV

TABLE 10.10

Simulation Parameters for Traps Placed within the GaN Buffer Layer

Simulation Parameters (GaN Buffer Layer)	Value
Donor type trap concentration	1E+16 cm^{-3}
Acceptor type trap concentration	1E+16 cm^{-3}
Capture cross section for both types of traps	1E-13 cm^2
Traps location (from conduction band)	1.5 eV

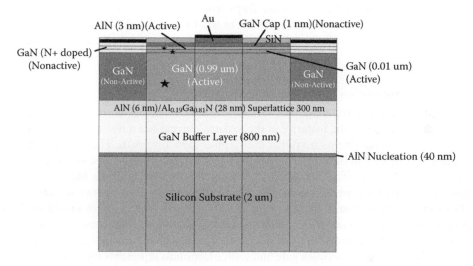

FIGURE 10.21 Simulation structure view of GaN HEMT for gate analysis.

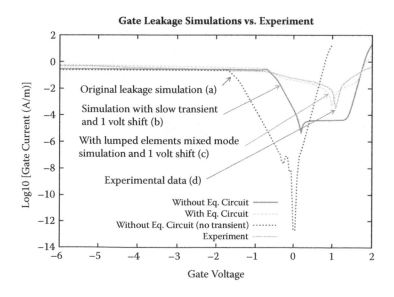

FIGURE 10.22 Comparison of gate leakage simulations (a–c) and experimental results from [217] (d).

level from simulation is that we assume the HEMT device and measurement technology are in ideal condition.

10.9.3.2 TCAD Simulation with Slow Transient

In fact, in real-world measurement, there is always a delay time for the voltage bias to ramp up, no matter how fast the equipment can be. We may consider this time delay to reach the desired voltage as a slow transient, which causes charging of the parasitic capacitance of device so that a larger background leakage current is present. In simulation, slow transient means that time is added as a second scan variable besides a voltage scan to mimic the real measurement environment. As plotted in Figure 10.22 line(b), with an arbitrary voltage ramp-up rate of 7 V/ms, we see a plateau at the bottom of the curve. This is similar to connecting a simple capacitor in parallel with the simulated device. Traps can be regarded as small capacitors because they are recombination centers that trap and release electrons after some time delay. Traps will create a background leakage current with a shape that has a plateau at the bottom, similar to what is seen in the slow transient simulation. Unfortunately, even with the inclusion of this effect, the leakage level is still well below the expected value.

10.9.3.3 TCAD Simulation with Equivalent Circuits

If we take a look at the experimental data, we can see that the leakage current has no plateau at the bottom, which suggests that the majority of the leakage current has a resistive path (or a combination of resistive and capacitive paths) rather than a pure capacitive path. The resistive leakage path may be caused by defects in the barrier region, like voids, that form high resistive leakage paths.

As we know, it is very difficult to accurately simulate the defects in the barrier with a TCAD simulator. A novel method, which combines the lumped elements like resistors and capacitors with a TCAD simulator, is proposed. Shown in Figure 10.23 is the setup view of mixed mode TCAD simulation with lumped elements. Three lumped elements are used. One series resistor connected to the gate and reflects the metal-semiconductor contact resistance (an indicator of metal contact quality). The parallel resistor between gate and source terminals represents the defects in the barrier

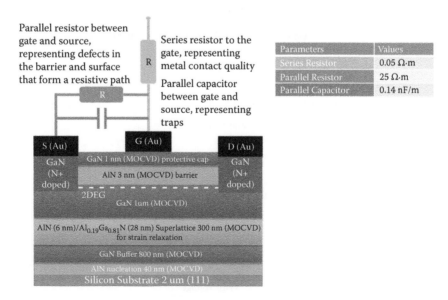

Parameters	Values
Series Resistor	0.05 Ω·m
Parallel Resistor	25 Ω·m
Parallel Capacitor	0.14 nF/m

FIGURE 10.23 Mixed mode TCAD simulation setup with equivalent parasitics and their corresponding values used in the simulation.

that form a resistive path. Finally, the parallel connected capacitor between gate and source terminals is used to simulate the traps. The values of these lumped elements are given in Figure 10.23 as
well [216].

As a result, the mixed-mode TCAD simulation (see Figure 10.22 line(c)) agrees quite well with
experimental data (Figure 10.22 line(d)). Please note that the leakage current minima from original
experimental data has a 1 V shift from 0, but the reason for this is not clear. Except the original
TCAD simulation (Figure 10.22 line(a)), the simulation results are shifted by 1 V to fit the experimental data.

10.10 MARKET PROSPECT OF COMPOUND SEMICONDUCTORS FOR POWER APPLICATIONS

Just like we mentioned in Chapter 3, the power electronics industry is very cost-sensitive. Customers
would prefer to have an old (silicon) technology at a lower cost than to accept a better and more
advanced technology. Even though the expected efficiency improvement of GaN and SiC devices is
impressive for DC/DC, AC/DC, and DC/AC converters, for GaN and SiC vendors, they must prove
their product's reliability, reach competitive costs, and educate the industry on how to best exploit
their advantages [218]. SiC and GaN are both good candidates for future power technology, each
with its own merits and drawbacks. SiC is more suitable for breakdown voltage above 600 V, with its
vertical current flow path similar to that of a VDMOS. GaN is currently offering breakdown ratings
equal or lower than 600 V [218]. GaN HEMTs are lateral devices similar to LDMOS, which limits
their current handling capability. This is one of the reasons that we include a discussion of GaN
in this book rather than SiC devices: a GaN HEMT is considered as a lateral rather than vertical
device, as are all the other devices we have talked about in this book.

In terms of market size, it has been estimated that the global market value for SiC power devices
in 2012 was about US$76 million with photovoltaic inverters as the major application. If the elec-
tric vehicle and hybrid electric vehicles use SiC devices instead of silicon, then by the year 2020
the market value can reach US$200 million [219]. At the time of the writing, GaN power devices
market value is estimated at less than US$10 million [220]. However, GaN devices are much less
expensive than SiC devices to produce: the slogan of GaN devices is to deliver SiC performance at
silicon costs.

This optimism comes from the existing GaN LED industry, which is becoming quite mature. It
is thought that since the LED industry has been able to grow by overcoming reliability challenges,
then so can GaN power devices. In fact, products such as a 600 V normally off HEMTs from
Transphorm, Inc., are already in the market [221]. The high-voltage GaN products may pick up the
momentum and eventually win the favor of the power electronics industry.

Appendix A: Carrier Statistics

The electron and hole concentrations in semiconductors are defined by Fermi–Dirac distributions and a parabolic density of states that, when integrated, yield [138]

$$n = N_c F_{1/2}\left(\frac{E_{fn} - E_c}{kT}\right) \tag{A.1}$$

$$p = N_v F_{1/2}\left(\frac{E_v - E_{fp}}{kT}\right) \tag{A.2}$$

where $F_{1/2}$ is the Fermi integral of order one-half. For the convenience of numerical evaluation, the approximation proposed by Bednarczyk and Bednarczyk is used [222]:

$$F_{1/2}(x) \approx (e^{-x} + \xi(x))^{-1} \tag{A.3}$$

$$\xi(x) = \frac{3}{4}\sqrt{\pi}[v(x)]^{\frac{3}{8}} \tag{A.4}$$

$$v(x) = x^4 + 50 + 33.6x\{1 - 0.68\exp[-0.17(x+1)^2]\} \tag{A.5}$$

This expression is accurate to within 0.4% of error in all ranges. In the limit of low carrier concentration, the previous equations are reduced to the familiar Boltzmann statistics:

$$n = N_c \exp\left(\frac{E_{fn} - E_c}{kT}\right) \tag{A.6}$$

$$p = N_v \exp\left(\frac{E_v - E_{fp}}{kT}\right) \tag{A.7}$$

For TCAD numerical simulations, the more general Fermi–Dirac statistics is used by default.

Appendix B: Process Simulation Source Code

The Suprem source code for a mock-up smart power IC process simulation flow

```
#- - - - - - - - - - - - - - - - - - - - - - - - - - - - - - - -
# This is a mock up BCD technology flow for 0.35 ~ 0.18 µm
#- - - - - - - - - - - - - - - - - - - - - - - - - - - - - - - -

include file = geo1.in

#- - - - - - - - - - - - - - - - - - - - - - - - - - - -
# Wafer Start
#- - - - - - - - - - - - - - - - - - - - - - - - - - - -
initialize boron orient = 100 conc = 1e17
struct outfile = 00_sub.str

#- - - - - - - - - - - - - - - - - - - - - - - - - - - -
# NBL screen oxide
#- - - - - - - - - - - - - - - - - - - - - - - - - - - -
diffuse time = 15 temp = 1000 dryo2
struct outfile = 01a_NBL_screen.str
#- - - - - - - - - - - - - - - - - - - - - - - - - - - -
# NBL PHOTO
#- - - - - - - - - - - - - - - - - - - - - - - - - - - -
include file = testflow1.msk
struct outfile = 01a_NBL_mask.str
implant arsenic dose = 1e14 energy = 15 angle = 0 rotation = 0
etch photoresist all
struct outfile = 01b_NBL_implant.str
etch oxide all

#- - - - - - - - - - - - - - - - - - - - - - - - - - - -
# NBL diff
#- - - - - - - - - - - - - - - - - - - - - - - - - - - -
```

```
# = = = = = = = = Temp Ramp Up = = = = = = = =
diffuse time = 30 temp = 800 final_temp = 1000 nitrogen
struct outf = 01c_NBL_ramp_up.str

# = = = = = = = = STABLE = = = = = = = =
diffuse time = 50 temp = 1000 nitrogen
struct outf = 01d_NBL_stable.str

# = = = = = = = = Temp Ramp Down = = = = = = = =
diffuse time = 30 temp = 1000 final_temp = 800 nitrogen
struct outf = 01e_NBL_ramp_down.str
#- - - - - - - - - - - - - - - - - - - - - - - - - - - - - -
# Epi growth #1
#- - - - - - - - - - - - - - - - - - - - - - - - - - - - - -
deposit silicon thick = 3 boron conc = 1e16 meshlayer = 30
structure outfile = 02a_epi_1.str


#- - - - - - - - - - - - - - - - - - - - - - - - - - - - - -
# Deep N+ sink #1
#- - - - - - - - - - - - - - - - - - - - - - - - - - - - - -
include file = testflow12.msk
structure outfile = 02b_nplus_sink_mask.str

implant phosphorus dose = 3e12 energy = 100 angle = 0 rotation = 0
implant phosphorus dose = 3e12 energy = 375 angle = 0 rotation = 0
implant phosphorus dose = 3e12 energy = 675 angle = 0 rotation = 0
implant phosphorus dose = 3e12 energy = 900 angle = 0 rotation = 0
etch photoresist all
structure outfile = 02b_nplus_sink_1.str


#- - - - - - - - - - - - - - - - - - - - - - - - - - - - - -
# Epi growth #2
#- - - - - - - - - - - - - - - - - - - - - - - - - - - - - -
deposit silicon thick = 1 boron conc = 1e16 meshlayer = 20
deposit silicon thick = 0.8 boron conc = 1e16 meshlayer = 30
deposit silicon thick = 0.2 boron conc = 1e16 meshlayer = 15
structure outfile = 02c_epi_2.str
```

```
#- - - - - - - - - - - - - - - - - - - - - - - - - - - - -
# Deep N+ sink #2
#- - - - - - - - - - - - - - - - - - - - - - - - - - - - -
include file = testflow12.msk
structure outfile = 02d_nplus_mask.str

implant phos dose = 2e12 energy = 150 angle = 0 rotation = 0
implant phos dose = 1e12 energy = 400 angle = 0 rotation = 0
implant phos dose = 1e12 energy = 550 angle = 0 rotation = 0
implant phos dose = 5e12 energy = 700 angle = 0 rotation = 0
etch photoresist all
structure outfile = 02d_nplus_sink_2.str

diffuse time = 120 temp = 1100
structure outfile = 02f_nplus_sink_3.str


#- - - - - - - - - - - - - - - - - - - - - - - - - - - - -
# Pad oxide for HVNW and HVPW
#- - - - - - - - - - - - - - - - - - - - - - - - - - - - -
deposit oxide thick = 0.01
struct outf = 03a_Pad_hvnw.str
#- - - - - - - - - - - - - - - - - - - - - - - - - - - - -
# n-drift PHOTO
#- - - - - - - - - - - - - - - - - - - - - - - - - - - - -
include file = testflow3.msk
structure outfile = 03b_hvnw_mask.str


#- - - - - - - - - - - - - - - - - - - - - - - - - - - - -
# HVNW Implant
#- - - - - - - - - - - - - - - - - - - - - - - - - - - - -
implant phosphor dose = 1e12 energy = 150 angle = 0 rotation = 0
implant phosphor dose = 1e12 energy = 350 angle = 0 rotation = 0
implant phosphor dose = 1e12 energy = 550 angle = 0 rotation = 0
implant phosphor dose = 1e12 energy = 750 angle = 0 rotation = 0
etch photoresis all
structure outfile = 03c_hvnw_imp.str
```

```
#- - - - - - - - - - - - - - - - - - - - - - - - - - - -
# HVPW PHOTO
#- - - - - - - - - - - - - - - - - - - - - - - - - - - -


#- - - - - - - - - - - - - - - - - - - - - - - - - - - -
# HVPW Implant
#- - - - - - - - - - - - - - - - - - - - - - - - - - - -


#- - - - - - - - - - - - - - - - - - - - - - - - - - - -
# Etch Pad oxide
#- - - - - - - - - - - - - - - - - - - - - - - - - - - -
etch oxide all
structure outfile = 03d_pad_hvnw_etch.str


#- - - - - - - - - - - - - - - - - - - - - - - - - - - -
# HV twin-well DRIVE-IN
#- - - - - - - - - - - - - - - - - - - - - - - - - - - -
diffuse time = 120 temp = 1100 nitrogen
structure outfile = 03e_hvw_drive_in.str


#- - - - - - - - - - - - - - - - - - - - - - - - - - - -
# Pbody for nLDMOS mask
#- - - - - - - - - - - - - - - - - - - - - - - - - - - -
include file = testflow6.msk
structure outfile = 04a_Pbody_mask.str


#- - - - - - - - - - - - - - - - - - - - - - - - - - - -
# Pbody implant
#- - - - - - - - - - - - - - - - - - - - - - - - - - - -
implant boron dose = 5e13 energy = 100 angle = 0 rotation = 0
etch photoresist all
structure outfile = 04b_Pbody_implant.str


#- - - - - - - - - - - - - - - - - - - - - - - - - - - -
# Pbody anneal
#- - - - - - - - - - - - - - - - - - - - - - - - - - - -
diffuse time = 30 temp = 1000 nitrogen
struct outf = 04c_Pbody_anneal.str
```

```
#- - - - - - - - - - - - - - - - - - - - - - - - - - - - - - -
# Pad oxide for Shallow Trench Isoltion
#- - - - - - - - - - - - - - - - - - - - - - - - - - - - - - -

# = = = = = = = Wet Oxide pad = = = = = = =
diffuse temp = 900 time = 15 weto2
struct outf = 05a_oxide_pad_STI.str

# = = = = = = = SiNi pad = = = = = = = =
deposit nitride thick = 0.08 temp = 800 meshlayer = 2
struct outf = 05b_nitride_pad_STI.str

# = = = = = = = Etch silicon = = = = = = =
include file = testflow2.msk
etch photoresist all
struct outf = 05c_STI_etch.str

# = = = = = = = STI liner oxide = = = = = = =
diffuse temp = 1000 time = 10 dryo2
struct outf = 05d_STI_liner.str
# = = = = = = = STI High Density Plasma fill = = = = = = =
deposit oxide thick = 0.5 meshlayer = 5 space = 0.01
struct outf = 05e_STI_fill.str

# = = = = = = = CMP stops at nitride = = = = = = =
etch start x = 0 y = -20
etch continue x = 0 y = -4.95
etch continue x = 34 y = -4.95
etch done x = 34 y = -20
structure outf = 05f_STI_CMP.str

# = = = = = = = oxide and nitride strip = = = = = = =
etch nitride all
structure outf = 05g_ON_strip.str

#- - - - - - - - - - - - - - - - - - - - - - - - - - - - - -
# NWELL PHOTO
#- - - - - - - - - - - - - - - - - - - - - - - - - - - - - - -
```

```
include file = testflow13.msk
structure outfile = 06a_Nwell_mask.str


#- - - - - - - - - - - - - - - - - - - - - - - - - - - - -
# NWELL implant
#- - - - - - - - - - - - - - - - - - - - - - - - - - - - -
implant phosphor dose = 1e13 energy = 50 angle = 0 rotation = 0
etch photoresist all
structure outfile = 06b_Nwell_implant.str


#- - - - - - - - - - - - - - - - - - - - - - - - - - - - -
# PWELL mask
#- - - - - - - - - - - - - - - - - - - - - - - - - - - - -
include file = testflow10.msk
structure outfile = 06c_Pwell_mask.str


#- - - - - - - - - - - - - - - - - - - - - - - - - - - - -
# PWELL implant
#- - - - - - - - - - - - - - - - - - - - - - - - - - - - -
implant boron dose = 1e13 energy = 30 angle = 0 rotation = 0
etch photoresist all
structure outfile = 06d_Pwell_implant.str


#- - - - - - - - - - - - - - - - - - - - - - - - - - - - -
# Nwell and Pwell Drive in
#- - - - - - - - - - - - - - - - - - - - - - - - - - - - -
# = = = = = = = = Remove OXIDE = = = = = = = =
etch oxide dry thick = 0.03
structure outf = 06e_oxide_strip.str

# = = = = = = = = Temp Ramp Up = = = = = = = =
diffuse time = 10 temp = 900 final_temp = 1000 nitrogen
struct outf = 06f_NW_ramp_up.str

# = = = = = = = = STABLE = = = = = = = =
diffuse time = 20 temp = 1000 nitrogen
struct outf = 06g_NW_stable.str

# = = = = = = = = Temp Ramp Down = = = = = = = =
```

```
diffuse time = 10 temp = 1000 final_temp = 900 nitrogen
struct outf = 06h_NW_ramp_down.str

#- - - - - - - - - - - - - - - - - - - - - - - - - - - - -
# LVNW for NPN emitter
#- - - - - - - - - - - - - - - - - - - - - - - - - - - - -
include file = testflow14.msk
structure outfile = 07a_LVNW_mask.str

#- - - - - - - - - - - - - - - - - - - - - - - - - - - - -
# LVNW implant
#- - - - - - - - - - - - - - - - - - - - - - - - - - - - -
implant phosphor dose = 5e14 energy = 20 angle = 0 rotation = 0
etch photoresist all
structure outfile = 07b_LVNW_implant.str

#- - - - - - - - - - - - - - - - - - - - - - - - - - - - -
# LVNW drive-in
#- - - - - - - - - - - - - - - - - - - - - - - - - - - - -
diffuse time = 10 temp = 1000 nitrogen
struct outf = 07c_LVNW_drive.str

#- - - - - - - - - - - - - - - - - - - - - - - - - - - - -
# Thick Gate Oxidation TGO 250A
#- - - - - - - - - - - - - - - - - - - - - - - - - - - - -
etch start x = 0 y = -20
etch continue x = 0 y = -4.95
etch continue x = 34 y = -4.95
etch done x = 34 y = -20
structure outfile = 08a_pre_TGO.str

diffuse time = 20 temp = 1000 dryo2
structure outfile = 08b_TGO.str

#- - - - - - - - - - - - - - - - - - - - - - - - - - - - -
# Low Voltage Gate Oxidation, LGO 10A
#- - - - - - - - - - - - - - - - - - - - - - - - - - - - -
include file = testflow11.msk
structure outfile = 08c_LGO_oxide_mask.str
etch oxide dry thick = 0.03
```

```
etch photoresist all
structure outfile = 08c_LGO_oxide_etch.str
diffuse time = 5 temp = 900 dryo2
structure outfile = 08d_LGO.str

#- - - - - - - - - - - - - - - - - - - - - - - - - - - - - - -
# Poly
#- - - - - - - - - - - - - - - - - - - - - - - - - - - - - - -
deposit poly thick = 0.25 meshlayer = 4 phosphorus conc = 1e20 space = 0.02
structure outfile = 09a_Poly_deposit.str
include file = testflow7.msk
structure outfile = 09b_Poly.str

# = = = = = = = = Poly anneal = = = = = = = =
diffuse time = 1 temp = 850 final_temp = 1000
diffuse time = 1 temp = 1000
diffuse time = 1 temp = 1000 final_temp = 850
structure outfile = 09c_Poly_anneal.str

#- - - - - - - - - - - - - - - - - - - - - - - - - - - - - - -
# NLDD and PLDD for logic devices
#- - - - - - - - - - - - - - - - - - - - - - - - - - - - - - -
include file = testflow16.msk
structure outf = 10a_nldd_mask.str
implant phosphorus dose = 9e12 energy = 20
etch photoresist all
diffuse time = 20/60 temp = 1000 nitrogen
etch oxide dry thick = 0.02
structure outf = 10b_nldd.str

# = = = = = = = = PLDD implant = = = = = = = =
include file = testflow15.msk
structure outf = 10c_pldd_mask.str
implant bf2 dose = 2e13 energy = 40
etch photoresist all
diffuse time = 30/60 temp = 1000 nitrogen
structure outf = 10d_pldd.str

#- - - - - - - - - - - - - - - - - - - - - - - - - - - - - - -
# nitride spacer
```

```
#- - - - - - - - - - - - - - - - - - - - - - - - - - - - - - -
deposit nitride thick = 0.25 meshlayer = 2 space = 0.05
structure outf = 11a_spacer_deposit.str
etch nitride dry thick = 0.255
structure outf = 11b_spacer.str

#- - - - - - - - - - - - - - - - - - - - - - - - - - - - - - -
# NSD
#- - - - - - - - - - - - - - - - - - - - - - - - - - - - - - -
include file = testflow5.msk
structure outf = 12a_nsd_mask.str
implant phos dose = 1e14 energy = 25
etch photoresist all
diffuse time = 15/60 temp = 950 nitrogen
etch oxide dry thick = 0.01
structure outf = 12b_nsd.str

#- - - - - - - - - - - - - - - - - - - - - - - - - - - - - - -
# PSD
#- - - - - - - - - - - - - - - - - - - - - - - - - - - - - - -
include file = testflow4.msk
structure outf = 12c_psd_mask.str
implant boron dose = 1e14 energy = 15
etch photoresist all
diffuse time = 5/60 temp = 950 nitrogen
structure outf = 12d_psd.str

#- - - - - - - - - - - - - - - - - - - - - - - - - - - - - - -
# Dual-Damascene BPSG flow ILD1a with etch stop omitted
#- - - - - - - - - - - - - - - - - - - - - - - - - - - - - - -
deposit oxide thick = 0.5 meshlayer = 5 space = 0.2
structure outfile = 13a_bpsg.str

#- - - - - - - - - - - - - - - - - - - - - - - - - - - - - - -
# Dual-Damascene BPSG flow ILD1b with etch stop omitted
#- - - - - - - - - - - - - - - - - - - - - - - - - - - - - - -
deposit oxide thick = 0.3 meshlayer = 2
structure outfile = 13b_bpsg_1b.str

#- - - - - - - - - - - - - - - - - - - - - - - - - - - - - - -
```

```
# BPSG CMP
#- - - - - - - - - - - - - - - - - - - - - - - - - - - - - -
etch start x = 0 y = -20
etch continue x = 0 y = -5.65
etch continue x = 34 y = -5.65
etch done x = 34 y = -20
structure outfile = 13c_bpsg_CMP.str

#- - - - - - - - - - - - - - - - - - - - - - - - - - - - - -
# Metal1 Etch
#- - - - - - - - - - - - - - - - - - - - - - - - - - - - - -
include file = testflow8.msk
structure outfile = 14a_Metal1_mask.str
etch oxide avoidmask depth = 0.3
etch photoresist all
structure outfile = 14a_Metal1_etch.str

#- - - - - - - - - - - - - - - - - - - - - - - - - - - - - -
# Contact Etch
#- - - - - - - - - - - - - - - - - - - - - - - - - - - - - -
include file = testflow9.msk
structure outfile = 14b_contacts_mask.str
etch oxide avoidmask depth = 1
etch photoresist all
structure outfile = 14b_contacts_etch.str

#- - - - - - - - - - - - - - - - - - - - - - - - - - - - - -
# Copper Electroplating
#- - - - - - - - - - - - - - - - - - - - - - - - - - - - - -
deposit cu thick = 1.2 meshlayer = 4
structure outfile = 14c_copper_depo.str

etch start x = 0 y = -15
etch continue x = 0 y = -5.55
etch continue x = 34 y = -5.55
```

```
etch done x = 34 y = -15

structure outfile = 14c_BEOL.str

#- - - - - - - - - - - - - - - - - - - - - - - - - - - - - -
# Export to device simulator
#- - - - - - - - - - - - - - - - - - - - - - - - - - - - - -

export outf = LDMOS.aps xpsize = 0.001
```

Appendix C: Trap Dynamics and AC Analysis

We start with the basic trap dynamic equation as follows

$$N_t \frac{df_t}{dt} = R_n - R_p \tag{C.1}$$

$$R_n = c_n n N_t (1 - f_t) - c_n n_1 N_t f_t \tag{C.2}$$

$$R_p = c_p p N_t f_t - c_p p_1 N_t (1 - f_t) \tag{C.3}$$

Consider a simple case that current is related to the trap rates by

$$I = q V_{tp} (R_n - R_p) \tag{C.4}$$

where V_{tp} is the volume of interests containing the traps. To solve the trap dynamic equations, we give details of the recombination rate as

$$\frac{R_n}{N_t} = c_n n (1 - f_t) - c_n n_1 f_t \tag{C.5}$$

$$\frac{R_p}{N_t} = c_p p f_t - c_p p_1 (1 - f_t) \tag{C.6}$$

In most situations, the carrier injection into the system is such that $n \gg n_1$ and $p \gg p_1$ so that the second terms above can be ignored. The simplified equation becomes

$$\frac{df_t}{dt} = c_n n (1 - f_t) - c_p p f_t \tag{C.7}$$

To study how the trap dynamic equation affects frequency response, we consider the system is under sinusoidal perturbation and all quantities vary with a factor $\exp(j\omega t)$

$$f_t = f_{t0} + \Delta f_t \exp(j\omega t) \tag{C.8}$$

$$n = n_0 + \Delta n \exp(j\omega t) \tag{C.9}$$

$$p = n_0 + \Delta p \exp(j\omega t) \tag{C.10}$$

The trap dynamic equation becomes

$$(\Delta f_t) j\omega = c_n (\Delta n) - \Delta(c_n n f_t) - \Delta(c_p p f_t) \tag{C.11}$$

We make the expansion in the first order:

$$\Delta(c_n n f_t) = c_n(\Delta n) f_{t0} + c_n n_0 (\Delta f_t) \tag{C.12}$$

$$\Delta(c_p p f_t) = c_p(\Delta p) f_{t0} + c_p p_0 (\Delta f_t) \tag{C.13}$$

We combine Equations (C.11) to (C.13) to solve for (Δf_t)

$$(\Delta f_t) j\omega = \Delta c_i - c_n n_0 (\Delta f_t) - c_p p_0 (\Delta f_t) \tag{C.14}$$

where Δc_i is a parameter depending on carrier injection condition and trap occupancy and is expressed as

$$\Delta c_i = c_n(\Delta n) - c_n(\Delta n) f_{t0} - c_p(\Delta p) f_{t0} \tag{C.15}$$

Simple algebra leads to the following solution for the trap AC response:

$$(\Delta f_t) = \Delta c_i / (j\omega + c_n n_0 + c_p p_0) \tag{C.16}$$

We rearrange to separate the real and imaginary parts:

$$(\Delta f_t) = -j \frac{\omega \Delta c_i}{\omega^2 + (c_n n_0 + c_p p_0)^2} + \frac{\Delta c_i (c_n n_0 + c_p p_0)}{\omega^2 + (c_n n_0 + c_p p_0)^2} \tag{C.17}$$

We combine Equations (C.1), (C.4), and (C.16) to obtain

$$(\Delta I) = \frac{V_{tp} N_t \omega^2 \Delta c_i + V_{tp} N_t j\omega \Delta c_i (c_n n_0 + c_p p_0)}{\omega^2 + (c_n n_0 + c_p p_0)^2} \tag{C.18}$$

The injection parameter Δc_i is a function of bias voltage V:

$$\Delta c_i = \frac{dc_i}{dV} \Delta V \tag{C.19}$$

and we are able to obtain the AC capacitance and conductance as

$$\Delta I = G_{tp} \Delta V + j\omega C_{tp} \Delta V \tag{C.20}$$

where the trap related conductance G_{tp} is given by

$$G_{tp} = \frac{V_{tp} N_t \omega^2 \left(\dfrac{dc_i}{dV}\right)}{\omega^2 + (c_n n_0 + c_p p_0)^2} \tag{C.21}$$

and the trap related capacitance is obtained as

$$C_{tp} = \frac{V_{tp} N_t \left(\dfrac{dc_i}{dV}\right)(c_n n_0 + c_p p_0)}{\omega^2 + (c_n n_0 + c_p p_0)^2} \tag{C.22}$$

Bibliography

[1] Ned Mohan, *Power Electronics, A First Course.* John Wiley & Sons, Inc., 2012.

[2] Argonne National Laboratory. [Online]. http//www.anl.gov/.

[3] Bimal K. Bose, *Power Electronics and Motor Drives, Advances and Trends.* Elsevier, 2006.

[4] Dragan Maksimovic, Regan Zane, and Robert Erickson, "Impact of digital control in power electronics," in *International Symposium on Power Semiconductor Devices and ICs,* Kitakyushu, 2004.

[5] Switched-mode power supply Wikipedia. [Online]. http://en.wikipedia.org/wiki/Switched-mode_power_supply.

[6] Marc Thompson, Introduction to power electronics. [Online]. http://www.thomp sonrd.com/NOTE S%2001%20INTRODUCTION%20TO% 20POWER%20ELECTRONICS.pdf.

[7] Rudy Severns, "History of soft switching," *Switching Power Magazine* by Ridley Engineering, Inc., 2001.

[8] Linear versus switch-mode power supplies. [Online]. http://digital.ni.com/public.nsf/ad0f28 2819902a1986256 f79005462b1/7438 e77138bddf1b862 56f660008e9cc/$FI LE/linear_versus_switching.pdf.

[9] Micheal Day, Understanding low drop out (LDO) regulators. [Online]. http://focus.ti.com/download/ trng/docs/seminar/Topic%209%20-%20Understanding%20LDO%20dropout.pdf.

[10] National Semiconductor. [Online]. https://www.national.com/ds/LM/LM340.pdf.

[11] Linear Technology LT3026. [Online]. http://www.linear.com/product/LTC3026.

[12] EE times Product How-To: Charge pumps can handle the Volts! [Online]. http://www.eetimes.com/ design/smart-energy-design/4394820/No-magnetics-required—charge-pumps-can-handle-the-Volts-.

[13] Vincent Wai-Shan Ng and Seth R. Sanders, August 2011. Switched capacitor DC-DC converter: superior where the buck converter has dominated. [Online]. http://www.eecs.berkeley.edu/Pubs/TechRpts/2011/ EECS-2011-94.pdf.

[14] Wikipedia: charge pump. [Online]. http://en.wikipedia.org/wiki/Charge_pump.

[15] Slobodan Ćuk and R. D. Middlebrook, "A general unified approach to modelling switching-converter power stages," in Proc. IEEE Power Elec. Spec. Conf. 1976, Cleveland, OH, 1976.

[16] ST Application Note. [Online]. http://www.st.com/internet/com/TECHNICAL_RESOURCES/ TECHNICAL_ LITERATURE/APPLICATION_NOTE/CD00003910.pdf.

[17] Robert W. Erickson, "DC-DC power converters," *Wiley Encyclopedia of Electrical and Electronics Engineering,* 2007, John Wiley & Sons.

[18] Full-bridge quarter brick DC/DC converter reference design using a dsPIC DSC. [Online]. http://ww1. microchip. com/downloads/en/AppNotes/01369A.pdf?from=rss.

[19] Bob Bell and Ajay Hari. February 2011. Topology key to power density in isolated DC-DC converters. [Online]. http://powerelectronics.com/power_systems/dc_dc_converters/topologies-used high-power-applications-201102/ index.html.

[20] Cadence PSPICE. [Online]. http://www.cadence.com/products/orcad/pspice_simulation/pages/default.aspx.

[21] Jeff Falin. November 2004. Cell phone power management needs specialized ICs. [Online]. http:// www.eetimes.com/design/power-management-design/4009520/Cell-phone-power-management-needs-specialized-ICs.

[22] Texas Instruments. 2008. Power Management Guide 3Q 2008 slvt145h.pdf. [Online]. http://www.ti.com/.

[23] Nazzareno Rossetti, *Managing Power Electronics, VLSl and DSP-Driven Computer Systems,* John Wiley & Sons, Inc., 2006.

[24] T. D. Burd, T. A. Pering, A. J. Stratakos, and R. W. Brodersen, "A dynamic voltage scaled microprocessor system," *IEEE J. Solid-St. Circ.,* vol. 35, no. 11, pp. 1571–1580, 2000.

[25] R. Gonzalez, B. M. Gordon, and A. Horowitz, "Supply and threshold voltage scaling for low power CMOS," *IEEE J. Solid-St. Circ.,* vol. 32, no. 8, pp. 1210–1216, 1997.

[26] R. Williams, W. Grabowski, A. Cowell, M. Darwish, and J. Berwick, "The dual-gate W-switched power MOSFET: A new concept for improving light load efficiency in DC–DC converters," in Proc. IEEE Int. Symp. Power Semicond. Devices ICs, Weimar, Germany, 1997.

[27] O. Trescases, A. Prodić, and W. T. Ng, "Digitally controlled current-mode DC-DC converter IC," *IEEE Trans. Circ. and Sys.,* vol. 58, no. 1, pp. 219–231, 2011.

[28] O. Trescases, W. T. Ng, H. Nishio, M. Edo, and T. Kawashima, "A digitally controlled DC-DC converter module with a segmented output stage for optimized efficiency," in 18th Int. Sym. on Power Semiconductor Devices and ICs (ISPSD '06), Naples, Italy, 2006.

[29] A. Yoo, M. Chang, O. Trescases, and W. T. Ng, "High performance low-voltage power MOSFETs with hybrid waffle layout structure in 0.25µm standard CMOS process," in 20th Int. Sym. on Power Semiconductor Devices and ICs (ISPSD '08), Orlando, Florida, 2008.

[30] Z. Zhao and A. Prodić, "Limit-cycle oscillations based auto-tuning system for digitally controlled DC-DC power supplies," IEEE Trans. Power Electron., vol. 24, no. 6, pp. 2211–2222, 2007.

[31] O. Trescases, W. T. Ng, M. Edo, H. Nishio, and T. Kawashima, "A digitally controlled DC-DC converter module with a segmented output stage for optimized efficiency," in Proc. Int. Symp. Power Semiconductor Devices and ICs, Naples, Italy, 2006.

[32] A. Zhao, A. A. Fomani, and W.T. Ng, "One-step digital dead-time correction for DC-DC converters," in Proc. Applied Power Electronics Conf., Palm Springs, CA, USA, 2010.

[33] A. Radic, Z. Lukic, A. Prodic, and R. de Nie, "Minimum deviation digital controller IC for single and two phase DC-DC switch-mode power supplies," in Proc. Applied Power Electronics Conf., Palm Springs, CA, USA, 2010.

[34] E. Meyer, Z. Zhang, and Y. F. Liu, "An optimal control method for buck converters using a practical capacitor charge balance technique," IEEE Trans. Power Electron., vol. 23, no. 4, pp. 1802–1812, 2008.

[35] J. Wang et al., "A digitally controlled integrated DC-DC converter with transient suppression," in 2010 22nd International Symposium on Power Semiconductor Devices & IC's (ISPSD), Hiroshima, Japan, 2010.

[36] J. Wang et al., "Integrated DC-DC converter with an auxiliary output stage for transient suppression," in Proc. Elec. Device and Solid State Circuits, Xi'an, China, 2009, pp. 380–383.

[37] Bo Lojek, History of Semiconductor Engineering, Springer, 2006.

[38] Semiconductor Industry Association. [Online]. http://www.sia-online.org/.

[39] Simon Li and Yue Fu, 3D TCAD Simulation for Semiconductor Processes, Devices and Optoelectronics, Springer, 2011.

[40] Xiao Hong. History of Semiconductor. [Online]. http://www.austincc.edu/HongXiao/overview/history-semi/.

[41] Traitorous eight, Wikipedia. [Online]. http://en.wikipedia.org/wiki/Traitorous_eight.

[42] M. Fischetti. 2009. A (partial, biased?) history of the MOSFET from a physicist's perspective. [Online]. www.ecs.umass.edu/ece/ece344/MOSFET_talk.ppt.

[43] Intel Corporation. 3-D, 22nm: new technology delivers an unprecedented combination of performance and power efficiency. [Online]. http://www.intel.com/content/www/us/en/silicon-innovations/intel-22nm-technology.html.

[44] Intel Corporation. Fun facts: exactly how small (cool) is 22 nanometers? [Online]. http://www.intel.com /content/dam/www/ public/us/en/documents/corporate-information/history-moores-law-fun-facts-factsheet.pdf.

[45] IC knowledge LLC. A simulation study of 450 mm wafer fabrication costs. [Online]. http://www.icknowledge.com/news/A%20Simulation%20Study%20of%20450mm%20Wafer%20Fabrication%20Costs%20revision%201.pdf.

[46] Texas Instruments. [Online]. http://www.ti.com/corp/docs/manufacturing/manufacturing.shtml.

[47] Soitec. [Online]. http://www.soitec.com/en/products-and-services/microelectronics/smart-power-soi/.

[48] Compound semiconductor. Turning 6-inch GaN LED manufacturing into reality. [Online]. http://www.compoundsemiconductor.net/csc/features-details/19733048/Turning-6-inch-GaN-LED-manufacturing-into-realit.html.

[49] Compound semiconductor. Cree's 150mm N-Type SiC wafers on the market. [Online]. http://www.compoundsemiconductor.net/csc/news-details/id/19735434/name/Cree's-150mm-n-type-SiC-wafers-on-the-marke.html.

[50] Peter Clarke. January 2, 2013. EDA market keeps growing. [Online]. http://www.eetimes.com/design/eda-design/4404248/EDA-market-growth.

[51] Electronic design automation. [Online]. http://en.wikipedia.org/wiki/Electronic_design_automation.

[52] BSIM Group. [Online]. http://www-device.eecs. berkeley .edu/bsim/.

[53] PSP. [Online]. http://pspmodel.asu.edu/.

[54] Professor Mitiko Miura. A Hiroshima University Model International Standard~HiSIM-LDMOS: A next generation, high-accuracy transistor model for semi-conductor development. [Online]. http://www.hiroshima-u.ac.jp/en/top/research_HU/researchnow/no2/index.html.

[55] NXP. [Online]. http://www.nxp.com/mode ls/high-voltage-models.html.

[56] Freescale RF high power models. [Online]. http://www.freescale.com/webapp/sps/site/overview.jsp?nodeId=0106B97520NFng.

[57] IC-CAP device modeling software—measurement control and parameter extraction. [Online]. http://www. home.agilent.com/agilent/product.jspx?cc=CA&lc=eng&ckey=1297149&nid=-34268.0.00&id=1297149.

[58] Jacek Korec, *Low Voltage Power MOSFETs*, Springer, 2011.

[59] Alireza Servati and Anthony R. Simon, *Introduction to Semiconductor Marketing*, Simon Publications, 2003.

[60] Rakesh Kumar, *Fabless Semiconductor Implementation*, McGraw-Hill, 2008.

[61] Peter Clarke, Qualcomm pits CMOS against GaAs for cellphone front ends. [Online]. http://www. eetimes.com/design/microwave -rf-design/4410008/Qualcomm-expected-to-ignite-CMOS-PA-market.

[62] Yole Development. June 2007. Semiconductor today, SiC power devices: if only we had a switch. [Online]. http://www.semiconductor-today.com /features/Semiconductor%20Today%20-%20SiC%20 power%20devices.pdf.

[63] Semiconductor sales leaders by year. [Online]. http://en.wikipedia.org/w iki/Semiconductor_sales_le aders_ by_year#Ranking_for_year_1987.

[64] Gartner, Inc. [Online]. http://www.gartner.com/technology/home.jsp.

[65] iSuppli. [Online]. http://www.isuppli.com/About/Pages/Contact.aspx.

[66] IC Insights. [Online]. http://www.icinsights.com/news/bulletins/PurePlay-Foundries-And-Fabless- Suppliers-Are-Star-Performers-In-Top-25-2012-Semiconductor-Supplier-Ranking/.

[67] EE times—Intel exec says fabless model 'collapsing'. [Online]. http://www.eetimes.com/ electronics-news/4371617/Intel-exec-says-fabless-model— collapsing-.

[68] Achronix semiconductor. [Online]. http://www.achronix.com/.

[69] EEtimes. [Online]. http://www.eetimes.com.

[70] IC Insights. Eleven companies move up in 1Q13 top 20 semi supplier ranking. [Online]. http://www.ic insights.com/news/bulletins/Eleven-Companies-Move-Up-In-1Q13-Top-20-Semi-Supplier-Ranking/.

[71] G. Roosmalen, Q. Zhang, and A. J. van Moore, *Creating High Value Micro/Nanoelectronics Systems*, Springer, 2009.

[72] Transistor count—Wikipedia. [Online]. http://en.wikipedia.org/wiki/Transistor_count.

[73] Wikipedia more than Moore. [Online]. http://en.wikipedia.org/wiki/Moore's_law.

[74] International Technology Roadmap for Semiconductors. [Online]. http://www.itrs.net/.

[75] The International Technology Roadmap for Semiconductors. 2011. [Online]. http://www.itrs.net/Links/2 011ITRS/2011Chapters/2011ExecSum.pdf.

[76] DC-DC converter ICs: power system in package, worldwide technology trends, forecasts and competitive environment, first edition. [Online]. http://www.reportlinker.com/p0685628-summary/DC-DC-Converter- ICs-Power-System-in-Package-Worldwide-Technology-Trends-Forecasts-and-Competitive- Environment-First-Edition.html.

[77] Jian Lu, Hongwei Jia, Andres Arias, Xun Gong, and Z. John Shen, "On-chip bondwire magnetics with ferrite-epoxy glob coating for power systems on chip," *International Journal of Power Management Electronics,* vol. 2008.

[78] Rongxiang Wu and J. K. O. Sin, "A novel silicon-embedded coreless inductor for high-frequency power management applications," *IEEE Electron Dev. Lett.,* vol. 32, no. 1, pp. 60–62, January 2011.

[79] R. Wu, J. K. O. Sin, and S. Y. (Ron) Hui, "A novel silicon-embedded coreless transformer for isolated DC-DC converter applications," in 23rd IEEE Int. Symp. on Power Semiconductor Devices & ICs, San Diego, CA, USA, 2011.

[80] X. Fang, R. Wu, L. Peng, and J. K. O. Sin, "A novel silicon-embedded toroidal power inductor with magnetic core," IEEE Electron Dev. Lett., vol. 34, no. 2, pp. 292–294, 2013.

[81] S. Gupta, J. C. Beckman, and S. L. Kosier, "Improved latch-up immunity in junction-isolated smart power ICs with unbiased guard ring," *IEEE Electron Dev. Lett.,* vol. 22, no. 12, pp. 600–602, December 2001.

[82] Bruno Murari. Smart Power Technology Evolves to Higher Levels of Complexity. [Online]. http://www. datasheetcatalog.org/datasheet/SGSThomsonMicroelectronics/mXxztx.pdf.

[83] Bruno Murari, Franco Bertotti, and Giovanni A. Vignola, *Smart Power ICs: Technologies and Applications,* Springer, 1995.

[84] Claudio Contiero, Antonio Andreini, and Paola Galbiati, Roadmap differentiation and emerging trends in BCD technology. [Online]. http://www.imec.be/essderc/ESSDERC2002/PDFs/D12_1.pdf.

[85] SMARTMOS Power Technology Brochure. [Online]. http://www.freescale.com/files/analog/doc/bro- chure/BR1567.pdf.

[86] Yong Liu, *Power Electronic Packaging,* Springer, 2012.

[87] Intel. [Online]. http://www.intel.com/technology/mooreslaw/.

[88] Wikipedia. [Online]. http://en.wikipedia.org/wiki/Tape-out#cite_note-1.

[89] (2006) Electronic Design. [Online]. http://electronicdesign.com/article/digital/ follow-heuristic-guide- lines-to-make-surface-mount-.aspx.

[90] JMP (statistical software)—Wikipedia. [Online]. https://en.wikipedia.org/wiki/JMP_(application_software).

[91] J. W. McPherson, R. B. Khamankar, and A. Shanware, "Complementary model for intrinsic time-dependent dielectric breakdown in SiO2 dielectrics," *J. Appl. Phy.,* vol. 88, no. 9, pp. 5351–5359, November 2000.

[92] S. Oussalah and B. Djezzar, "Field acceleration model for TDDB: still a valid tool to study the reliability of thick SiO2-based dielectric layers?," *IEEE Trans. Electron. Dev.,* vol. 54, no. 7, pp. 1713–1717, July 2007.

[93] Yee-Chia Yeo, Qiang Lu, and Chenming Hu, "MOSFET gate oxide reliability: anode hole injection model and its applications," *Int. J. High Speed Electron. and Sys.,* vol. 11, no. 3, pp. 849–886, 2001.

[94] Hot-carrier injection. [Online]. https://en.wikipedia.org/wiki/Hot-carrier_injection.

[95] Jone F. Chen, Kuen-Shiuan Tian, Shiang-Yu Chen, Kuo-Ming Wu, and C. M. Liu, "On-resistance degradation induced by hot-carrier injection in LDMOS transistors with STI in the drift region," *IEEE Electron. Dev. Lett.,* vol. 29, no. 9, pp. 1071–1073, September 2008.

[96] Electromigration. [Online]. http://en.wikipedia.org/wiki/Electromigration.

[97] J. R. Lloyd and J. J. Clement, "Electromigration in copper conductors," *Thin Solid Films,* vol. 262, no. 1–2, pp. 135–141, June 1995.

[98] C. Cismaru and V. Ramanathan, "Plasma induced damage testing methodology for the 0.13 μm CMOS technology," in 6th International Symposium on Plasma- and Process-Induced Damage, Monterey, CA, 2001, pp. 48–51.

[99] TSMC. Power IC. [Online]. http://www.tsmc.com/english/dedicatedFoundry/technology/power_ic.htm.

[100] UMC. Power Management IC. [Online]. http://www.umc.com/chinese/pdf/PMIC_DM.pdf.

[101] Global foundries. (2011) BCDlite™—The Ideal Green Power Management Platform. [Online]. http://www.globalfoundries.com/newsletter/2011/ 2/bcdlite.aspx.

[102] SMIC. Analog/Power. [Online]. http://www.smics.com/eng/foundry/technology/analog_power.php.

[103] M. Darwish and K. Board, "Lateral resurfed COMFET," *Electron. Lett.,* vol. 20, no. 12, pp. 519–520, June 1984.

[104] Johnny K. O. Sin and C. A. T. Salama, "The SINFET: A new high conductance, hing switching speed mos-gated transistor," *Electron. Lett.,* vol. 21, no. 24, pp. 1134–1136, November 1985.

[105] M. R. Simpson, P. A. Gough, F. I. Hshieh, and V. Rumennik, "Analysis of the lateral insulated gate transistor," in International Electron Devices Meeting, Washington DC, December 1985.

[106] Deva N. Pattanayak et al., "n-Channel lateral insulated gate trasistors: part I—steady-state characteristics," *IEEE Trans. Electron. Dev.,* vol. 33, no. 12, pp. 1956–1963, 1986.

[107] T. P. Chow et al., "Interaction between monolithic, junction-isolated lateral insulated-gate bipolar transistors," *IEEE Trans. Electron. Dev.,* vol. 38, no. 2, pp. 310–315, 1991.

[108] W. W. T. Chan, J. K. O. Sin, P. K. T. Mok, and S. S. Wong, "CMOS latchup characterization for LDMOS/LIGBT power integrated circuits," in IEEE International Conference on Semiconductor Electronics, Malaysia, 1996.

[109] I. Wacyk, R. Jayaraman, L. Casey, and J. K. O. Sin, "Fast LIGBT switching due to plasma confinement through pulse width control," in 3rd International Symposium on Power Semiconductor Devices & ICs, Baltimore, 1991.

[110] W. W. T. Chan, J. K. O. Sin, and S. S. Wong, "A novel cross-talk isolation structure for bulk CMOS power ICs," *IEEE Trans. Electron. Dev.,* vol. 45, no. 7, pp. 1580–1586, 1998.

[111] Zener diode—Wikipedia. [Online]. http://en.wikipedia.org/wiki/Zener_diode.

[112] Donald T. Comer, "Zener Zap Anti-Fuse Trim in VLSI circuits," *VLSI Design,* vol. 5, no. 1, pp. 89–100, 1996.

[113] SmartMOS technology. [Online]. http://www.freescale.com/webapp/sps/site/overview.jsp?code=TM_SMARTMOS.

[114] L. Clavelier, B. Charlet, B. Giffard, and M. Roy, "Deep trench isolation for 600V SOI power devices," in 33rd European Solid-State Circuits Conference, Estoril, Portugal, 2003.

[115] R. Bashir S. Lee, "Modeling and characterization of deep trench isolation structures," *Microelectron. J.,* vol. 32, pp. 295–300, 2001.

[116] Crosslight Software, *CSuprem Manual,* Burnaby, BC, Canada, 2009.

[117] P. Helmut, "Advanced process modeling for VLSI technology," Ph.D. thesis, Technical University of Wien, 1996.

[118] M. Radi, "Three-dimensional simulation of thermal oxidation," Ph.D. thesis, Technical University of Wien, 1998.

[119] W. Bohmayr, G. Schrom, and S. Selberherr, "Investigation of channeling in field oxide corners by three-dimensional Monte Carlo simulation of ion implantation," in 4th Int. Conf. on Solid-State and Integrated-Circuits Technology, Beijing, China, 1995.

[120] H. Puchner, "A transient activation model for phosphorus after sub-amorphizing channeling implants," in Proc. 26th Euro. Solid State Device Res. Conf., Bologna, Italy, 1996, pp. 157–160.

[121] Robert W. Dutton and Zhiping Yu, *Technology CAD Computer Simulation of IC Processes and Devices,* Kluwer Academic Publishers, 1993.

[122] Axcelis 2D Implant Simulator. [Online]. http://www.axcelis.com/2d-implant-simulator.

[123] S. E. Hansen and M. D. Deal, Suprem-IV.GS, *Two-Dimensional Process Simulation for Silicon and Galium Arsenide*, Stanford University, 1993.

[124] D. Chin, S. Y. Oh, S. M. Hu, R. W. Dutton, and J. L. Moll, "Two-dimensional oxidation," *IEEE Trans. Electron. Dev.*, vol. 30, no. 7, pp. 744–749, 1983.

[125] J. P. Peng, D. Chidambarrao, and G. R. Srinivasan, "Viscoelestic modeling of thermal oxidation," in 177th Electrochem Society Meeting, 1990.

[126] M. Radi, "Three-dimensional simulation of thermal oxidation," Ph.D. thesis, University of Wien, 1998.

[127] S. R. Vinay, *On the Numerical Modeling of Thermal Oxidation in Silicon*, Stanford University, 1997.

[128] C. Stephan, "Multidimensional viscoelestic model modeling of silicon oxidation and titanium silicidation," Ph.D. thesis, University of Florida, 1996.

[129] J. Lorenz, *3-Dimensional Process Simulation*, Springer, 1995.

[130] W. A. Tiler, "On the kinetics of thermal oxidation of silicon I. A theoretical perspective," *J. Electrochem. Soc.*, vol. 127, no. 3, pp. 619–624, 1980.

[131] M. E. Law, "Grid adaptation near moving boundaries in two dimensions for IC process simulation," *IEEE Trans*. CAD, vol. 14, no. 10, pp. 1223–1230, 1995.

[132] H. Takato et al., "Impact of surrounding gate transistor (SGT) for Ultra-High-Density LS's," *IEEE Trans. on Electron. Dev.*, vol. 38, no. 3, pp. 573–578, 1991.

[133] B. E. Deal and A. S. Grove, "General relationship for the thermal oxidation of silicon," *J. Appli. Phys.*, vol. 36, no. 12, May 1965.

[134] S. Selberherr, *Analysis and Simulation of Semiconductor Devices,* Springer-Verlag, 1984.

[135] James D. Plummer, Michael D. Deal, and Peter B. Griffin, *Silicon VLSI Technology, Fundamentals, Practice and Modeling,* Prentice Hall, 2000.

[136] H. Mastumoto and M. Fukuma, "Numerical modeling of nonuniform Si thermal oxidation," *IEEE Trans. Electron. Dev.*, vol. 32, no. 2, pp. 132–140, 1985.

[137] A. Poncet, "Finite-Element simulation of local oxidation of silicon," *IEEE Trans. CAD,* vol. 4, no. 1, pp. 41–53, 1985.

[138] S. M. Sze, *Physics of Semiconductor Devices,* 2nd edition, John Wiley & Sons, 1981.

[139] P. M. Fahey, P. B. Griffin, and J. D. Plummer, "Point defects and dopant diffusion in silicon," *Rev. Mod. Phys.*, vol. 61, no. 2, pp. 289–384, 1989.

[140] W. Taylor, U. Gosele, and T. Y. Tan, "Present understanding of point defect parameters and diffusion in silicon: An overview," in Pro. Phy. and Model. in Semi. Tech., the Electrochemical Society. Proceedings, Vol. 93-6. Honolulu, HI., 1993.

[141] C. F. Machala et al., "A strategy for enabling predictive TCAD in development of sub-100nm CMOS technologies," in IEEE Int. Conf. on Simu. of Semi. Proc. and Dev. (SISPAD), Kobe, Japan, 2002.

[142] M. A. Ismail et al., "Calibration parameters in TCAD for predictive MOSFET device simulations," in IEEE Int. Conf. Semi. Elec. (ICSE), Kuala Lumpur, Malaysia, 2012.

[143] M. Lorenzini et al., "Simulation of 0.35um/0.25um CMOS Technology Doping Profiles," *VLSI Design*, vol. 13, pp. 459–463, 2001.

[144] H. Park et al., "Systematic calibration of process simulators for predictive TCAD," in IEEE Int. Conf. Simul. of Semi. Proc. and Dev. (SISPAD), Cambridge, MA, USA, 1997.

[145] J.-H. Lee et al., "Systematic global calibration of a process simulator," in International Conference on Modeling and Simulation of Microsystems, Cambridge, MA, USA, 2000.

[146] Synopsys Inc., Taurus TSUPREM-4 User Guide, 2005.

[147] Yue Fu and Simon Li, "3D simulation for power semiconductor devices," in CMOS ET 2012, Vancouver, 2012.

[148] HEXFET, International Rectifier. [Online]. http://www.irf.com/technical-info/guide/device.html.

[149] Nvidia. CUDA. [Online]. http://www.nvidia.ca/object/cuda_home_new.html.

[150] Central processing unit—Wikipedia. [Online]. https://en.wikipedia.org/wiki/Central_processing_unit.

[151] Graphic processing unit—Wikipedia. [Online]. http://en.wikipedia.org/wiki/Graphics_processing_unit.

[152] Florin Manolache. Introduction to Parallel Computing and Scientific Computation. [Online]. http://www.math.cmu.edu/~florin/M21-765 /slides/35/index.html.

[153] Amdahl's law—Wikipedia. [Online]. http://en.wikipedia.org/wiki/Amdahl's_law.

[154] Nvidia. What Is GPU Computing? [Online]. http://www.nvidia.ca/object/what-is-gpu-computing.html.

[155] R.S. Varga, *Matrix Iterative Analysis*, Prentice-Hall, 1962.

[156] Crosslight Software, *APSYS Manual*. Burnaby, BC, Canada, 2006.

[157] E. F. Schubert, "Chapter 16—High doping effects," in *Physical Foundations of Solid-State Devices*, Rensselaer Polytechnic Institute, 2009.

[158] B. Jayant Baliga, *Fundamentals of Power Semiconductor Devices*, Springer, 2010.

[159] A. G. Chynoweth, "Ionization rates for electrons and holes in silicon," *Phys. Rev.,* vol. 109, no. 5, pp. 1537–1540, 1958.

[160] Josef Lutz, Heinrich Schlangenotto, Uwe Scheuermann, and Rik De Doncker, *Semiconductor Power Devices Physics, Characteristics, Reliability,* Springer, 2011.

[161] C. A. Lee, R. A. Logan, R. L. Batdorf, J. J. Kleimack, and W. Wiegmann, "Ionization rates of holes and electrons in silicon," *Phys. Rev.,* vol. 134, pp. A761–A773, 1964.

[162] T. Ogawa, "Avalanche breakdown and multiplication in silicon pin junctions," *Japanese J. Applied Physics,* vol. 4, pp. 473–484, 1965.

[163] R. Van Overstraeten and H. De Man, "Measurement of the ionization rates in diffused silicon p-n junctions," *Solid-St. Electron.,* vol. 13, no. 5, pp. 583–608, May 1970.

[164] C. R. Crowell and S. M. Sze, "Temperature dependence of avalanche multiplication in semiconductors," *Appl. Phys. Lett.,* vol. 9, no. 6, pp. 242–244, 1966.

[165] Y. Okuto and C. R. Crowell, "Threshold energy effects on avalanche breakdown voltage in semiconductor junctions," *Solid-St. Electron.,* vol. 18, pp. 161–168, 1975.

[166] S. M. Sze and K. Ng, *Physics of Semiconductor Devices,* Third edition, John Wiley & Sons, Inc., 2007.

[167] Calvin Hu Chenming, *Modern Semiconductor Devices for Integrated Circuits*, Pearson, 2012.

[168] Valentin O. Turin, "A modified transferred-electron high-field mobility model," *Solid-St. Electron.* vol. 49, no. 10, 2005, pp. 1678–1682.

[169] Software and Manuals. [Online]. http://www-tcad.stanford.edu/tcad/programs/oldftpable.html.

[170] C. Lombardi, S. Manzini, A. Saporito, and M. Vanzi, "A physically based mobility model for numerical simulation of non-planar devices," IEEE Trans. Comp. Aided Des. Int. Circ. Sys. vol. 7, no. 11, pp. 1164–1171, November 1988.

[171] M. Farahmand et al., "Monte Carlo simulation of electron transport in the III-nitride wurtzite phase material system: binaries and ternaries," *IEEE Trans. Electron. Dev.,* vol 48, no.3, 2001, pp. 535–542.

[172] A. Reklaitis and L. Reggiani, "Monte Carlo study of hot-carrier transport in bulk wurtzite GaN and modeling of a near-terahertz impact avalanche transit time diode," *J. Appl. Phy.*, vol. 95, no. 12, pp. 7925–7935, June 2004.

[173] S. E. Laux, "Techniques for small-signal analysis of semiconductor devices," *IEEE Trans. Electron Dev.*, vol. ED-32, p. 2028, 1985.

[174] Q. Li and R. W. Dutton, "Numerical small-signal AC modeling of deep-level-trap related frequency-dependent output conductance and capacitance for GaAs MESFET's on semi-insulating substrates," *IEEE Trans. Electron. Dev.,* vol. 38, pp. 1285–1288, 1991.

[175] A. A. Grinberg, M. S. Shur, R. J. Fischer, and H. Morkoc, "An investigation of the effect of graded layers and tunneling on the performance of AlGaAs/GaAs heterojunction bipolar transistors," *IEEE Trans. Electron. Dev.,* vol. ED-31, pp. 1758–1765, 1984.

[176] Sysnosys Inc., *Taurus Medici User Guide,* 2003.

[177] J. H. Lee et al., "Circuit model parameter generation with TCAD simulation," in IEEE Int. Conf. Solid-State Int. Circ. Tech. (ICSICT), 2004.

[178] Il-Yong Park et al., "BD180—a new 0.18 um BCD (Bipolar-CMOS-DMOS) Technology from 7V to 60V," in Proc. 20th Int. Sym. on Power Semi. Dev. & ICs, Orlando, FL, 2008.

[179] Seok-Woo Lee et al., "Gate oxide thinning effects at the edge of shallow trench isolation in the dual gate oxide process," in 6th International Conference on VLSI and CAD, 1999. ICVC '99, Seoul, 1999.

[180] Guizhong Pan, *MOS Int. Circ. Pro. and Manu. Tech. (in Chinese).* Shanghai: Shanghai Scientific & Technical Publishers, 2012.

[181] J. A. Appels and H. M. J. Vaes, "High Voltage Thin Layer Devices," in Proc. IEDM, Washington, DC, 1979.

[182] A. W. Ludikhuize, "High-Voltage DMOS and PMOS in Analog IC's," in Proc. IEDM, San Francisco, CA, 1982.

[183] Vladimir Rumennik, Donald R. Disney, and Janardhanan S. Ajit, "High-voltage transistor with multi-layer conduction region," US patent. US 6,207,994 B1, March 27, 2001.

[184] A. W. Ludikhuize, "A Review of RESURF Technology," in IEEE ISPSD, Toulouse, France, 2000.

[185] D. R. Disney, A. K. Paul, M. Darwish, R. Basecki, and V. Rumennik, "A new 800 V lateral MOSFET with dual conduction paths," in IEEE ISPSD, Osaka, Japan, 2001.

[186] B. Jayant Baliga, Advanced Power MOSFET Concepts, Springer, 2010.

[187] Geant4—Wikipedia. [Online]. http://en.wikipedia.org/wiki/Geant4.

[188] EKV MOSFET Model. [Online]. http://ekv.epfl.ch/.

[189] VBIC. [Online]. http://www.designers-guide.org/VBIC/.

[190] Gennady Gildenblat, Compact Modeling—Principles, Techniques and Applications, Springer, 2010.

[191] E. C. Griffith et al., "Capacitance modelling of LDMOS transistors," in IEEE Solid-State Dev. Res. Conf., 2000.

[192] P. A. Gough, M. R. Simpson, and V. Rumennik, "Fast switching lateral insulated gate transistor," IEDM Tech. Dig., pp. 218–221, 1986.

[193] Johnny K. O. Sin and Satyen Mukherjee, "Analysis and characterization of the segmented anode LIGBT," IEEE Trans. on Electron. Dev., vol. 40, no. 7, 1993.

[194] J. K. O. Sin, C. A. T. Salama, and L. Z. Hou, "The SINFET—A Schottky injection MOS-gated power transistor," IEEE Trans. Electron. Dev., vol. 33, no. 12, pp. 1940–1947, December 1986.

[195] T. Laska, M. Munzer, F. Pfirsch, C. Schaeffer, and T. Schmidt, "The field stop IGBT (FS IGBT)—a new power device concept with a great improvement potential," in IEEE ISPSD, Toulouse, France, 2000.

[196] M. Kitagawa, I. Omura, S. Hasegawa, T. Inoue, and A. Nakagawa, "A 4500V injection enhanced insulated gate bipolar transistor (IEGT) operating in a mode similar to a thyristor," in IEEE IEDM, Washington, DC, 1993.

[197] S. G. Nassif-Khalil, Li Zhang Hou, and C. Andre T. Salama, "SJ/RESURF LDMOST," IEEE Trans. Electron. Dev., vol. 51, no. 7, pp. 1185–1191, June 2004.

[198] Abraham Yoo, Jacky C. W. Ng, Johnny K. O. Sin, and Wai Tung Ng, "High performance CMOS-compatible Super Junction FinFETs for Sub-100V applications," in Inter. Ele. Dev. Meet. (IEDM'10), San Francisco, CA, USA, 2010.

[199] Kevin Jing Chen, "GaN smart power chip technology," in IEEE International Conference of Electron Devices and Solid-State Circuits, EDSSC 2009, Xi'an, 2009.

[200] King-Yuen Wong, Wanjun Chen, Xiaosen Liu, Chunhua Zhou, and Kevin J. Chen, "GaN smart power IC technology," Phy. Stat. Solidi (B), vol. 247, no. 7, pp. 1732–1734, 2010.

[201] Michael A. Briere, December 2008, GaN-based power devices offer game-changing potential in power-conversion electronics. [Online]. http://www.eetimes.com/design/power-management-design /4010344/ GaN-based-power-devices-offer-game-changing-potential-in-power-conversion-electronics.

[202] DigiKey. [Online]. http://www.digikcy.com/Web%20Export/Supplier%20Content/Microsemi_278/PDF/ Microsemi_GalliumNitride_VS_Silicon Carbide.pdf?redirected=1.

[203] Avogy. Semiconductor Materials for Power Devices. [Online]. http://www.avogy.com/index.php/ semiconductor-materials-for-power-devices.

[204] Avogy, Inc. GaN on GaN Compared to GaN on Si/SiC or Sapphire. [Online]. http://www.avogy.com/ index.php/gan-and-gan-compared-to-gan-on-si-sic-or-sapphire.

[205] S. J. Pearton, GaN and ZnO-Based Materials and Devices, Springer, 2012.

[206] Compound Semiconductor. GaN substrates to challenge silicon. [Online]. http://www.compoundsemi-conductor.net/ csc/indepth-details/19735741/GaN-substrat es-to-challenge-silico.html.

[207] Vincenzo Fiorentini, Fabio Bernardini, and Oliver Ambacher, "Evidence for nonlinear macroscopic polarization in IIIV nitride alloy heterostructures," Appl. Phys. Lett., vol. 80, no. 7, p. 1204, 2002.

[208] V. Fiorentini, F. Bernardini, and O. Ambacher, "Ambacher. Evidence for nonlinear macroscopic polarization in iii-v nitride alloy heterostructures," Appl. Phys. Lett., vol. 80, no. 7, pp. 1204–1206, 2002.

[209] K. Burak Üçer. Epitaxy—Wake Forest University. [Online]. http://users.wfu.edu/ucerkb/Nan242/L12-Epitaxy.pdf.

[210] Weiwei Kuang, TCAD Simulation and Modeling of AlGaN/GaN HFETs, North Carolina State University, 2008.

[211] Hadis Morko, Handbook of Nitride Semiconductors and Devices, Materials Properties, Physics and Growth, Wiley-VCH, 2008.

[212] Gang Xie, Bo Zhang, Fred Y. Fu, and W. T. Ng, "Breakdown voltage enhancement for GaN high electron mobility transistors," in Proc. 22nd Int. Symp. Power Semi. Dev. & ICs, Hiroshima, Japan, 2010.

[213] V. O. Turin and A. A. Balandin, "Performance degradation of GaN field-effect transistors due to thermal boundary resistance at GaN = substrate interface," *Electron. Lett.,* vol. 40, no. 1, pp. 81–83, January 2004.

[214] Gang Xie et al., "Breakdown voltage enhancement for power AlGaN/GaN HEMTs with air-bridge field plate," *IEEE Electron. Dev. Lett.,* vol. 33, no. 5, pp. 670–672, 2012.

[215] Ayse Merve Ozbek, "Measurement of impact ionization coefficients in GaN," Ph.D. thesis, North Carolina State University, 2012.

[216] Simon Li and Yue Fu, "Gate leakage current analysis of high electron mobility transistors," in IEEE Int. Conf. Ele. Dev. S. S. Cir. (EDSSC 2012), Bangkok, Thailand, 2012.

[217] Tongde Huang, Xueliang Zhu, Ka Ming Wong, and Kei May Lau, "Low-leakage-current AlN/GaN MOSHFETs using Al2O3 for increased 2DEG," *IEEE Electron. Dev. Lett.,* vol. 33, no. 2, pp. 212–214, February 2012.

[218] Industry review: SiC and GaN power devices jostle to grow their role, April 2013. [Online]. http://www.bluetoad.com/publicatio n/repo30/16981/157700/9ed9015ea3adbd2919e2aa605de5271663fcce5a.pdf.

[219] Brain Coppa. SiC Power Inversion Chips for PV & EVs. [Online]. http://www.planetanalog.com/author.asp?section_id=3056&doc_id=559910.

[220] Yole Development. March 2012. Power Gan Report 2012. [Online]. http://www.researchandmarkets.com/reports/2115657/power_gan_report_2012.

[221] Transphorm. Products. [Online]. https://www.transphormusa.com/products.

[222] D. Bednarczyk and J. Bednarczyk, "The approximation of the fermi–dirac integral F1/2(η)," *Phys. Lett. A,* vol. 64, no. 4, pp. 409–410, January 1978.

Index

Note: Page numbers ending in "e" refer to equations. Page numbers ending in "f" refer to figures. Page numbers ending in "t" refer to tables.